Quantum Probability and Related Topics

Proceedings of the 30th Conference

QP–PQ: Quantum Probability and White Noise Analysis*

QP–PQ: Quantum Probability and White Noise Analysis

Vol. 27: Quantum Probability and Related Topics
eds. R. Rebolledo and M. Orszag

Vol. 26: Quantum Bio-Informatics III
From Quantum Information to Bio-Informatics
eds. L. Accardi, W. Freudenberg and M. Ohya

Vol. 25: Quantum Probability and Infinite Dimensional Analysis
Proceedings of the 29th Conference
eds. H. Ouerdiane and A. Barhoumi

Vol. 24: Quantum Bio-Informatics II
From Quantum Information to Bio-informatics
eds. L. Accardi, W. Freudenberg and M. Ohya

Vol. 23: Quantum Probability and Related Topics
eds. J. C. García, R. Quezada and S. B. Sontz

Vol. 22: Infinite Dimensional Stochastic Analysis
eds. A. N. Sengupta and P. Sundar

Vol. 21: Quantum Bio-Informatics
From Quantum Information to Bio-Informatics
eds. L. Accardi, W. Freudenberg and M. Ohya

Vol. 20: Quantum Probability and Infinite Dimensional Analysis
eds. L. Accardi, W. Freudenberg and M. Schürmann

Vol. 19: Quantum Information and Computing
eds. L. Accardi, M. Ohya and N. Watanabe

Vol. 18: Quantum Probability and Infinite-Dimensional Analysis
From Foundations to Applications
eds. M. Schürmann and U. Franz

Vol. 17: Fundamental Aspects of Quantum Physics
eds. L. Accardi and S. Tasaki

Vol. 16: Non-Commutativity, Infinite-Dimensionality, and Probability at the Crossroads
eds. N. Obata, T. Matsui and A. Hora

Vol. 15: Quantum Probability and Infinite-Dimensional Analysis
ed. W. Freudenberg

Vol. 14: Quantum Interacting Particle Systems
eds. L. Accardi and F. Fagnola

*For the complete list of the published titles in this series, please visit:
www.worldscibooks.com/series/qqpwna_series.shtml

QP–PQ
Quantum Probability and White Noise Analysis
Volume XXVII

Quantum Probability and Related Topics

Proceedings of the 30th Conference

Santiago, Chile 23 – 28 November 2009

Editors

Rolando Rebolledo
Universidad Católica de Chile, Chile

Miguel Orszag
Universidad Católica de Chile, Chile

NEW JERSEY • LONDON • SINGAPORE • BEIJING • SHANGHAI • HONG KONG • TAIPEI • CHENNAI

Published by

World Scientific Publishing Co. Pte. Ltd.
5 Toh Tuck Link, Singapore 596224
USA office: 27 Warren Street, Suite 401-402, Hackensack, NJ 07601
UK office: 57 Shelton Street, Covent Garden, London WC2H 9HE

British Library Cataloguing-in-Publication Data
A catalogue record for this book is available from the British Library.

QP–PQ: Quantum Probability and White Noise Analysis — Vol. 27
QUANTUM PROBABILITY AND RELATED TOPICS
Proceedings of the 30th Conference

ISBN-13 978-981-4338-73-8
ISBN-10 981-4338-73-7

Printed in Singapore by World Scientific Printers.

FOREWORD

This volume collects a number of contributions presented at the 30th conference on Quantum Probability and Related Topics, held in Santiago, Chile on November 23–28, 2009.

Current works at the frontiers of research in various fields were discussed at the meeting. Particularly, we refer to quantum probability, infinite dimensional analysis, spectral theory, quantum information and statistics. Included in this volume are expository papers, which will help increase communication between researchers working in these areas. All the contributions have been refereed by at least one expert and revised, some extensively, for publication. No other divisions have been made; the material has been arranged purely in an alphabetical manner. We owe thanks to all the authors for their contribution to the volume. We are in debt with all the referees for their careful reading and evaluation of articles.

The Conference was organised by Marco Corgini, Franco Fagnola, Claudio Fernández, Carlos Lizama, Carlos Mora, Miguel Orszag, Humberto Prado, Carlos Saavedra, Soledad Torres, and Rolando Rebolledo. Carla Yovane chaired the administrative team composed by María Eliana Altamirano and Elizabeth Llanquitur. We gratefully acknowledge their continued and efficient assistance.

Several institutions and organisations have contributed with their financial support to the success of the Conference. We are particularly grateful for the support of this conference to the Faculty of Mathematics of the Catholic University and the Bicentennial Programme on Science and Technology of the Chilean government, through the grant PBCT-ADI 13.

Rolando Rebolledo and Miguel Orszag

CONTENTS

EXISTENCE OF THE FOCK REPRESENTATION FOR CURRENT ALGEBRAS OF THE GALILEI ALGEBRA

L. ACCARDI*, A. BOUKAS and J. MISIEWICZ

Centro Vito Volterra, Università di Roma Tor Vergata
via Columbia 2, 00133 Roma, Italy
**E-mail: accardi@volterra.mat.uniroma2.it*
http://volterra.mat.uniroma2.it

Department of Mathematics, American College of Greece
Aghia Paraskevi, Athens 15342, Greece
E-mail: andreasboukas@acg.edu

Department of Mathematics, University of Zielona Góra, Poland
E-mail: J.Misiewicz@wmie.uz.zgora.pl

Keywords: Galilei algebra; Heisenberg algebra; Central extension; Fock representation; Positive definite kernel; Infinitely divisible distribution

1. Introduction and statement of the problem

The investigation of the 3 problems stated below has led, in the past 10 years, to a multiplicity of new results and to the discovery of several unexpected connections between different fields of mathematics and physics:

Problem I: construct a *continuous analogue* of the $*$–Lie algebra (and associative $*$–algebra) of differential operators in d variables with polynomial coefficients

$$DOPC(\mathbb{R}^d) := \left\{ \sum_{n\in\mathbb{N}^d} P_n(x)\partial_x^n \ ; \ x \in \mathbb{R}^d \ ; \ P_n \text{ complex polynomials in } d \text{ real variables} \right\}$$

where *continuous* means that the space

$$\mathbb{R}^d \equiv \{\text{functions } \{1, \ldots, d\} \to \mathbb{R}\}$$

is replaced by some function space

$$\{\text{functions } \mathbb{R} \to \mathbb{R}\}$$

Since, for $d = 1$, this algebra can be canonically identified to the universal enveloping algebra of the one–mode Heisenberg algebra $Heis_{\mathbb{C}}(1)$, this problem is equivalent to the old standing problem of constructing a theory of nonlinear quantum (boson) fields: hence its connections with the renormalization problem.

Problem II: construct $*$–representations of this algebra (typically a generalization of the Fock representation) as operators acting on some domain in a Hilbert space $\mathcal{H}$

Problem III: prove the unitarity of these representations, i.e. that the skew symmetric elements of this $*$–algebra can be exponentiated, leading to strongly continuous 1–parameter unitary groups.

The combination of the constructive results obtained in this direction with the no–go theorems have made clear since the early developments of this programme, that the algebra $DOPC(\mathbb{R}^d)$ is *too large* to allow a nontrivial realization of this programme and that one has to restrict one's attention to appropriately chosen subalgebras of it (see the survey paper[7]).

The investigation of the connections of the renormalization problem with the problem of central extensions of $*$–Lie algebras has led to the discovery that the one–mode Heisenberg algebra $Heis_{\mathbb{C}}(1)$ admits a unique non trivial central extension and to its boson representation.

More precisely: let a, $a^\dagger$, h (central element) denote the generators of the one–mode Heisenberg algebra $Heis_{\mathbb{C}}(1)$, with relations

$$(a)^* = a^\dagger \ ; \ h^* = h$$

$$[a, a^\dagger]_{Heis} = h \quad ; \quad [a, h]_{Heis} = [h, a^\dagger]_{Heis} = 0$$

Here and in what follows all omitted commutators are assumed to be equal to zero. We define p and q by the equations

$$a^\dagger = p + i\,q \quad ; \quad a = p - i\,q \tag{1}$$

and we suppose that $a^\dagger, a, h, p, q$ are realized in the Schroedinger representation, where the powers of all these operators are well defined. The following result was proved in[6] and,[2] see also[8] and.[9]

Theorem 1.1. *Up to isomorphisms there exists a unique nontrivial central extension of the Heisenberg algebra $Heis_{\mathbb{C}}(1)$, denoted $CeHeis_{\mathbb{C}}(1)$, with generators $a, a^\dagger, h$, E (central element) and relations*

$$[a, a^\dagger]_{CeHeis} = h \quad ; \quad [h, a^\dagger]_{CeHeis} = E \quad ; \quad [a, h]_{CeHeis} = E$$

$$(a)^* = a^\dagger \quad , \quad h^* = h \quad , \quad E^* = E$$

Furthermore (boson representation of $CeHeis_{\mathbb{C}}(1)$):

(i) one can choose constants $A, B, C, D \in \mathbb{C}$ such that the elements

$$p^2_{CH} := \frac{B\,a^\dagger - \bar{B}\,a}{A\bar{B} - \bar{A}B}$$

$$q_{CH} := \frac{A\,a^\dagger - A\,a}{A\bar{B} - \bar{A}B}$$

$$p_{CH} := h/2iD$$

and the central element form a new set of generators of $CeHeis_{\mathbb{C}}(1)$

(ii) the map

$$(p^2_{CH}, p_{CH}, q_{CH}, E) \mapsto (p^2, p, q, 1) \equiv (q^2, q, p, 1)$$

extends to a $*$–Lie algebra isomorphism between $CeHeis_{\mathbb{C}}(1)$ and the Galilei algebra $(q^2, q, p, 1)$.

Various realizations of the Galilei algebra were well known and studied both in the Mathematical and Physics literature (Bourbaki,[12] Feinsilver and Schott,[14] Franz,[16] Ovando,[18] ...). The apparently new point in the above theorem was the identification of this algebra with the unique nontrivial central extension of the Heisenberg algebra.

The continuous extension of $CeHeis_{\mathbb{C}}(1)$, i.e. the current algebra of this $*$–Lie algebra over $\mathbb{R}$ (all what we say in the following remains valid for $\mathbb{R}^d$ for any $d \in \mathbb{N}$), has generators (in the sense of operator valued distributions, see[11])

$$\{q^2_s \ , \ q_s \ , \ p_t \ : \ s, t \in \mathbb{R}\}$$

with brackets and involution well defined and deduced from the corresponding brackets and involution of the usual (linear) free boson field in momentum representation:

$$[a_t, a^\dagger_s] = \delta(t-s) \ ; \ [a^\dagger_t, a^\dagger_s] = [a_t, a_s] = 0 \tag{2}$$

$$(a_s)^* = a^\dagger_s \tag{3}$$

Therefore, for the Galilei algebra, Problem (I) above is easily solved.

The solution of Problem (II) for this algebra, namely:

Can one construct a generalization of the Fock $*$–representations for the current algebra of $CeHeis_{\mathbb{C}}(1)$ over $\mathbb{R}$?

was open and, by a result of Accardi and Boukas obtained in,[2] was reduced to the proof of the infinite divisibility of the vacuum distribution of the self–adjoint elements of $CeHeis_{\mathbb{C}}(1)$.

The proof of this result for a large set of parameters has been recently obtained in a joint paper (cf.[10]) by L. Accardi, A. Boukas and J. Misiewicz. These results strongly support the conjecture that the statement is true for all values of the parameters involved. The validity of this conjecture is related to some deep results of classical probability. In the following we outline the main steps of the argument.

The first step is to compute the vacuum characteristic function of the self–adjoint elements of $CeHeis_{\mathbb{C}}(1)$ (see section 2 below). Then we split this characteristic function as a product of two factors, one of which is the characteristic function of the Gamma distribution and, by a scaling argument we reduce the proof of infinite divisibility to the proof that the remaining factor is also the characteristic function of some probability distribution i.e., by Bochner's theorem, that it is positive definite and continuous at 0. This last statement is proved by adapting to our goals the Fock space characterization of infinitely divisible measures on Lie groups as developed by Araki, Woods, Parthasarathy, Schmidt, Guichardet (see the monographs,[19][17] and the bibliography therein).

In the Appendix of the present paper we give a synthetic re–formulation, with a few integrations, of the part of the theory that is needed to deduce the above mentioned theorem.

2. Random variables in $CeHeis_{\mathbb{C}}(1)$

As explained in section (1) we suppose that p, q are realized in the one mode Schroedinger representation where we know that the operators of the form

$$X := Lq^2 + Bq + Cp \quad ; \quad L, B, C \in \mathbb{R}$$

are essentially self–adjoint, so that their vacuum characteristic functions

$$\langle \Phi, e^{is\,X} \Phi \rangle \quad ; \quad s \in \mathbb{R}$$

are well defined. In terms of $\{b^{+2}, b^2, b^+b, b^+, b, 1\}$, where $[b, b^+] = 1$, using $p = b^+ + b$ and $q = i(b - b^+)$, and with the notation

$$M := C + iB$$

we obtain

$$X = -Lb^{+2} - Lb^2 + 2Lb^+b + \bar{M}b^\dagger + Mb + L$$

Our first goal is to compute the characteristic function of X. To achieve this goal we use the following splitting formula proved in.[2]

Lemma 2.1. *(Splitting formula) Let $L \in \mathbb{R}$ and $M, N \in \mathbb{C}$. Then for all $s \in \mathbb{R}$ such that $2\,L\,s + 1 > 0$*

$$e^{s\,(L\,b^2 + L\,b^{\dagger 2} - 2\,L\,b^\dagger\,b - L + M\,b + N\,b^\dagger)}\,\Phi = e^{w_1(s)\,b^{\dagger 2}}\,e^{w_2(s)\,b^\dagger}\,e^{w_3(s)}\,\Phi \tag{4}$$

where

$$w_1(s) = \frac{L\,s}{2\,L\,s + 1} \tag{5}$$

$$w_2(s) = \frac{L\,(M+N)\,s^2 + N\,s}{2\,L\,s + 1} \tag{6}$$

$$w_3(s) = \frac{(M+N)^2\,(L^2\,s^4 + 2\,L\,s^3) + 3\,M\,N\,s^2}{6\,(2\,L\,s + 1)} - \frac{\ln(2\,L\,s + 1)}{2} \tag{7}$$

Therefore, putting $N := \bar{M}$ in (7) and replacing L by $-L$ in Lemma 1 we obtain

$$e^{sX}\,\Phi = e^{s\,(-L\,b^2 - L\,b^{\dagger 2} + 2\,L\,b^\dagger\,b + L + M\,b + \bar{M}\,b^\dagger)}\,\Phi = e^{w_1(s)\,b^{\dagger 2}}\,e^{w_2(s)\,b^\dagger}\,e^{w_3(s)}\,\Phi \tag{8}$$

where

$$w_1(s) = \frac{L\,s}{2\,L\,s - 1} \tag{9}$$

$$w_2(s) = \frac{2\,C\,L\,s^2 - \bar{M}\,s}{2\,L\,s - 1} \tag{10}$$

$$w_3(s) = \frac{4\,C^2(L^2\,s^4 - 2\,L\,s^3) + 3|M|^2\,s^2}{6\,(1 - 2\,L\,s)} - \frac{\ln(1 - 2\,L\,s)}{2} \tag{11}$$

By analytic continuation (in s)

$$e^{w_3(is)} = (1 - 2\,L\,i\,s)^{-1/2}e^{\frac{4\,C^2(L^2\,s^4 + 2\,i\,L\,s^3) - 3\,|M|^2\,s^2}{6\,(1 - 2\,i\,L\,s)}} \tag{12}$$

In conclusion: the characteristic function of the vacuum distribution of the operator random variable

$$X = Lq^2 + Bq + Cp$$

is

$$\langle \Phi, e^{is\,(Lq^2 + Bq + Cp)}\,\Phi \rangle = (1 - 2\,L\,is)^{-1/2}e^{\frac{4\,C^2(L^2\,s^4 + 2i\,L\,s^3) - 3|M|^2\,s^2}{6\,(1 - 2\,i\,L\,s)}} \tag{13}$$

$$=: (1 - 2\,L\,is)^{-1/2}e^{\psi_L(s,M)}$$

3. Infinite divisibility

Recall, from equation (13), the definition

$$\psi_L(s,M) := \frac{4Re(M)^2 Ls^3\,(Ls+2i) - 3|M|^2 s^2}{6\,(1-2iLs)} \tag{14}$$

and notice two important remarks

(i) (scaling property) for every $t \in \mathbb{R}_+$

$$\psi_L(s,\sqrt{t}M) := t\frac{4Re(M)^2 Ls^3\,(Ls+2i) - 3|M|^2 s^2}{6\,(1-2iLs)} = t\psi_L(s,M) \tag{15}$$

(ii) for every $L \in \mathbb{R}$ and $M \in \mathbb{C}$,

$$\varphi_{L,M}(s) := (1 - 2\,L\,is)^{-1/2} e^{\psi_L(s,M)} \tag{16}$$

where $\varphi_L(s,M)$, given by the left hand side of (13), is a characteristic function for every $t \in \mathbb{R}_+$ and for every $L \in \mathbb{R}$ and $M \in \mathbb{C}$.

Denoting

$$\varphi(s) := (1 - 2\,L\,is)^{-1/2}$$

we have that

$$\varphi_{L,M}(s) := \varphi(s) e^{\psi_L(s,M)} \tag{17}$$

Suppose that one can prove that, for every $L \in \mathbb{R}$ and $M \in \mathbb{C}$, the function

$$e^{\psi_L(s,M)}$$

is a characteristic function. Then this characteristic function is infinitely divisible because the scaling property implies that, for any $t \geq 0$:

$$(e^{\psi_L(s,M)})^t = e^{t\psi_L(s,M)} = e^{\psi_L(s,\sqrt{t}M)}$$

which is a characteristic function because of our assumption.

Therefore the left hand side of (16), which is already known to be a characteristic function, is infinitely divisible being the product of two infinitely divisible characteristic functions.

3.1. *A particular case*

Theorem 3.1. *If*

$$C = Re(M) = 0 \Leftrightarrow M \in i\mathbb{C}$$

the vacuum characteristic functions given by (13) are infinitely divisible with cumulant generating function given by

$$-i\frac{|M|^2}{4L}s - \int_0^\infty (1 - e^{isu})\frac{1/2}{u}e^{-2Lu}du - \frac{|M|^2}{8L^2}\int_0^\infty (1 - e^{isu})e^{-2Lu}du \quad (18)$$

Remark: Formula (18) gives also the explicit form of the associated Levy measure.

Appendix: Infinitely divisible kernels

4. Kernels and matrices

Definition 4.1. Let S be a set, a function $K : (x, y) \in S \times S \to k(x, y) \in \mathbb{C}$ is called a $\mathbb{C}$*–valued kernel* on S (only kernel if no ambiguity is possible). If S is a finite set, then there exists a $d \in \mathbb{N}$ such that, up to relabeling the elements of S can be identified to the subset $\{1, \ldots, d\} \subseteq \mathbb{N}$. With this identification a kernel on S is identified to the $d \times d$ complex matrix

$$k(i, j) =: k_{i,j} \quad ; \quad i, j \in \{1, \ldots, d\}$$

4.1. *Positive definite matrices*

In the following $\mathbb{C}$ will be considered as a Hilbert space with scalar product $\langle x, y\rangle := \bar{x}y$ so that $\mathcal{B}(\mathbb{C})$ is identified to $\mathbb{C}$ acting on itself by multiplication. For $d \in \mathbb{N}$ and $k = (k_{ij}) \in M_d(\mathbb{C})$ and $x_1, \cdots, x_d \in \mathbb{C}$ we will use the notation

$$\bar{x}_i k_{ij} x_j := \sum_{i,j=1}^{d} \bar{x}_i k_{ij} x_j = \langle x, ky\rangle$$

(i.e. we assume summation over repeated indices) where $\langle \cdot, \cdot \rangle$ denotes the hermitian scalar product on $\mathbb{C}^d$

$$\langle x, y\rangle := \sum_{j=1}^{d} \bar{x}_j y_j \quad ; \quad x = (x_1, \cdots, x_d)\ ,\ y = (y_1, \cdots, y_d) \in \mathbb{C}^d$$

$\mathbb{R}^d$ is identified to the subspace of $\mathbb{C}^d$ obtained as range of the projection

$$x = (x_1, \cdots, x_d) \in \mathbb{C}^d \mapsto Re(x) := (Re(x_1), \cdots, Re(x_d))$$

The restriction on $\mathbb{R}^d$ of the hermitian scalar product on $\mathbb{C}^d$ is real valued. The $d \times d$ complex matrices $M_d(\mathbb{C})$ act naturally on $\mathbb{C}^d$; this action induces an action of the $d \times d$ real matrices $M_d(\mathbb{R})$ on $\mathbb{C}^d$.

The adjoint k^* of a matrix $k = (k_{ij})$ is defined by

$$\langle kx, y\rangle = \langle x, k^*y\rangle \quad ; \quad x = (x_1, \cdots, x_d)\ ,\ y = (y_1, \cdots, y_d) \in \mathbb{C}^d$$

Thus

$$(k^*)_{ij} = \overline{k_{ji}}$$

Definition 4.2. A matrix $k = (k_{ij}) \in M_d(\mathbb{C})$ is called:

(i) Hermitian if

$$\langle x, kx\rangle = \bar{x}_i k_{ij} x_j \in \mathbb{R} \qquad ; \qquad \forall x_1, \cdots, x_d \in \mathbb{C}$$

(ii) skew–Hermitian if

$$\langle x, kx\rangle = \bar{x}_i k_{ij} x_j = 0 \qquad ; \qquad \forall x_1, \cdots, x_d \in \mathbb{C}$$

(iii) symmetric if it is Hermitian and with real coefficients, i.e.

$$k = (k_{ij}) \in M_d(\mathbb{R})$$

(iv) symplectic if it is skew–Hermitian and with real coefficients, i.e.

$$k = (k_{ij}) \in M_d(\mathbb{R}) \quad ; \quad \langle x, kx\rangle = \bar{x}_i k_{ij} x_j = 0 \quad ; \quad \forall x_1, \cdots, x_d \in \mathbb{C}$$

Lemma 4.1. *A matrix $k = (k_{ij}) \in M_d(\mathbb{C})$ is:*

(i) Hermitian if and only if

$$k_{ji} = \overline{k_{ij}} \Leftrightarrow k = k^* \qquad ; \qquad \forall x_1, \cdots, x_d \in \mathbb{C} \tag{19}$$

In particular $k = (k_{ij}) \in M_d(\mathbb{C})$ is Hermitian if and only if its restriction on $\mathbb{R}^d$ is Hermitian.

(ii) skew–Hermitian if and only if its diagonal coefficients are zero and

$$k_{ji} = -\overline{k_{ij}} \Leftrightarrow k = -k^* \quad ; \quad \forall x_1, \cdots, x_d \in \mathbb{C}\ ;\ i \neq j \tag{20}$$

(iii) symplectic if and only if $\forall i, j \in \{1, \cdots, d\}$

$$k_{ij} = -k_{ji}$$

(iv) skew–Hermitian and Hermitian if and only if its diagonal coefficients are zero and its off–diagonal coefficients are purely imaginary and satisfy (19).

Proof. Let $k = (k_{ij}) \in M_d(\mathbb{C})$. k is Hermitian if and only if $\forall x_1, \cdots, x_d \in \mathbb{C}$

$$\langle x, kx\rangle = \bar{x}_i k_{ij} x_j = (\bar{x}_i k_{ij} x_j)^- = x_i \bar{k}_{ij} \bar{x}_j = \bar{x}_i \bar{k}_{ji} x_j$$

Since $x_1, \cdots, x_d \in \mathbb{C}$ are arbitrary, this is equivalent to (19). The restriction to $\mathbb{R}^d$ of the above identity gives

$$x_i k_{ij} x_j = x_i \bar{k}_{ji} x_j$$

From this (19) follows by choosing $\forall i_0, j_0 \in \{1, \cdots, d\}$

$$x_j = \delta_{i_0,i} \quad ; \quad y_i = \delta_{j_0,j}$$

k is skew–Hermitian if and only if $\forall x_1, \cdots, x_d \in \mathbb{C}$

$$0 = \langle x + y, k(x + y)\rangle = \langle x, ky\rangle + \langle y, kx\rangle \Leftrightarrow \langle x, ky\rangle = -\langle y, kx\rangle$$

$$\Leftrightarrow \bar{x}_i k_{ij} y_j = -\bar{y}_j k_{ji} x_i$$

Fixing $i_0, j_0 \in \{1, \cdots, d\}$ and choosing

$$x_j = \delta_{i_0,i} \quad ; \quad y_i = \delta_{j_0,j}$$

one finds

$$k_{i_0 j_0} = -k_{j_0 i_0}$$

Finally $k = (k_{ij}) \in M_d(\mathbb{C})$ is skew–Hermitian and Hermitian if and only if

$$k_{ji} = -k_{ij} = -\bar{k}_{ji} \quad ; \quad \forall i, j \in \{1, \cdots, d\}$$

i.e. $k_{ii} = 0$ and k_{ji} is purely imaginary for $i \neq j$. □

Lemma 4.2. *A matrix $h = (h_{ij}) \in M_d(\mathbb{C})$ is Hermitian if and only if the matrix $Re(h) := (Re(h_{ij})) \in M_d(\mathbb{R})$ is symmetric, i.e.*

$$Re(h_{ij}) = Re(h_{ji}) \tag{21}$$

and the matrix $Im(h) := (Im(h_{ij})) \in M_d(\mathbb{R})$ is symplectic, i.e.

$$Im(h_{ij}) = -Im(h_{ji}) \tag{22}$$

Proof. $h = (h_{ij})$ is Hermitian if and only if

$$\mathrm{Re}(h_{ij}) - i\mathrm{Im}(h_{ij}) = \bar{h}_{ij} = h_{ji} = \mathrm{Re}(h_{ji}) + i\mathrm{Im}(h_{ji})$$

and this is equivalent to the identities (21) and (22). □

Definition 4.3. A matrix $(k_{ij}) \in M_d(\mathbb{C})$ is said to be positive definite if

$$\sum_{i,j=1}^{d} x_i^* k_{ij} x_j \geq 0 \tag{23}$$

$\forall\, x_1, \ldots, x_d \in \mathbb{C}$. If in addition

$$\sum_{i,j=1}^{d} x_i^* k_{ij} x_j = 0 \Leftrightarrow x_1 = \cdots = x_d = 0 \tag{24}$$

then the matrix (k_{ij}) $(i, j = 1, \ldots, d)$ is called strictly positive definite.

Remark: Note that the diagonal elements of a positive definite matrix are positive because $\forall\, x \in \mathbb{C}$

$$x^* k_{ii} x \geq 0$$

Moreover, considering (4.3) with $d = 2$, one sees that a positive definite kernel on S always satisfies the Schwarz inequality

$$|K(x_1, x_2)|^2 \leq K(x_1, x_1) K(x_2, x_2) \tag{25}$$

Lemma 4.3. *$k \in M_d(\mathbb{C})$ is PD if and only if for every skew–Hermitian matrix $S \in M_d(\mathbb{C})$, $k + S$ is PD*

Proof. Clear from the definition of PD. □

Remark: If k is positive definite and Hermitian and $S \in M_d(\mathbb{C})$ is skew–Hermitian, then $k + S$ is PD but it is Hermitian if and only if S is. If this is the case, being also skew–Hermitian, by Lemma (4.1) S must have zero diagonal coefficients and off–diagonal coefficients purely imaginary and satisfying (19).
Thus there exist matrices which are positive definite but not Hermitian.

Lemma 4.4. *A matrix $k \in M_d(\mathbb{C})$ is positive definite and Hermitian if and only if there exist two Hermitian matrices $a, b \in Herm_d(\mathbb{C})$ such that*

$$k = (a^2 + b^2) + i(ab - ba) \tag{26}$$

Remark: In particular any such k has the form

$$k = k_{H,+} + ik_S \tag{27}$$

with $k_{H,+}$ PDH and k_S skew adjoint. The decomposition (26) is highly non unique, reflecting the non uniqueness of the square root.

Proof. k is positive definite and Hermitian if and only if there exists a matrix $x \in M_d(\mathbb{C})$ such that $k = x^*x$. Writing

$$x = \frac{1}{2}(x + x^*) + \frac{1}{2}(x - x^*) =: a + ib \tag{28}$$

by construction a and b are Hermitian. Therefore k is positive definite if and only if it has the form

$$k = (a + ib)^*(a + ib) = (a - ib)(a + ib) = (a^2 + b^2) + i(ab - ba) \qquad \square$$

Remark: Using the identity (28) any matrix k can be written in the form

$$k = k_H + ik_S = \mathrm{Re}(k_H) + i\mathrm{Im}(k_H) + i\mathrm{Re}(k_S) - \mathrm{Im}(k_S) \tag{29}$$

where

$$k_H = k_H^* \qquad ; \qquad k_S = k_S^*$$

$\mathrm{Re}(k_H)$, $\mathrm{Re}(k_S)$ are real Hermitian (symmetric) and $\mathrm{Im}(k_H)$, $\mathrm{Im}(k_S)$ are real skew–symmetric (symplectic).

Lemma 4.5. *$k = (k_{ij})$ is positive definite if and only if, in the notation (29):*

$$Re(k_S) = 0 \tag{30}$$

$$Re(k_H) \quad \text{is positive definite Hermitian} \tag{31}$$

$k = (k_{ij})$ is positive definite and Hermitian if and only if

$$k_S = 0 \tag{32}$$

Remark: In particular any positive definite Hermitian $k = (k_{ij})$ has the form

$$k = k_R + ik_I \tag{33}$$

where $k_R = \mathrm{Re}(k_H)$ is real PDH and $k_I = \mathrm{Im}(k_H)$ real Hermitian.

Proof. The decomposition (29) and the positive definiteness of k imply that

$$0 \leq \langle x, kx\rangle = \langle x, \mathrm{Re}(k_H)x\rangle + \langle x, i\mathrm{Im}(k_H)x\rangle + \langle x, i\mathrm{Re}(k_S)x\rangle - \langle x, \mathrm{Im}(k_S)x\rangle$$

$$= \langle x, \mathrm{Re}(k_H)x\rangle + i\langle x, \mathrm{Re}(k_S)x\rangle$$

In particular the right hand side is real and, by Lemma (4.2), $\mathrm{Re}(k_H) \in M_d(\mathbb{R})$ is symmetric, so that:

$$\begin{aligned}\langle x, \mathrm{Re}(k_H)x\rangle + i\langle x, \mathrm{Re}(k_S)x\rangle &= \langle \mathrm{Re}(k_H)x, x\rangle - i\langle \mathrm{Re}(k_S)x, x\rangle \\ &= \langle x, \mathrm{Re}(k_H)x\rangle - i\langle x, \mathrm{Re}(k_S)x\rangle \\ \Leftrightarrow \langle x, \mathrm{Re}(k_S)x\rangle &= -\langle x, \mathrm{Re}(k_S)x\rangle \Leftrightarrow \langle x, \mathrm{Re}(k_S)x\rangle = 0\end{aligned}$$

But, since $\mathrm{Re}(k_S)$ is real Hermitian, it can be symplectic if and only if it is zero, i.e. (30) holds. In this case

$$0 \leq \langle x, \mathrm{Re}(k_H)x\rangle$$

which is (31). Finally $k = (k_{ij})$ is positive definite and Hermitian if and only if

$$\mathrm{Re}(k_H) + i\mathrm{Im}(k_H) - \mathrm{Im}(k_S) = k = k^* = \mathrm{Re}(k_H)^* - i\mathrm{Im}(k_H)^* - \mathrm{Im}(k_S)^* =$$

$$= \mathrm{Re}(k_H) + i\mathrm{Im}(k_H) + \mathrm{Im}(k_S) \Leftrightarrow 2\mathrm{Im}(k_S) = 0 \Leftrightarrow \mathrm{Im}(k_S) = 0 \qquad \square$$

Lemma 4.6. *Any positive definite matrix k can be written in the form*

$$k = Re(k_H) + iIm(k_H) - Im(k_S) \tag{34}$$

where $Re(k_H)$ is a real positive definite matrix and $Im(k_H)$ $Im(k_S)$ are real skew–symmetric. Conversely if $Re(k_H)$, $Im(k_H)$ and $Im(k_S)$ have these properties, then the right hand side of (34) is a positive definite matrix.

Proof. Clear from Lemma (4.5). $\square$

5. Positive definite kernels

Definition 5.1. A kernel $k(\cdot,\cdot)$ on S is called positive definite if $\forall n \in \mathbb{N}$, $\forall x_1, \ldots, x_n \in S$, $\forall \xi_1, \ldots, \xi_n \in \mathbb{N}$

$$\xi_k^* k(x_k, x_j)\xi_j \geq 0 \tag{35}$$

i.e. if the matrix $(k(x_i, x_j))$ is positive definite.

Remark: From (35) it is clear that PD kernels on S are a cone, closed under pointwise convergence.

Lemma 5.1. *Given a kernel k on S, define the kernels $\bar{k}$, k^* respectively by*

$$\bar{k}(x, y) := k(x, y)^* \tag{36}$$

$$k^*(x,y) := k(y,x)^* \tag{37}$$

Then one of these kernels is PD (resp. CPD) if and only if the other two are.

Proof.

$$\xi_j^* k^*(x_j,x_k)\xi_k = \xi_j^* k(x_k,x_j)^*\xi_k = (\xi_k^* k(x_k,x_j)\xi_j)^* \geq 0 \Leftrightarrow \xi_k^* k(x_k,x_j)\xi_j \geq 0$$

Thus k is PD (resp. CPD) if and only if k^* is. Similarly

$$\xi_j^* \bar{k}(x_j,x_k)\xi_k = \xi_j^* k(x_j,x_k)^*\xi_k = (\xi_k^* k(x_j,x_k,)\xi_j)^* = (\xi_j k(x_j,x_k,)\xi_k^*)^* \geq 0 \Leftrightarrow$$

$$\xi_j k(x_j,x_k,)\xi_k^* \geq 0$$

Thus k is PD (resp. CPD) if and only if $\bar{k}$ is. □

Corollary 5.1. *If k is a PD (resp. CPD) kernel then its real part defined by*

$$Re(k)(x,y) := \frac{1}{2}(k(x,y) + k(y,x)^*)$$

is PD.

Corollary 5.2. *If $k(f,g)$ is a positive definite kernel on S then, for any family $t_f \in \mathbb{C}$, also the kernel*

$$t_f^* k(f,g) t_g$$

is positive definite. Conversely, if there exists a map

$$c : f \in G \to c_f \in \mathbb{C} \setminus \{0\}$$

(i.e. with all c_f's invertible) such that the kernel

$$H(f,g) := c_f^* k(f,g) c_g \tag{38}$$

is PD, then the kernel $k(f,g)$ itself is PD.

Proof. Let $k(f,g)$ be positive definite. Then it is clear that, for any family $t_f \in \mathbb{C}$, also the kernel

$$k_t(f,g) := t_f^* k(f,g) t_g = k(f,g) t_f^* t_g$$

is PD. Conversely, if the kernel $H(f,g)$, defined by (38) is PD then, by the first part of the Lemma, also the kernel

$$k(f,g) = (c_f^*)^{-1} H(f,g) c_g^{-1} = (c_f^{-1})^* H(f,g) c_g^{-1}$$

is PD. □

Definition 5.2. Two positive definite kernels on S, k_1, k_2 are called equivalent if for some function
$t : S \to \mathbb{C} \setminus \{0\}$

$$k_1(f,g) = t_f^* k_2(f,g) t_g$$

5.1. *Schur's lemma*

Lemma 5.2. *(Schur) If $H = (H_{jk}), k = (k_{jk})$ are PD then their Schur product*

$$H \circ k := (H_{jk} k_{jk})$$

is PD. In particular the cone of PD kernels on S is a semigroup under Schur multiplication with identity given by

$$(1)_{ij} := (1^*)_{ij} := 1 \qquad ; \qquad \forall i,j \in S$$

Proof By the spectral theorem for matrices

$$a = x^* x \qquad : \qquad b = y^* y$$

$$\overline{\lambda}_i a_{ik} b_{ik} \lambda_k = \overline{\lambda}_i x_{ij}^* x_{jk} y_{il}^* y_{lk} \lambda_k = \overline{\lambda}_i \overline{x_{ji} y_{li}} x_{jk} y_{lk} \lambda_k = |x_{jk} y_{lk} \lambda_k|^2 \geq 0$$

Lemma 5.3. *The family of $\mathbb{C}$–valued positive definite matrices (kernels on S) is closed under:*
- pointwise multiplication (Schur's Lemma)
- pointwise addition
- pointwise multiplication by a positive scalar
- pointwise limits
- integrals

Proof. From the definition of positive-definiteness. □

Definition 5.3. A Borel function $f : \mathbb{C} \to \mathbb{C}$ is called *completely positive (Hermitian)* if $\forall d \in \mathbb{N}$ and for any positive definite (Hermitian) matrix $A = (a_{ij})$ the matrix

$$f \circ A := (f(a_{ij}))$$

is positive definite (Hermitian)

Remark: The above definition is equivalent to say that, for any set S and for any kernel k on S, the composite kernel on S

$$f \circ k =: f(k)$$

is positive definite (Hermitian).

Remark: An interesting open problem is to characterize the completely positive (Hermitian) Borel functions $f : \mathbb{C} \to \mathbb{C}$. The following is an example.

Corollary 5.3. *The exponential function on $\mathbb{C}$ is completely positive and Hermitian.*

Proof. We have to prove that, if k is a positive definite (Hermitian) kernel on S, then its exponential kernel has these properties. By Schur's Lemma $\forall n \in \mathbb{N}$ $k^{\circ n}$ (Schur's power) is positive definite. Lemma (5.3) implies that e^k is positive definite. Moreover, if k is Hermitian then $\forall x, y \in S$

$$(e^{k(x,y)})^* = e^{\overline{k(x,y)}} = e^{k(y,x)}$$

Therefore e^k is Hermitian. □

Definition 5.4. A kernel h is called exponential if it is of the form $h = e^k$ where k is a kernel on S. In this case h is called the exponential kernel of k.

6. Functions of positive definite matrices and kernels

Some PD kernels have the property that their Schur inverses are also PD. This however is not a general property. In the present section we investigate this problem.

Theorem 6.1. *Let $A = (a_{ij})$ be any Hermitian complex matrix. Then there exists an $\varepsilon_A > 0$ such that $\forall \varepsilon \in [-\varepsilon_A, \varepsilon_A]$*

$$1 + \varepsilon A$$

is invertible and both $1 + \varepsilon A$ and its inverse are PD and Hermitian.

Proof. Since, for any unitary matrix U, the properties of being PDH are invariant under the transformation $A \mapsto U^* A U$, we can suppose that A is diagonal, say $A = diag(s_1, \ldots, s_n)$, with the s_j real. Then

$$1 + \varepsilon A = diag(1 + \varepsilon s_1, \ldots, 1 + \varepsilon s_n)$$

is Hermitean, PD and, for $|\varepsilon|$ small enough, invertible. For the same ε:

$$(1+\varepsilon A)^{-1}_{ij} = \delta_{ij}(1+\varepsilon s_j)^{-1} \geq 0$$

Thus $(1+\varepsilon A)^{-1}$ is also positive and this proves the thesis. □

Theorem 6.2. *Let $A \equiv (a_{ij})$ be an Hermitian matrix. Then there exists an $\varepsilon_A > 0$ such that $\forall t \in (-\varepsilon_A, \varepsilon_A)$ the matrix $e^{\circ(1+tA)} \equiv (e^{\delta_{ij}} e^{ta_{ij}})$ is positive definite and Hermitian.*

Proof Let (a_{ij}) be as in the statement and let $\varepsilon_A > 0$ be as in Theorem (6.1). Then, $\forall t \in (-\varepsilon_A, \varepsilon_A)$, the matrix $(1+tA) \equiv (\delta_{ij} + ta_{ij})$ is Hermitean PD hence, by Schur's lemma the same is true for

$$\frac{(1+tA)^{\circ n}}{n!} \equiv \frac{((\delta_{ij} + ta_{ij})^n}{n!} \qquad ; \qquad \forall n \in \mathbb{N}$$

Summing the exponential series one obtains the thesis.
Remark. As shown by the following lemma, the matrix $(e^{\delta_{ij}})$ is a typical example of a matrix which is PD and Schur–invertible but whose Schur–inverse is not PD.

Lemma 6.1. *Let $C > 1$ be a real number. Then the matrix $(C^{\delta_{ij}})$ is PD and Schur–invertible but its Schur–inverse is not PD. If $C = 1$, then $(C^{\delta_{ij}})$ is the Schur identity, which coincides with its inverse.*

Proof. The entries of $(C^{\delta_{ij}})$ are either 1 or $C \neq 0$. So $(C^{\delta_{ij}})$ is Schur–invertible. Moreover

$$\bar{x}_i(C^{\delta_{ij}})x_j = \sum_{i\neq j} \bar{x}_i x_j + C\sum_i |x_i|^2 = |\sum_i x_i|^2 + (C-1)\sum_i |x_i|^2 \geq 0$$

so it is PD because $C > 1$. If $C < 1$, then for $\sum_i x_i = 0$ the above sum takes generically negative values. Thus the inverse of $(C^{\delta_{ij}})$ is not PD.

Corollary 6.1. *If K is a positive definite Hermitian kernel on S then, for any $0 \leq a < b \in \mathbb{R}_+$, the kernel on S*

$$e^{bK(x,y)} - e^{aK(x,y)}$$

is positive definite Hermitian.

Remark: In particular the kernels

$$e^{K(x,y)} - 1$$

$$\sinh(tK(x,y)) \qquad ; \qquad \forall t \in \mathbb{R}_+$$

are PD Hermitian (the PD and Hermitianity of $\cosh(tK(x,y))$ follows from Lemma (5.3).

Proof. The kernels $K(x,y)$ and ($\forall t \in \mathbb{R}$) are positive definite Hermitian. From this it follows that

$$\int_a^b K(x,y)e^{tK(x,y)}dt = e^{bK(x,y)} - e^{aK(x,y)}$$

is PD Hermitian. $\square$

7. Conditionally positive definite kernels

Lemma 7.1. *If, $\forall t \in \mathbb{R}_+$ the kernel $(e^{tk_{ij}})$ is PD, then the kernel (k_{ij}) is such that for any finite family of complex numbers (λ_j) satisfying*

$$\sum_j \lambda_j = 0 \tag{39}$$

one has:

$$\sum_{j,k} \overline{\lambda}_i k_{ij} \lambda_j \geq 0 \tag{40}$$

Such a k admits a unique representation of the form

$$k = k_0 + S \tag{41}$$

with $k_0 = k_0^*$ is positive definite and S real symplectic.

In the representation (41), $S = 0$ if and only if $\forall t \in \mathbb{R}_+$ the kernel $(e^{tk_{ij}})$ is PD and Hermitian,

Proof. We begin proving (40). Suppose that $\forall t \in \mathbb{R}_+$ the kernel $(e^{tk_{ij}})$ is PD, i.e. that $\forall t \in [0,T]$

$$\overline{\lambda}_i e^{tk_{ij}} \lambda_j \geq 0$$

then, whenever condition (39) is satisfied, one has, $\forall t \in [0,T]$, assuming summation over repeated indices:

$$0 \leq \overline{\lambda}_i e^{tk_{ij}} \lambda_j = \overline{\lambda}_i e^{tk_{ij}} \lambda_j - \overline{\lambda}_i \lambda_j$$

therefore, for $t \geq 0$:

$$0 \leq \overline{\lambda}_i \frac{(e^{tk_{ij}} - 1)}{t} \lambda_j \tag{42}$$

and, letting $t \downarrow 0$, (40) follows.

Finally if, $\forall t \in \mathbb{R}_+$ the kernel $(e^{tk_{ij}})$ is Hermitian, then

$$\frac{d}{dt}|_{t=0}(e^{tk_{ij}})^* = (\frac{d}{dt}|_{t=0}e^{tk_{ij}})^* = \overline{k_{i,j}} = \frac{d}{dt}|_{t=0}e^{tk_{j,i}} = k_{j,i}$$

i.e. k is self–adjoint, which implies $S = 0$.

To prove the hermitianity of the kernel $k = (k_{ij})$ denote $\forall i, j \in S$

$$k_{ij} =: Re(k_{ij}) + iIm(k_{ij}) =: k_{H,ij} + ik_{S,ij} =: k_H(ij) + ik_S(ij) \qquad (43)$$

with $k_{H,jj}, k_{S,jj} \in \mathbb{R}$. Since $\forall t \in \mathbb{R}_+$ the kernel $(e^{tk_{ij}})$ is PD, by a previous Remark,

$$e^{tk_{jj}} = e^{tk_{H,jj}}(\cos(tk_{S,jj}) + i\sin(tk_{S,jj}) \geq 0 \quad ; \quad \forall j \in S \; ; \; \forall t \geq 0$$

which is possible if and only if $\forall t \geq 0$ and $\forall j \in X$

$$\sin(tk_{S,jj}) = 0 \quad \Leftrightarrow \quad k_{S,jj} = 0$$

and this proves that

$$k_{ii} \in \mathbb{R} \quad ; \quad \forall i \in S \qquad (44)$$

Finally condition (40) implies in particular that $\forall \lambda \in \mathbb{C}$ and $\forall i, j \in X$

$$0 \leq \bar{\lambda}k(i,i)\lambda + \lambda k(j,j)\bar{\lambda} - \bar{\lambda}k(i,j)\lambda - \bar{\lambda}k(j,i)\lambda$$

and, since $k_S(i,i) = 0$ because of (44), putting $\lambda = 1$, this becomes

$$\begin{aligned} 0 &\leq k(i,i) + k(j,j) - k(i,j) - k(j,i) \\ &= k_H(i,i) + k_H(j,j) - k_H(i,j) - k_H(i,j) - i(k_S(i,j) + k_S(j,i)) \end{aligned}$$

Thus

$$k_S(i,j) + k_S(j,i) = 0 \Leftrightarrow k_S(i,j) = -k_S(j,i) \qquad (45)$$

Thus we can write

$$k_{ij} = k_{H,i,j} + ik_{S,i,j} = \frac{1}{2}(k_{H,i,j} + k_{H,j,i}) + \frac{1}{2}(k_{H,i,j} - k_{H,j,i}) + ik_{S,i,j}$$

and defining

$$k_{0,i,j} := \frac{1}{2}(k_{H,i,j} + k_{H,j,i}) + ik_{S,i,j} = \overline{k_{0,j,i}}$$

$$S_{i,j} := \frac{1}{2}(k_{H,i,j} - k_{H,j,i})$$

we see that k_0 is self–adjoint, S real symplectic and (41) is satisfied. If $k = k_0' + S'$ is another representation of the form (41), then

$$k + S = k_0' + S' \Leftrightarrow k - k_0' = S' - S$$

This implies that

$$Im(k - k_0') = 0 \Leftrightarrow Im(k) = Im(k_0')$$

$$Re(k - k_0') = S' - S$$

But the left hand side is symmetric and the right hand side is anti–symmetric. Hence both sides must be zero and the representation (41) is unique. □

Definition 7.1. In the notations of Lemma (7.1) a complex matrix $(k_{ij}) \in M_d(\mathbb{C})$ is said to be conditionally positive definite (CPD) if it satisfies the necessary conditions of Lemma (7.1), i.e. it has the form (41) and

$$\sum_{j,k} \overline{\lambda}_i k_{ij} \lambda_j \geq 0 \tag{46}$$

for any finite family of complex numbers (λ_j) such that

$$\sum_j \lambda_j = 0 \tag{47}$$

If in addition k is Hermitian, we say that k is CPD Hermitian.

Remark: From (41) it follows that, if k is CPD then its diagonal elements are real.

Remark: If $f : S \to \mathbb{C}$ is any function, the kernels

$$k(x, y) := f(x) + f(y) \quad ; \quad f(x) - f(y)$$

are CPD and the kernel

$$k(x, y) := f(x) + \overline{f(y)}$$

is CPD Hermitian. All these kernels are called *trivial.*

Definition 7.2. A kernel $k(\cdot, \cdot)$ on E is called conditionally positive definite if $\forall\, n \in \mathbb{N}$, $\forall\, x_1, \ldots, x_n$, the matrix $(k(x_i, x_j))$ is conditionally positive definite.

Lemma 7.2. *Let k be a CPD kernel on X and, for any pair of functions $\alpha, \beta : X \to \mathbb{C}$, define the kernel*

$$h(x,y) := k(x,y) + \alpha(x) + \beta(y) \tag{48}$$

Then the kernel (48):

(i) always satisfies condition (46) for any finite family of complex numbers (λ_j) satisfying (47)

(ii) is CPD if and only if there exist real valued functions

$$a, b, \gamma : x \in S \to a(x), b(x), \gamma(x) \in \mathbb{R} \tag{49}$$

such that

$$\alpha(x) = a(x) - i\gamma(x) \quad ; \quad \beta(x) = b(x) + i\gamma(x) \tag{50}$$

(iii) is CPD Hermitian if and only if, in the decomposition (41) of k,S is trivial

(iv) if k is Hermitian, h is Hermitian if and only if for some constant $c \in \mathbb{R}$ one has

$$\beta(x) = \overline{\alpha}(x) + c \tag{51}$$

In this case, denoting

$$\alpha_0(x) := \frac{1}{2}c + \alpha(x)$$

one has

$$h(x,y) := k(x,y) + \alpha_0(x) + \overline{\alpha_0}(y) \tag{52}$$

and h is CPD if and only if k is.

Proof. (i) It is clear that, if k is CPD, then for any choice of the functions α and β, the right hand side of (48) satisfies condition (46) for any finite family of complex numbers (λ_j) satisfying (47).

(ii) From the decomposition (41) of k one deduces that, for h to be CPD, one must have the identity

$$h(x,y) = k_0(x,y) + S(x,y) + \alpha(x) + \beta(y) = h_0(x,y) + S_0(x,y) \tag{53}$$

for some $h_0 = h_0^*$ and S_0 real symplectic. This is equivalent to

$$k_0(x,y) - h_0(x,y) = S_0(x,y) - S(x,y) - \alpha(x) - \beta(y)$$

Taking adjoints this gives

$$\overline{k_0(x,y)} - \overline{h_0(x,y)} = S_0(x,y) - S(x,y) - \overline{\alpha(x)} - \overline{\beta(y)}$$

Since h_0 and k_0 are Hermitian, this implies that

$$S_0(x,y) - S(x,y) - \overline{\alpha(x)} - \overline{\beta(y)} = \overline{k_0(x,y)} - \overline{h_0(x,y)} = k_0(y,x) - h_0(y,x) =$$

$$= S_0(y,x) - S(y,x) - \alpha(y) - \beta(x) = -S_0(x,y) + S(x,y) - \alpha(y) - \beta(x) \Leftrightarrow$$

$$\Leftrightarrow 2S_0(x,y) - 2S(x,y) = \overline{\alpha(x)} - \alpha(y) + \overline{\beta(y)} - \beta(x)$$

exchanging the roles of x and y one has

$$2S_0(y,x) - 2S(y,x) = 2S(x,y) - 2S_0(x,y) = \overline{\alpha(y)} - \alpha(x) + \overline{\beta(x)} - \beta(y)$$

Thus the right hand sides of the two equations must be opposite:

$$\overline{\alpha(x)} - \alpha(y) + \overline{\beta(y)} - \beta(x) = -\overline{\alpha(y)} + \alpha(x) - \overline{\beta(x)} + \beta(y) \Leftrightarrow$$

$$\Leftrightarrow 0 = (\alpha(y) - \overline{\alpha(y)}) + (\alpha(x) - \overline{\alpha(x)}) + (\beta(x) - \overline{\beta(x)}) + (\beta(y) - \overline{\beta(y)})$$

Putting $x = y$ gives

$$0 = 2(\alpha(x) - \overline{\alpha(x)}) + 2(\beta(x) - \overline{\beta(x)}) = 4iIm(\alpha(x)) + 4iIm(\beta(x)) \Leftrightarrow$$

$$\Leftrightarrow Im(\beta(x)) = -Im(\alpha(x)) =: -\gamma(x)$$

$$\Leftrightarrow \alpha(x)) = a(x) - i\gamma(x) \quad ; \quad \beta(x) = b(x) - i\gamma(x) \quad ; \quad a(x), b(x), \gamma(x) \in \mathbb{R}$$

and this proves (50). Conversely, if (49) and (50) are satisfied, then

$$\begin{aligned} h(x,y) &= k_0(x,y) + a(x) + b(y) + i(\gamma(y) - \gamma(x)) + S(x,y) \\ &= k_0(x,y) + \frac{1}{2}(a(x) + a(y)) + \frac{1}{2}(b(y) + b(x)) \\ &\quad + i(\gamma(y) - \gamma(x)) + \frac{1}{2}(a(x) - a(y)) + \frac{1}{2}(b(y) - b(x)) + S(x,y) \end{aligned}$$

Thus defining

$$h_0(x,y) := k_0(x,y) + \frac{1}{2}(a(x) + a(y)) + \frac{1}{2}(b(y) + b(x)) + i(\gamma(y) - \gamma(x))$$

$$S_0(x,y) := \frac{1}{2}(a(x) - a(y)) + \frac{1}{2}(b(y) - b(x)) + S(x,y)$$

one has

$$\overline{h_0(x,y)} := \overline{k_0(x,y)} + \frac{1}{2}(a(x) + a(y)) + \frac{1}{2}(b(y) + b(x)) - i(\gamma(y) - \gamma(x))$$

$$= k_0(y,x) + \frac{1}{2}(a(y)+a(x)) + \frac{1}{2}(b(x)+b(y)) + i(\gamma(x)-\gamma(y)) = h_0(y,x)$$

$$S_0(y,x) := \frac{1}{2}(a(y)-a(x)) + \frac{1}{2}(b(x)-b(y)) + S(y,x)$$

$$= -\frac{1}{2}(a(x)-a(y)) - \frac{1}{2}(b(y)-b(x)) - S(x,y) = -S_0(x,y)$$

hence h_0 is Hermitian and S_0 is symplectic, so that h is CPD.

(iii) is CPD Hermitian if and only if

$$0 = S_0(x,y) := \frac{1}{2}(a(x)-a(y)) + \frac{1}{2}(b(y)-b(x)) + S(x,y) \Leftrightarrow$$

$$\Leftrightarrow S(x,y) = \frac{1}{2}(a(x)-b(x)) - \frac{1}{2}(a(y)-b(y)) =: f(x) - f(y)$$

i.e. S is trivial.

(iv) h is Hermitian if and only if

$$\overline{h(x,y)} = h(y,x)$$

Given (48) this is equivalent to

$$\overline{k(x,y)} + \overline{\alpha}(x) + \overline{\beta}(y) = k(y,x) + \alpha(y) + \beta(x)$$

If k is Hermitian, this is equivalent to

$$\overline{\alpha}(x) + \overline{\beta}(y) = \alpha(y) + \beta(x) \Leftrightarrow \overline{\alpha}(x) - \beta(x) = \alpha(y) - \overline{\beta}(y) \tag{54}$$

Since $x, y \in S$ are arbitrary, there exists a constant $a \in \mathbb{C}$ such that

$$\overline{\alpha}(x) - \beta(x) =: a$$

is independent of $x \in S$. Moreover, from (54) with $x = y$, we see that a must be real. Therefore

$$\beta(x) = \overline{\alpha}(x) - a \tag{55}$$

which is (51) with $c = -a$. Therefore h can be written in the form (52) and from this it is clear that h is CPD if and only if k is. □

Theorem 7.1. *Let X be a set, k a kernel on X and $x_0 \in X$ an element of X. Then the following statements are equivalent:*

i) k is conditionally positive definite and

$$k(x_0, x_0) \leq 0 \tag{56}$$

ii) the kernel

$$\langle x, y \rangle := k(x,y) - k(x,x_0) - k(x_0,y) \tag{57}$$

is positive definite on X.

Proof. For any finite set $F \subseteq S$ and any choice of $\lambda_x \in \mathbb{C}$ $(x \in F)$ define

$$\lambda_{x_0} := -\sum_{x \subset F} \lambda_x \tag{58}$$

so that

$$\sum_{x \in F \cup \{x_0\}} \lambda_x = 0$$

Now consider the quantity

$$\begin{aligned}
&\sum_{x,y \in F} \bar{\lambda}_x (k(x,y) - k(x,x_0) - k(x_0,y))\lambda_y \\
&= \sum_{x,y \in F} \bar{\lambda}_x k(x,y)\lambda_y - (\sum_x \lambda_x)^{-} \sum_{y \in F} k(x_0,y)\lambda_y - \sum_{x \in F} \bar{\lambda}_x k(x,x_0)(\sum_x \lambda_y) \\
&= \sum_{x,y \in F} \bar{\lambda}_x k(x,y)\lambda_y + \bar{\lambda}_{x_0} \sum_{y \in F} k(x_0,y)\lambda_y + \sum_{x \in F} \bar{\lambda}_x k(x,x_0)\lambda_{x_0} \\
&= \sum_{x,y \in F} \bar{\lambda}_x k(x,y)\lambda_y + \bar{\lambda}_{x_0} \sum_{y \in F} k(x_0,y)\lambda_y \\
&\quad + \sum_{x \in F} \bar{\lambda}_x k(x,x_0)\lambda_{x_0} + \bar{\lambda}_{x_0} k(x_0,x_0)\lambda_{x_0} - \bar{\lambda}_{x_0} k(x_0,x_0)\lambda_{x_0} \\
&= \sum_{x,y \in F \cup \{x_0\}} \bar{\lambda}_x k(x,y)\lambda_y - \bar{\lambda}_{x_0} k(x_0,x_0)\lambda_{x_0}
\end{aligned} \tag{59}$$

If k is conditionally positive definite, then this is

$$\geq -\bar{\lambda}_{x_0} k(x_0,x_0)\lambda_{x_0}$$

and if (56) holds, this is ≥ 0.

Conversely, if the kernel (57) is PD, then the quantity (59) is positive for any finite set $F \subseteq S$ and any choice of $\lambda_x \in \mathbb{C}$ $(x \in F)$. In particular, if the $\lambda_x \in \mathbb{C}$ are chosen so that

$$\sum_{x \in F} \lambda_x = 0$$

then the identity (59) gives

$$0 \leq \sum_{x,y \in F} \bar{\lambda}_x (k(x,y) - k(x,x_0) - k(x_0,y))\lambda_y$$

$$= \sum_{x,y\in F} \bar{\lambda}_x k(x,y)\lambda_y - (\sum_x \lambda_x)^- \sum_{y\in F} k(x_0,y)\lambda_y - \sum_{x\in F} \bar{\lambda}_x k(x,x_0)(\sum_x \lambda_y)$$

$$= \sum_{x,y\in F} \bar{\lambda}_x k(x,y)\lambda_y$$

i.e. the kernel $k(x,y)$ is CPD.

Finally, if the kernel (57) is PD, then choosing $x_0 = x = y$ one finds:

$$0 \leq \langle x_0, x_0 \rangle = k(x_0,x_0) - k(x_0,x_0) - k(x_0,x_0) = -k(x_0,x_0)$$

which is equivalent to (56). □

Theorem 7.2. *The following statements are equivalent:*

(i) There exists an interval $[0,T]$ such that $\forall t \in [0,T]$ the kernel $(e^{tk_{ij}})$ is PD

(ii) $\forall t \in \mathbb{R}_+$ the kernel $(e^{tk_{ij}})$ is PD

(iii) the kernel (k_{ij}) is CPD

Proof. (i) $\Rightarrow$ (ii). By Schur's theorem, $\forall s,t \in [0,T]$ the kernel

$$e^{sk_{ij}} e^{tk_{ij}} = e^{(s+t)k_{ij}}$$

is PD and from this (ii) follows.

Since clearly (ii) $\Rightarrow$ (i), it follows that (i) $\Leftrightarrow$ (ii).

(ii) $\Rightarrow$ (iii). This is the content of Lemma (7.1).

(iii) $\Rightarrow$ (i) (in fact we will prove that (iii) $\Rightarrow$ (ii)). Suppose that (k_{ij}) is CPD. Then, by a previous Remark, its diagonal elements are real. Therefore one can find a function $\alpha, : i \in X \to \alpha_i \in \mathbb{C}$ such that the kernel

$$h_{ij} := k_{ij} - \alpha_i - \overline{\alpha}_j$$

which is CPD by Lemma (7.2), satisfies

$$h_{ii} = k_{ii} - \alpha_i - \overline{\alpha}_i = k_{ii} - 2Re(\alpha_i) \leq 0 \qquad ; \qquad \forall i \in S$$

By Theorem (7.1) for any $i_0 \in X$ the kernel

$$h_{ij} - h_{ii_0} - h_{i_0j} = k_{ij} - \alpha_i - \overline{\alpha}_j - k_{ii_0} + \alpha_i + \overline{\alpha}_{i_0} - k_{i_0j} + \alpha_{i_0} + \overline{\alpha}_j$$

$$= k_{ij} - k_{ii_0} - k_{i_0j} + \alpha_{i_0} + \overline{\alpha}_{i_0} =: k_{ij} - \beta_i + \overline{\beta}_j$$

with

$$\beta_i := k_{ii_0} - \alpha_{i_0}$$

is PD. Therefore, $\forall t \geq 0$

$$e^{t(k_{ij}-\beta_i-\overline{\beta}_j)} = e^{-t\beta_i} e^{tk_{ij}} e^{-t\overline{\beta}_j} = e^{-t\beta_i} e^{tk_{ij}} (e^{-t\beta_j})^*$$

is PD hence, by Corollary (5.2), also

$$e^{tk_{ij}}$$

is PD $\forall t \geq 0$ and this proves (ii) hence (i). □

Lemma 7.3. *For two Hermitian Kernels h, k the following are equivalent:*
(i) $\forall u_0 \in X$ the positive definite kernels

$$k(u,v) - k(u,u_0) - k(u_0,v) \quad ; \quad h(u,v) - h(u,u_0) - h(u_0,v) \quad (60)$$

differ by an additive real constant c.

(ii) There exists a function $\alpha : X \to \mathbb{C}$ such that

$$h(u,v) = k(u,v) + \alpha(u) + \overline{\alpha(v)} + c$$

Proof. If the kernels (60) differ by an additive real constant c, then $\forall u, u_0, v \in X$

$$k(u,v) - k(u,u_0) - k(u_0,v) = h(u,v) - h(u,u_0) - h(u_0,v) + c$$

$$\Leftrightarrow (k-h)(u,v) = (k-h)(u,u_0) + (k-h)(u_0,v) + c$$

defining

$$\alpha(u) := (k-h)(u,u_0)$$

the Hermitianity of h and k implies that

$$\overline{\alpha(v)} := (k-h)(u_0,v)$$

Conversely let h be of the form

$$h(u,v) = k(u,v) + \alpha(u) + \overline{\alpha(v)} + c$$

then

$$h(u,v) - h(u,u_0) - h(u_0,v) =$$

$$k(u,v) + \alpha(u) + \overline{\alpha(v)} + c - k(u,u_0) - \alpha(u) - \overline{\alpha(u_0)} - c - k(u_0,v) - \alpha(u_0) - \overline{\alpha(v)} - c =$$

$$= k(u,v) - k(u,u_0) - k(u_0,v) - \overline{\alpha(u_0)} - \alpha(u_0) - c$$

$$= k(u,v) + k(u,u_0) + k(u_0,v) - \mathrm{Re}(\alpha(u_0)) + c$$

i.e. the positive definite kernels (60) differ by an additive real constant c.□

8. Infinitely divisible kernels

Definition 8.1. A (positive definite) kernel

$$k : X \times X \to \mathbb{C}$$

is called infinitely divisible, if $\forall n \in \mathbb{N}$ there exists a (positive definite) kernel k_n such that

$$k_n^n = k$$

where k_n^n denotes the n–th Schur power of k_n.

Lemma 8.1. *For a PD kernel the following statements are equivalent:*

(i) k is infinitely divisible

(ii) $\forall n \in \mathbb{N}$ and $\forall x, y \in X$ there exists a choice of the n–th root of $k(x, y)$ such that the kernel

$$k^{1/n}(x, y) := (k(x, y))^{1/n}$$

is PD (iii) For each $t \geq 0$ the kernel

$$k^t(x, y) := k(x, y)^t$$

is PD.

Proof. By definition if k is infinitely divisible then for each n and for some PD kernel k_n^n,

$$k_n^n = k \Leftrightarrow k_n(i, j)^n = k(i, j) \qquad ; \qquad \forall i, j$$

Therefore, denoting $\log(k(i, j))$ the principal determination of the logarithm of $k(i, j)$ one has

$$\log(k_n(i, j)) = (1/n) \log(k(i, j))$$

Thus, defining

$$k(i, j)^{1/n} := e^{(1/n) \log(k(i,j))}$$

one has

$$k_n(i, j) = k(i, j)^{1/n}$$

hence $k(i, j)^{1/n}$ is PD. Therefore for each $k \in \mathbb{N}$

$$k(i, j)^{k/n}$$

is also PD. Finally, defining k^t by continuity, one has that k^t is PD $\forall t \geq 0$.

$\square$

9. The Kolmogorov decomposition theorem for $\mathbb{C}$–valued PD kernels

Remark: The following theorem shows that every positive definite kernel can be seen as a scalar product in some Hilbert space.

Definition 9.1. A pre–Hilbert space is a vector space $\mathcal{H}$, over the real or complex numbers, with an antilinear embedding into its algebraic dual $\mathcal{H}^*$

$$\xi \in \mathcal{H} \mapsto \xi^* \in \mathcal{H}^* \tag{61}$$

such that the kernel on $\mathcal{H}$ defined by

$$\langle \xi, \eta \rangle := \xi^*(\eta)$$

is positive definite hermitean.

Remark. With small modifications, the discussion below can be extended to general positive definite kernels.

Theorem 9.1. *(Kolmogorov decomposition for positive definite hermitean kernels) Let X be a set. k is a $\mathbb{C}$–valued positive definite hermitean kernel on X if and only if then there exists an Hilbert space $\mathcal{H}$ and a map*

$$v : X \to \mathcal{H}$$

such that, in the notation (61):

$$k(x,y) = \langle v(x), v(y) \rangle = v(x)^*(v(y)) \quad ; \quad \forall x,y \in X \tag{62}$$

$$\{v(x) : x \in X\} \quad \textit{is total in } \mathcal{H}$$

If $\{v_0, \mathcal{H}_0\}$ is another pair with these properties, then there exists a unitary operator $U : \mathcal{H} \to \mathcal{H}_0$ such that $Uv = v_0$.

Proof. The atomic measures $\{\delta_x : x \in X\}$ are linearly independent probability measures on X. Denote $\tilde{X}$ the complex vector space generated by them. Since k is positive definite and hermitean, the map

$$(\delta_x, \delta_y) \mapsto k(x,y)$$

uniquely extends, by sesquilinearity, to a sesquilinear map on $\tilde{X} \times \tilde{X}$ which is PDH by construction. Denote $\mathcal{H}$ the completion of the pre–Hilbert space obtained in this way and

$$\eta : \tilde{X} \to \mathcal{H}$$

the quotient map. Denote v the restriction of η on X. Then the map $v : X \to \mathcal{H}$ satisfies

$$\langle v(x), v(y)\rangle = \langle \eta(\delta_x), \eta(\delta_y)\rangle = k(x, y)$$

The remaining statements are clear. □

Definition 9.2. A pair $\{v, H\}$, defined by Theorem (9.1), is called a Kolmogorov representation of the PDH kernel $K : X \times X \to \mathbb{C}$.

10. Boson Fock spaces

Let $\mathcal{H}$ be an Hilbert space. Its scalar product $\langle \cdot , \cdot \rangle$ is a positive definite hermitean kernel on $\mathcal{H}$ hence, by Corollary (5.3), also the kernel $\exp\langle \cdot , \cdot \rangle$ is.

Definition 10.1. The Kolmogorov decomposition of the PD kernel $\exp\langle \cdot , \cdot \rangle$ on $\mathcal{H}$ is denoted

$$\{e^{\mathcal{H}}, v =: e^{\cdot}\}$$

and called the exponential space (or Boson Fock space over $\mathcal{H}$).

The total set

$$\exp(\mathcal{H}) := v(\mathcal{H}) := \{e^f = v_f \in e^{\mathcal{H}} \ : \ f \in \mathcal{H}\}$$

is called the set of exponential vectors of $e^{\mathcal{H}}$.

Remark: Recall that by definition of Kolmogorov decomposition the characterizing property of the exponential vectors is

$$\langle e^f, e^g\rangle = \langle v_f, v_g\rangle = e^{\langle f,g\rangle} \quad ; \quad \forall f, g \in \mathcal{H} \tag{63}$$

11. Infinitely divisible kernels and Boson Fock spaces

Definition 11.1. Let S be a set. A kernel k on S is called exponentially PDH if there exists a PDH kernel q such that k is equivalent to e^q, i.e. if there exists a function $c : f \in S \to c_f \in \mathbb{C} \setminus \{0\}$ such that

$$c_f^* k(f, g) c_g = e^{q(f,g)} \quad ; \quad f, g \in S \tag{64}$$

Theorem 11.1. *For a kernel k on a set S the following statements are equivalent.*

(i) $k(f,g)$ is exponentially PDH

(ii) there exists a CPD hermitean kernel q_0 such that k has the form

$$k(f,g) = e^{q_0(f,g)} \quad ; \quad f,g \in S \tag{65}$$

(iii) $k(f,g)$ is an infinitely divisible PDH kernel, i.e. $k(f,g)^t$ is PDH, $\forall t \geq 0$

(iv) there exists a PDH kernel q on S and a map $f \in S \mapsto \kappa_f \in \mathbb{C}$ such that, denoting $\{\mathcal{H}, v\}$ (resp. $\{\mathcal{K}, u\}$) the Kolmogorov decomposition of k (resp. q), then the map

$$e^{\kappa_f} v_f \in \mathcal{H} \mapsto e^{u_f} \in \Gamma(\mathcal{K}) \tag{66}$$

extends to a unitary isomorphism between $\mathcal{H}$ and the Fock space $\Gamma(\mathcal{K})$ over $\mathcal{K}$.

Proof. (i)$\Rightarrow$ (ii). If k is an exponentially PDH kernel then it has the form (64) with $q(\,\cdot\,,\,\cdot\,)$ PDH. Therefore, defining

$$\kappa_f := Logc_f$$

where $Logc_f$ denotes the principal determination of the logarithm of c_f one has

$$c_f =: e^{\kappa_f}$$

and the identity (69) holds with

$$q_0(f,g) := q(f,g) - \kappa_f^* - \kappa_g \tag{67}$$

From Lemma (7.2) we know that the kernel $q_0(f,g)$ is CPDH because $q(f,g)$ is PDH hence a fortiori CPDH.

(ii)$\Leftrightarrow$(iii). $k(f,g)$ has the form (69), for some CPD kernel q_0, if and only if $\forall t \geq 0$

$$k(f,g)^t = e^{tq_0(f,g)} \tag{68}$$

and, since $q_0(f,g)$ is CPDH, the right and side is PDH by Theorem (7.2). This is equivalent to say that $k(f,g)$ is an infinitely divisible PDH kernel.

(iii)$\Rightarrow$ (iv) If k is infinitely divisible then, by Theorem (7.2), it has the form (69) for some CPDH kernel q_0. Then by Theorem (7.1), and up to replacing q_0 by an equivalent CPD hermitian kernel, $\forall f_0 \in S$ the kernel

$$q_1(f,g) := q_0(f,g) - q_0(f,f_0) - q_0(f_0,g)$$

is PDH. By construction the Kolmogorov decomposition $\{\mathcal{H}, v\}$, of k, satisfies

$$\langle v_f, v_g \rangle = k(f,g) = e^{q_0(f,g)}$$

Therefore

$$\langle e^{-q_0(f,f_0)^*} v_f, e^{-q_0(f_0,g)} v_g \rangle = \langle e^{-q_0(f_0,f)} v_f, e^{-q_0(f_0,g)} v_g \rangle =$$

$$= e^{-q_0(f,f_0)} e^{q_0(f,g)} e^{-q_0(f_0,g)} = e^{-q_1(f,g)}$$

and (iv) follows by choosing

$$\kappa_f := q_0(f_0, f)$$

(iv) $\Leftrightarrow$ (i). Suppose that (iv) holds and define

$$c_f := e^{\kappa_f}$$

the unitary isomorphism (66) implies that

$$c_f^* k(f,g) c_g = \langle e^{\kappa_f} v_f, e^{\kappa_g} v_g \rangle = \langle e^{u_f}, e^{u_g} \rangle = e^{\langle u_f, u_g \rangle} = e^{q(u_f, u_g)}$$

this means that $k(f,g)$ is exponentially PDH, i.e. (i) holds. Since the converse statement, i.e. that (i) implies (iv) is clear, this ends the proof. □

Theorem 11.2. *Let S be a set. The infinitely divisible PD hermitean kernels on S, under pointwise multiplication, are a semi–group whose only invertible elements are the kernels equivalent to the identity kernel.*

Proof. According to Theorem (11.1) (ii), k is an infinitely divisible hermitian kernel on S if and only if on S there exists a CPDH kernel q_0 such that k has the form

$$k(f,g) = e^{q_0(f,g)} \quad ; \quad f,g \in S \tag{69}$$

Up to replacing q_0 by an equivalent CPDH kernel one can assume that there exists $f_0 \in S$ such that

$$q_0(f_0, f_0) = 0$$

In fact, for any $f_0 \in S$, if $\alpha : \mathbb{C} \to \mathbb{C}$ is any function such that

$$q_0(f_0, f_0) = 2\text{Re}(\alpha_{f_0})$$

then the CPDH kernel defined by

$$q_0'(f,g) := q_0(f,g) - \alpha_f - \overline{\alpha_g}$$

by construction is equivalent to q_0 and such that $q_0'(f_0, f_0) = 0$.
Finally, if k has the form (69) and has an infinitely divisible PD (hence also hermitean) inverse, then $\forall t \geq 0$

$$k(f,g)^{-t} = e^{-tq_0(f,g)} \quad ; \qquad f,g \in S$$

is CPDH. But since $q_0(f,g)$ is CPD, $-q_0(f,g)$ can be CPD if and only if for any finite set F and any λ_f $(f \in F)$ such that $\sum_{f\in F} \lambda_f = 0$, one has

$$\sum_{f,g\in F} \lambda_f q_0(f,g)\lambda_g = 0$$

By hemiteanity $q_0(f,f)$ is real for all $f \in S$. Therefore, possibly replacing $q_0(f,g)$ by the equivalent CPDH kernel $q_0(f,g) - \frac{1}{2}q_0(f,f) - \frac{1}{2}q_0(g,g)$, one can assume that

$$q_0(f,f) = 0 \quad ; \quad \forall f \in S$$

With this assumption, choosing $F = \{f,g\}$ and $\lambda_f = -\lambda_g = 1$ and using hermiteanity, we obtain

$$0 = q_0(f,f) + q_0(g,g) - q_0(f,g) - q_0(g,f) = -2Re(q_0(f,g))$$

i.e. $q_0(f,g)$ is purely imaginary and we can write

$$q_0(f,g) = iq_1(f,g)$$

with $q_1(f,g)$ real. Hermiteanity then implies that

$$q_1(f,g) = -q_1(g,f)$$

Fixing $f_0 \in S$, the PDH kernel

$$q(f,g) := q_0(f,g) - q_0(f_0,f) - q_0(g,f_0)$$

is identically zero beacuse of the Schwarz inequality and

$$\begin{aligned} q(f,f) &:= q_0(f,f) - q_0(f_0,f) - q_0(f,f_0) = -iq_1(f_0,f) - iq_1(f,f_0) \\ &= -iq_1(f_0,f) + iq_1(f_0,f) = 0 \end{aligned}$$

Thus $k(f,g)$ is equivalent to the identity kernel and this ends the proof. □

References

1. Accardi, L., Boukas, A.: Random variables and positive definite kernels associated with the Schroedinger algebra, *Proceedings of the VIII International Workshop Lie Theory and its Applications in Physics*, Varna, Bulgaria, June 16-21, 2009, pages 126-137, American Institute of Physics , AIP Conference Proceedings 1243.

2. ———: The centrally extended Heisenberg algebra and its connection with the Schrödinger, Galilei and Renormalized Higher Powers of Quantum White Noise Lie algebras, with Luigi Accardi, *Proceedings of the VIII International Workshop Lie Theory and its Applications in Physics*, Varna, Bulgaria, June 16-21, 2009, pages 115-125, American Institute of Physics , AIP Conference Proceedings 1243.
3. ———: Renormalized higher powers of white noise (RHPWN) and conformal field theory, *Infinite Dimensional Analysis, Quantum Probability, and Related Topics* **9**, No. 3, (2006) 353-360.
4. ———: The emergence of the Virasoro and w_∞ Lie algebras through the renormalized higher powers of quantum white noise , *International Journal of Mathematics and Computer Science*, **1**, No.3, (2006) 315–342.
5. ———: Fock representation of the renormalized higher powers of white noise and the Virasoro–Zamolodchikov–w_∞ $*$–Lie algebra, *J. Phys. A: Math. Theor.*, 41 (2008).
6. ———: Cohomology of the Virasoro–Zamolodchikov and Renormalized Higher Powers of White Noise $*$–Lie algebras, *Infinite Dimensional Anal. Quantum Probab. Related Topics*, Vol. 12, No. 2 (2009) 120.
7. ———: Quantum probability, renormalization and infinite dimensional $*$–Lie algebras, *SIGMA (Symmetry, Integrability and Geometry: Methods and Applications)*, 5 (2009), 056, 31 pages.
8. ———: Central extensions and stochastic processes associated with the Lie algebra of the renormalized higher powers of white noise, *Proceedings of the 11th workshop: non-commutative harmonic analysis with applications to probability*, Bedlewo, Poland, August 2008, Banach Center Publ. 89 (2010), 13-43.
9. ———: On the central extensions of the Heisenberg algebra, *QP-PQ: Quantum Probability and White Noise Analysis - vol. 25, Quantum Probability and Infinite Dimensional Analysis.* Proceedings of the 29th conference, Hammamet,Tunisia,13-18, October 2008, pages 1-12.
10. Accardi, L., Boukas, A., Misiewicz, J. : Renormalization and central extensions of Lie algebras, submitted.
11. Accardi L., Lu Y. G., Volovich I. V.: White noise approach to classical and quantum stochastic calculi, Lecture Notes of the Volterra International School of the same title, Trento, Italy (1999), Volterra Center preprint 375, Università di Roma Tor Vergata.
12. Bourbaki, N.: *Groupes et Algebres de Lie*, 2nd ed., Hermann, Paris, 1971.
13. Feinsilver, P. J., Schott, R.: *Algebraic structures and operator calculus. Volumes I and III*, Kluwer, 1993.
14. ———: Differential relations and recurrence formulas for representations of Lie groups, *Stud. Appl. Math.*, 96 (1996), no. 4, 387–406.
15. Feinsilver, P. J., Kocik, J., Schott, R.: Representations of the Schroedinger algebra and Appell systems, *Fortschr. Phys.* 52 (2004), no. 4, 343–359.
16. Franz U.: *Representations et processus stochastiques sur les groupes de Lie et sur les groupes quantiques*, Memoire de D.E.A., Université Henri Poincaré – Nancy I (1994)
17. Guichardet A., Symmetric Hilbert spaces and related topics, Lect. Notes

Math. 261, Springer, Berlin, 1972.
18. Ovando, G.: Four dimensional symplectic Lie algebras, *Beitrage Algebra Geom.* 47 (2006), no. 2, 419–434.
19. Parthasarathy K.R., Schmidt K., Positive definite kernels continuous tensor products and central limit theorems of probability theory, Springer Lecture Notes in Mathematics no. 272, 1972.

MODULAR STRUCTURES AND LANDAU LEVELS

F. BAGARELLO*

Dipartimento di Metodi e Modelli Matematici, Università di Palermo, Palermo, I-90128, Italy
**E-mail: bagarell@unipa.it*
www.unipa.it/~ bagarell

We review some recent results concerning Landau levels and Tomita-Takesaki modular theory. We also extend the general framework behind this to quasi *-algebras, to take into account the possible appearance of unbounded observables.

Keywords: Tomita-Takesaki modular structure. Landau levels.

1. Introduction and mathematical theory

In a recent paper,[1] together with Ali and Honnouvo we have constructed an explicit application of the Tomita-Takesaki's modular theory to the so-called Landau levels. Here we review these results and we further extend the general settings to consider the presence also of unbounded operators.

The motion of a quantum electron in a constant electromagnetic field produces energy (Landau) levels which are linearly spaced and infinitely degenerate. The Hamiltonian of the system, using "clever" variables, can be written as the hamiltonian of a single harmonic oscillator. Moreover, if the sense of the magnetic field is reversed, one obtains a second Hamiltonian, similar to the first one, but commuting with it. Both these Hamiltonians can be written in terms of two pairs of mutually commuting oscillator type creation and annihilation operators, which then generate two von Neumann algebras which mutually commute, and in fact are commutants of each other. This leads to the existence of a modular structure, in the sense of the Tomita-Takesaki theory[2–4] .

1.1. *Summary of the mathematical theory*

We will now quickly review some basic facts of the Tomita-Takesaki modular theory of von Neumann algebras. Details and proofs of statements may be found, for example, in[4–6] . Let $\mathfrak{A}$ be a von Neumann algebra on a Hilbert space $\mathfrak{H}$ and $\mathfrak{A}'$ its commutant. Let $\mathbf{\Phi} \in \mathfrak{H}$ be a unit vector which is cyclic and separating for $\mathfrak{A}$. Then the corresponding state φ on the algebra, $\langle \varphi \,;\, A\rangle = \langle \mathbf{\Phi} \mid A\mathbf{\Phi}\rangle$, $A \in \mathfrak{A}$, is faithful and normal. Consider the antilinear map,

$$S : \mathfrak{H} \longmapsto \mathfrak{H} \,, \qquad SA\mathbf{\Phi} = A^*\mathbf{\Phi} \,,\ \forall A \in \mathfrak{A} \,. \tag{1}$$

Since $\mathbf{\Phi}$ is cyclic, this map is densely defined and in fact it can be shown that it is closable. We denote its closure again by S and write its polar decomposition as

$$S = J\Delta^{\frac{1}{2}} = \Delta^{-\frac{1}{2}}J \,, \quad \text{with} \quad \Delta = S^*S \,. \tag{2}$$

The operator Δ, called the *modular operator*, is positive and self-adjoint. The operator J, called the *modular conjugation operator*, is antiunitary and satisfies $J = J^*$, $J^2 = I_{\mathfrak{H}}$. Note that the antiunitarity of J implies that $\langle J\phi \mid J\psi\rangle = \langle \psi \mid \phi\rangle$, $\forall \phi, \psi \in \mathfrak{H}$.

Since Δ is self-adjoint, using its spectral representation we see that for $t \in \mathbb{R}$ the family of operators $\Delta^{-\frac{it}{\beta}}$, for some fixed $\beta > 0$, defines a unitary family of automorphisms of the algebra $\mathfrak{A}$. Denoting these automorphisms by $\alpha_\varphi(t)$, we may write

$$\alpha_\varphi(t)[A] = \Delta^{\frac{it}{\beta}} A \Delta^{-\frac{it}{\beta}} \,, \quad \forall A \in \mathfrak{A} \,. \tag{3}$$

Thus, they constitute a strongly continuous one-parameter group of automorphisms, called the *modular automorphism group*. Denoting the generator of this one-parameter group by $\mathbf{H}_\varphi$, we get

$$\Delta^{-\frac{it}{\beta}} = e^{it\mathbf{H}_\varphi} \quad \text{and} \quad \Delta = e^{-\beta\mathbf{H}_\varphi} \,. \tag{4}$$

It can then be shown that the state φ is invariant under this automorphism group, that

$$e^{-\beta\mathbf{H}_\varphi}\mathbf{\Phi} = \mathbf{\Phi} \,, \qquad \Delta^{\frac{it}{\beta}}\,\mathfrak{A}\,\Delta^{-\frac{it}{\beta}} = \mathfrak{A} \,, \tag{5}$$

and that the antilinear map J interchanges $\mathfrak{A}$ with its commutant $\mathfrak{A}'$:

$$J\mathfrak{A}J = \mathfrak{A}' \,. \tag{6}$$

Finally, the state φ can be shown to satisfy the *KMS (Kubo-Martin-Schwinger) condition*, with respect to the automorphism group $\alpha_\varphi(t)$, $t \in \mathbf{R}$, in the following sense. For any two $A, B \in \mathfrak{A}$, the function

$$F_{A,B}(t) = \langle \varphi \,;\, A\,\alpha_\varphi(t)[B]\rangle \,, \tag{7}$$

has an extension to the strip $\{z = t + iy \mid t \in \mathbb{R},\ y \in [0, \beta]\} \subset \mathbb{C}$ such that $F_{A,B}(z)$ is analytic in the open strip $(0, \beta)$ and continuous on its boundaries. Moreover, it also satisfies the boundary condition (*at inverse temperature* β),

$$\langle \varphi \,;\, A\, \alpha_\varphi(t + i\beta)[B] \rangle = \langle \varphi \,;\, \alpha_\varphi(t)[B]\, A \rangle \,, \quad t \in \mathbb{R} \,. \tag{8}$$

1.1.1. *The role of the Hilbert-Schmidt operators*

A simple example of the Tomita-Takesaki theory and its related KMS states can be built on the space of Hilbert-Schmidt operators on a Hilbert space. The set of Hilbert-Schmidt operators is by itself a Hilbert space, and there are two preferred algebras of operators on it, which carry the modular structure. The presentation here follows that in[7] (Chapter 8, Section 4).

Again, let $\mathfrak{H}$ be a (complex, separable) Hilbert space of dimension N (finite or infinite) and $\{\zeta_i\}_{i=1}^N$ an orthonormal basis of it ($\langle \zeta_i \mid \zeta_j \rangle = \delta_{ij}$). We denote by $\mathcal{B}_2(\mathfrak{H})$ the space of all Hilbert-Schmidt operators on $\mathfrak{H}$. This is a Hilbert space with scalar product

$$\langle X \mid Y \rangle_2 = \mathrm{Tr}[X^* Y] \,.$$

The vectors

$$\{X_{ij} = |\zeta_i\rangle\langle\zeta_j| \mid i, j = 1, 2, \ldots, N\} \,, \tag{9}$$

form an orthonormal basis of $\mathcal{B}_2(\mathfrak{H})$:

$$\langle X_{ij} \mid X_{k\ell} \rangle_2 = \delta_{ik}\delta_{\ell j} \,.$$

In particular, the vectors

$$\mathbb{P}_i = X_{ii} = |\zeta_i\rangle\langle\zeta_i| \,, \tag{10}$$

are one-dimensional projection operators on $\mathfrak{H}$. In what follows I will denote the identity operator on $\mathfrak{H}$ and I_2 that on $\mathcal{B}_2(\mathfrak{H})$.

We identify a special class of linear operators on $\mathcal{B}_2(\mathfrak{H})$, denoted by $A \vee B,\ A, B \in \mathcal{L}(\mathfrak{H})$, which act on a vector $X \in \mathcal{B}_2(\mathfrak{H})$ in the manner:

$$(A \vee B)(X) = AXB^* \,.$$

Using the scalar product in $\mathcal{B}_2(\mathfrak{H})$, we see that

$$\mathrm{Tr}[\mathrm{X}^*(\mathrm{AYB}^*)] = \mathrm{Tr}[(\mathrm{A}^*\mathrm{XB})^*\mathrm{Y})] \Longrightarrow (\mathrm{A} \vee \mathrm{B})^* = \mathrm{A}^* \vee \mathrm{B}^* \,,$$

and since for any $X \in \mathcal{B}_2(\mathfrak{H})$

$$(A_1 \vee B_1)(A_2 \vee B_2)(X) = A_1[(A_2 \vee B_2)(X)]B_1^* = A_1 A_2 X B_2^* B_1^* \,,$$

we have,

$$(A_1 \vee B_1)(A_2 \vee B_2) = (A_1A_2) \vee (B_1B_2)\ . \tag{11}$$

There are two special von Neumann algebras which can be built out of these operators. These are

$$\mathfrak{A}_\ell = \{A_\ell = A \vee I \mid A \in \mathcal{L}(\mathfrak{H})\}\ , \qquad \mathfrak{A}_r = \{A_r = I \vee A \mid A \in \mathcal{L}(\mathfrak{H})\}\ . \tag{12}$$

They are mutual commutants and both are factors:

$$(\mathfrak{A}_\ell)' = \mathfrak{A}_r\ , \qquad (\mathfrak{A}_r)' = \mathfrak{A}_\ell\ , \qquad \mathfrak{A}_\ell \cap \mathfrak{A}_r = \mathbb{C}I_2\ . \tag{13}$$

Consider now the operator $J : \mathcal{B}_2(\mathfrak{H}) \longrightarrow \mathcal{B}_2(\mathfrak{H})$, whose action on the vectors X_{ij} in (9) is given by

$$JX_{ij} = X_{ji} \Longrightarrow J^2 = I_2 \quad \text{and} \quad J(|\phi\rangle\langle\psi|) = |\psi\rangle\langle\phi|\ , \quad \forall\phi,\psi \in \mathfrak{H}\ . \tag{14}$$

This operator is antiunitary, and since

$$[J(A \vee I)J]X_{ij} = J(A \vee I)X_{ji} = J(AX_{ji}) = J(A|\zeta_j\rangle\langle\zeta_i|) =$$

$$= |\zeta_i\rangle\langle\zeta_j|A^* = (I \vee A)X_{ij}\ ,$$

we immediately get

$$J\mathfrak{A}_\ell J = \mathfrak{A}_r\ . \tag{15}$$

A KMS state can be introduced starting from a set of non-zero, positive numbers $\alpha_i\ ,\ \ i = 1, 2, \ldots, N$, satisfying $\sum_{i=1}^N \alpha_i = 1$. Indeed, let

$$\mathbf{\Phi} = \sum_{i=1}^{N} \alpha_i^{\frac{1}{2}}\ \mathbb{P}_i = \sum_{i=1}^{N} \alpha_i^{\frac{1}{2}}\ X_{ii} \in \mathcal{B}_2(\mathfrak{H})\ . \tag{16}$$

Then $\mathbf{\Phi}$ defines a vector state φ on the von Neumann algebra $\mathfrak{A}_\ell$: for any $A \vee I \in \mathfrak{A}_\ell$, we put

$$\langle\varphi\ ;\ A \vee I\rangle = \langle\mathbf{\Phi} \mid (A \vee I)(\mathbf{\Phi})\rangle_2 = \mathrm{Tr}[\mathbf{\Phi}^* A \mathbf{\Phi}] = \mathrm{Tr}[\rho_\varphi A], \tag{17}$$

with $\rho_\varphi = \sum_{i=1}^N \alpha_i \mathbb{P}_i$. Moreover $\mathbf{\Phi}$ is cyclic and separating for $\mathfrak{A}_\ell$, while φ is faithful and normal[1] .

It is also possible to show that the state φ is indeed a KMS state for a particular choice of α_i. For that, we first need to construct a time evolution $\alpha_\varphi(t)$, $t \in \mathbb{R}$, on the algebra $\mathfrak{A}_\ell$, using the state φ, with respect to which it has the KMS property, for fixed $\beta > 0$,

$$\langle\varphi\ ;\ A_\ell\, \alpha_\varphi(t + i\beta)[B_\ell]\rangle = \langle\varphi\ ;\ \alpha_\varphi(t)[B_\ell]\ A_\ell\rangle\ , \quad \forall A_\ell, B_\ell \in \mathfrak{A}_\ell\ , \tag{18}$$

and moreover the function,

$$F_{A_\ell,B_\ell}(z) = \langle \varphi\,;\, A_\ell\, \alpha_\varphi(z)[B_\ell] \rangle\,, \tag{19}$$

is analytic in the strip $\{\Im(z) \in (0,\beta)\}$ and continuous on its boundaries.
We start by defining the operators,

$$\mathbf{P}_{ij} = \mathbb{P}_i \vee \mathbb{P}_j\,, \qquad i,j = 1,2,\ldots,N \tag{20}$$

where the $\mathbb{P}_i$ are the projection operators on $\mathfrak{H}$ defined in (10). Clearly, the $\mathbf{P}_{ij}$ are projection operators on the Hilbert space $\mathcal{B}_2(\mathfrak{H})$.

Using ρ_φ in (17) and for a fixed $\beta > 0$, we define the operator H_φ as:

$$\rho_\varphi = e^{-\beta H_\varphi} \Longrightarrow H_\varphi = -\frac{1}{\beta}\sum_{i=1}^{N} (\ln \alpha_i)\, \mathbb{P}_i\,. \tag{21}$$

Next we define the operators:

$$H^\ell_\varphi = H_\varphi \vee I\,, \qquad H^{\mathrm{r}}_\varphi = I \vee H_\varphi\,, \qquad \mathbf{H}_\varphi = H^\ell_\varphi - H^{\mathrm{r}}_\varphi\,, \tag{22}$$

Since $\sum_{i=1}^N \mathbb{P}_i = I$, we may also write

$$H^\ell_\varphi = -\frac{1}{\beta}\sum_{i,j=1}^{N} \ln \alpha_i \mathbf{P}_{ij}\,, \quad \text{and} \quad H^{\mathrm{r}}_\varphi = -\frac{1}{\beta}\sum_{i,j=1}^{N} \ln \alpha_j \mathbf{P}_{ij}\,.$$

Thus,

$$\mathbf{H}_\varphi = -\frac{1}{\beta}\sum_{i,j=1}^{N} \ln\left[\frac{\alpha_i}{\alpha_j}\right] \mathbf{P}_{ij}\,. \tag{23}$$

Using the operator.

$$\Delta_\varphi := \sum_{i,j=1}^{N} \left[\frac{\alpha_i}{\alpha_j}\right] \mathbf{P}_{ij} = e^{-\beta \mathbf{H}_\varphi}\,, \tag{24}$$

we further define a time evolution operator on $\mathcal{B}_2(\mathfrak{H})$:

$$e^{i\mathbf{H}_\varphi t} = [\Delta_\varphi]^{-\frac{it}{\beta}}\,. \qquad t \in \mathbb{R}\,, \tag{25}$$

and we note that, for any $X \in \mathcal{B}_2(\mathfrak{H})$,

$$\begin{aligned} e^{i\mathbf{H}_\varphi t}(X) &= \sum_{i,j=1}^{N} \left[\frac{\alpha_i}{\alpha_j}\right]^{-\frac{it}{\beta}} \mathbf{P}_{ij}(X) = \left[\sum_{i=1}^{N} (\alpha_i)^{-\frac{it}{\beta}} \mathbb{P}_i\right] \vee \left[\sum_{j=1}^{N} (\alpha_j)^{-\frac{it}{\beta}} \mathbb{P}_j(X)\right] \\ &= e^{iH_\varphi t}(X) e^{-iH_\varphi t}\,, \end{aligned}$$

so that

$$e^{i\mathbf{H}_\varphi t} = e^{iH_\varphi t} \vee e^{iH_\varphi t}\,, \tag{26}$$

where H_φ is the operator introduced in (21). From the definition of the vector $\mathbf{\Phi}$ in (16), it is clear that it commutes with H_φ and hence that it is invariant under this time evolution:

$$e^{i\mathbf{H}_\varphi t}(\mathbf{\Phi}) = e^{iH_\varphi t}\,\mathbf{\Phi}\,e^{-iH_\varphi t} = \mathbf{\Phi}\,. \tag{27}$$

Finally, using $e^{i\mathbf{H}_\varphi t}$ we define the time evolution α_φ on the algebra $\mathfrak{A}_\ell$, in the manner (see (3)):

$$\alpha_\varphi(t)[A_\ell] = e^{i\mathbf{H}_\varphi t}\,A_\ell\,e^{-i\mathbf{H}_\varphi t} \qquad \forall A_\ell \in \mathfrak{A}_\ell\,. \tag{28}$$

Writing $A_\ell = A \vee I\,,\ \ A \in \mathcal{L}(\mathfrak{H})$, and using the composition law (11), we see that

$$e^{i\mathbf{H}_\varphi t}\,A_\ell\,e^{-i\mathbf{H}_\varphi t} = \left[e^{iH_\varphi t}\,A\,e^{-iH_\varphi t}\right] \vee I\,, \tag{29}$$

so that, by virtue of (17),

$$\langle \varphi\,;\,\alpha_\varphi(t)[A_\ell]\rangle = \mathrm{Tr}\left[\rho_\varphi\,e^{iH_\varphi t}\,A\,e^{-iH_\varphi t}\right] = \langle \varphi\,;\,A_\ell\rangle\,, \tag{30}$$

since ρ_φ and H_φ commute. Thus, the state φ is invariant under the time evolution α_φ. We finally refer to[1] for the proof of the KMS condition.

We now analyze the antilinear operator $S_\varphi : \mathcal{B}_2(\mathfrak{H}) \longrightarrow \mathcal{B}_2(\mathfrak{H})$, which acts as (see (1))

$$S_\varphi(A_\ell\mathbf{\Phi}) = A_\ell^*\mathbf{\Phi}\,, \qquad \forall A_\ell \in \mathfrak{A}_\ell\,. \tag{31}$$

Taking $A_\ell = A \vee I$,

$$S_\varphi(A_\ell\mathbf{\Phi}) = A_\ell^*\mathbf{\Phi}\,,\quad \forall A_\ell \in \mathfrak{A}_\ell \quad \Longleftrightarrow \quad S_\varphi(A\mathbf{\Phi}) = A^*\mathbf{\Phi}\,,\quad \forall A \in \mathcal{L}(\mathfrak{H})\,.$$

Using (16) we may write,

$$S_\varphi(A\mathbf{\Phi}) = A^*\mathbf{\Phi} \quad \Longrightarrow \quad \sum_{i=1}^{N} \alpha_i^{\frac{1}{2}}\,S_\varphi(A\mathbb{P}_i) = \sum_{i=1}^{N} \alpha_i^{\frac{1}{2}}\,A^*\mathbb{P}_i\,.$$

Taking $A = X_{k\ell}$ (see (9)) and using $X_{k\ell}\mathbb{P}_i = \delta_{\ell i}X_{ki}$, we then get

$$\alpha_\ell^{\frac{1}{2}}\,S_\varphi(X_{k\ell}) = \alpha_k^{\frac{1}{2}}\,S_\varphi(X_{\ell k}) \quad \Longrightarrow \quad S_\varphi(X_{k\ell}) = \left[\frac{\alpha_k}{\alpha_\ell}\right]^{\frac{1}{2}} X_{\ell k}\,. \tag{32}$$

Since any $A \in \mathcal{L}(\mathfrak{H})$ can be written as $A = \sum_{i,j=1}^{N} a_{ij}X_{ij}$, where $a_{ij} = \langle \zeta_i \mid A\zeta_j\rangle$, and furthermore, since $\mathbf{P}_{ij}(X_{k\ell}) = X_{ij}\delta_{ik}\delta_{j\ell}$, we obtain using (14) and (24),

$$S_\varphi = J[\Delta_\varphi]^{\frac{1}{2}}\,, \tag{33}$$

which in fact, also gives the polar decomposition of S_φ.

2. Application to Landau levels

We now show how the above setup, based on $\mathcal{B}_2(\mathfrak{H})$, can be applied to a specific physical situation namely, to the case of an electron subject to a constant magnetic field, as discussed in[8].

In that case, $\mathfrak{H} = L^2(\mathbb{R})$ and the mapping $\mathcal{W} : \mathcal{B}_2(\mathfrak{H}) \longrightarrow L^2(\mathbb{R}^2, dx\, dy)$, with

$$(\mathcal{W}X)(x,y) = \frac{1}{(2\pi)^{\frac{1}{2}}} \mathrm{Tr}[U(x,y)^* X], \quad \text{where} \quad U(x,y) = e^{-i(xQ+yP)}, \tag{34}$$

Q, P being the usual position and momentum operators ($[Q,P] = iI$), transfers the whole modular structure unitarily to the Hilbert space $\widetilde{\mathfrak{H}} = L^2(\mathbb{R}^2, dx\, dy)$. The mapping $\mathcal{W}$ is often referred to as the *Wigner transform* in the physical literature.

To work this out in some detail, we start by constructing the Hamiltonian H_φ (see (21)), using the oscillator Hamiltonian $H_{\text{osc}} = \frac{1}{2}(P^2 + Q^2)$ on $\mathfrak{H}$. Let us choose the orthonormal basis set of vectors ζ_n , $n = 0, 1, 2, \ldots \infty$, to be the eigenvectors of H_{osc}:

$$H_{\text{osc}} \zeta_n = \left(n + \frac{1}{2}\right) \zeta_n \, . \tag{35}$$

As it is well known, the ζ_n are the Hermite functions,

$$\zeta_n(x) = \frac{1}{\pi^{\frac{1}{4}}} \frac{1}{\sqrt{2^n\, n!}}\, e^{-\frac{x^2}{2}} h_n(x) \, , \tag{36}$$

the h_n being the Hermite polynomials, obtainable as:

$$h_n(x) = (-1)^n\, e^{x^2} \partial_x^n\, e^{-x^2} \, . \tag{37}$$

Consider now the operator $e^{-\beta H_{\text{osc}}}$, for some fixed $\beta > 0$. We have

$$e^{-\beta H_{\text{osc}}} = \sum_{n=0}^{\infty} e^{-(n+\frac{1}{2})\beta} \mathbb{P}_n \quad \text{and} \quad \mathrm{Tr}\left[e^{-\beta H_{\text{osc}}}\right] = \frac{e^{-\frac{\beta}{2}}}{1 - e^{-\beta}} \, .$$

Thus we take

$$\rho_\varphi = \frac{e^{-\beta H_{\text{osc}}}}{\mathrm{Tr}\left[e^{-\beta H_\varphi}\right]} = (1 - e^{-\beta}) \sum_{n=0}^{\infty} e^{-n\beta} \mathbb{P}_n, \ \text{and}\ \mathbf{\Phi} = \left[1 - e^{-\beta}\right]^{\frac{1}{2}} \sum_{n=0}^{\infty} e^{-\frac{n}{2}\beta} \mathbb{P}_n \, . \tag{38}$$

Following (17) and (21) we write

$$\rho_\varphi = \sum_{n=0}^{\infty} \alpha_n \mathbb{P}_n \, , \qquad \alpha_n = (1 - e^{-\beta}) e^{-n\beta} \, ,$$

and

$$H_\varphi = -\frac{1}{\beta}\sum_{n=0}^{\infty} \ln\left[(1-e^{-\beta})e^{-n\beta}\right]\mathbb{P}_n = \sum_{n=0}^{\infty}\left[n - \frac{\ln(1-e^{-\beta})}{\beta}\right]\mathbb{P}_n$$
$$= H_{\rm osc} - \left[\frac{1}{2} + \frac{\ln(1-e^{-\beta})}{\beta}\right] I\,, \tag{39}$$

which is the Hamiltonian giving the time evolution $\alpha_\varphi(t)$, with respect to which the above ρ_φ defines the KMS state φ. Since the difference between H_φ and $H_{\rm osc}$ is just a constant, we shall identify these two Hamiltonians in the sequel.

As stated earlier, the dynamical model that we consider is that of a single electron of unit charge, placed in the xy-plane and subjected to a constant magnetic field, pointing along the *positive z-direction*. The classical Hamiltonian of the system, in convenient units, is

$$H_{\rm elec} = \frac{1}{2}(\vec{p} - \vec{A})^2 = \frac{1}{2}\left(p_x + \frac{y}{2}\right)^2 + \frac{1}{2}\left(p_y - \frac{x}{2}\right)^2\,, \tag{40}$$

where we have chosen the magnetic vector potential to be $\vec{A}^\uparrow := \vec{A} = \frac{1}{2}(-y, x, 0)$ (so that the magnetic field is $\vec{B} = \nabla \times \vec{A}^\uparrow = (0,0,1)$).

Next, on $\widetilde{\mathfrak{H}} = L^2(\mathbb{R}^2, dxdy)$, we introduce the quantized observables,

$$p_x + \frac{y}{2} \longrightarrow Q_- = -i\frac{\partial}{\partial x} + \frac{y}{2}\,, \qquad p_y - \frac{x}{2} \longrightarrow P_- = -i\frac{\partial}{\partial y} - \frac{x}{2}\,, \tag{41}$$

which satisfy $[Q_-, P_-] = iI_{\widetilde{\mathfrak{H}}}$ and in terms of which the quantum Hamiltonian corresponding to $H_{\rm elec}$ becomes

$$H^\uparrow = \frac{1}{2}\left(P_-^2 + Q_-^2\right)\,. \tag{42}$$

This is the same as the oscillator Hamiltonian in one dimension, $H_{\rm osc}$, given above (and hence the same as H_φ in (39), with our convention of identifying these two). The eigenvalues of this Hamiltonian are then $E_\ell = (\ell + \frac{1}{2})$, $\ell = 0, 1, 2, \ldots \infty$. However, this time each level is infinitely degenerate, and we will denote the corresponding normalized eigenvectors by $\Psi_{n\ell}$, with $\ell = 0, 1, 2, \ldots, \infty$, indexing the energy level and $n = 0, 1, 2, \ldots, \infty$, the degeneracy at each level. If the magnetic field were aligned along the *negative z-axis* (with $\vec{A}^\downarrow = \frac{1}{2}(y, -x, 0)$ and $\vec{B} = \nabla \times \vec{A}^\downarrow = (0, 0, -1)$), the corresponding quantum Hamiltonian would have been

$$H^\downarrow = \frac{1}{2}\left(P_+^2 + Q_+^2\right)\,, \tag{43}$$

with

$$Q_+ = -i\frac{\partial}{\partial y} + \frac{x}{2}\,, \qquad P_+ = -i\frac{\partial}{\partial x} - \frac{y}{2}\,, \tag{44}$$

and $[Q_+, P_+] = iI_{\widetilde{\mathfrak{H}}}$. The two sets of operators $\{Q_\pm, P_\pm\}$, mutually commute:

$$[Q_+, Q_-] = [P_+, Q_-] = [Q_+, P_-] = [P_+, P_-] = 0\,. \tag{45}$$

Thus, $[H^\downarrow, H^\uparrow] = 0$ and the eigenvectors $\Psi_{n\ell}$ of $H^\uparrow$ can be chosen so that they are also the eigenvectors of $H^\downarrow$ in the manner

$$H^\downarrow\Psi_{n\ell} = \left(n + \frac{1}{2}\right)\Psi_{n\ell}\,, \qquad H^\uparrow\Psi_{n\ell} = \left(\ell + \frac{1}{2}\right)\Psi_{n\ell}\,, \tag{46}$$

so that $H^\downarrow$ lifts the degeneracy of $H^\uparrow$ and vice versa.

Then, it is well known (see, for example,[7]) that the map $\mathcal{W}$ in (34) is unitary and straightforward computations (see, for example[8]) yield,

$$\mathcal{W}\begin{pmatrix} Q \vee I_{\mathfrak{H}} \\ P \vee I_{\mathfrak{H}} \end{pmatrix}\mathcal{W}^{-1} = \begin{pmatrix} Q_+ \\ P_+ \end{pmatrix}, \qquad \mathcal{W}\begin{pmatrix} I_{\mathfrak{H}} \vee Q \\ I_{\mathfrak{H}} \vee P \end{pmatrix}\mathcal{W}^{-1} = \begin{pmatrix} Q_- \\ P_- \end{pmatrix}, \tag{47}$$

and

$$\mathcal{W}\begin{pmatrix} H_{\text{osc}} \vee I_{\mathfrak{H}} \\ I_{\mathfrak{H}} \vee H_{\text{osc}} \end{pmatrix}\mathcal{W}^{-1} = \begin{pmatrix} H^\downarrow \\ H^\uparrow \end{pmatrix}, \qquad \mathcal{W}X_{n\ell} = \Psi_{n\ell}, \tag{48}$$

where the $X_{n\ell}$ are the basis vectors defined in (9) and the $\Psi_{n\ell}$ are the normalized eigenvectors defined in (46). This also means that these latter vectors form a basis of $\widetilde{\mathfrak{H}} = L^2(\mathbb{R}^2, dxdy)$. Finally, note that the two sets of operators, $\{Q_+, P_+\}$ and $\{Q_-, P_-\}$, generate the two von Neumann algebras $\mathfrak{A}_+$ and $\mathfrak{A}_-$, respectively, with $\mathcal{W}\mathfrak{A}_\ell\mathcal{W}^{-1} = \mathfrak{A}_+$ and $\mathcal{W}\mathfrak{A}_{\mathrm{r}}\mathcal{W}^{-1} = \mathfrak{A}_-$. Thus physically, the two commuting algebras correspond to the two directions of the magnetic field. The KMS state $\mathbf{\Psi} = \mathcal{W}\mathbf{\Phi}$, with $\mathbf{\Phi}$ given by (38) is just the *Gibbs equilibrium state* for this physical system.

2.1. *A second representation*

It is interesting to pursue this example a bit further using different variables in the description of the system. As before, let us consider the electron in a uniform magnetic field oriented in the positive z-direction, with vector potential $\vec{A}^\uparrow = \frac{1}{2}(-y, x, 0)$ and magnetic field $\vec{B} = \nabla \times \vec{A}^\uparrow = (0, 0, 1))$. The classical Hamiltonian is given by $H^\uparrow = \frac{1}{2}\left(\vec{p} - \vec{A}^\uparrow\right)^2$. There are several possible ways to write this Hamiltonian, which are more convenient than using the coordinates x, y and z. One such representation was used above and we indicate below a second possibility. Note that the quantized Hamiltonian may be splitted into a free part H_0 and an interaction or angular momentum part, $H^{\uparrow}_{\rm int}$:

$$\begin{cases} H^{\uparrow} = H_0 + H^{\uparrow}_{\rm int}, \\ H_0 = H_{0,x} + H_{0,y} = \frac{1}{2}\left(\widehat{p}_x^2 + \frac{\widehat{x}^2}{4}\right) + \frac{1}{2}\left(\widehat{p}_y^2 + \frac{\widehat{y}^2}{4}\right), \\ H^{\uparrow}_{\rm int} = -\frac{1}{2}(\widehat{x}\widehat{p}_y - \widehat{y}\widehat{p}_x) = -\widehat{l}_z\,. \end{cases} \tag{49}$$

with the usual definitions of $\widehat{x}, \widehat{p}_x$, etc. Of course, $[\widehat{x}, \widehat{p}_x] = [\widehat{y}, \widehat{p}_y] = iI_{\widetilde{\mathfrak{H}}}$, while all the other commutators are zero. Introducing the corresponding annihilation operators

$$a_x = \frac{1}{\sqrt{2}}[\widehat{x} + i\widehat{p}_x]\,, \qquad a_y = \frac{1}{\sqrt{2}}[\widehat{y} + i\widehat{p}_y]\,, \tag{50}$$

and their adjoints

$$a_x^* = \frac{1}{\sqrt{2}}[\widehat{x} - i\widehat{p}_x]\,, \qquad a_y^* = \frac{1}{\sqrt{2}}[\widehat{y} - i\widehat{p}_y]\,, \tag{51}$$

which satisfy the canonical commutation rules $[a_x, a_x^*] = [a_y, a_y^*] = I_{\widetilde{\mathfrak{H}}}$, while all the other commutators are zero, the hamiltonian $H^{\uparrow}$ can be written as $H^{\uparrow} = H_0 + H^{\uparrow}_{\rm int}$, with $H_0 = \left(a_x^* a_x + a_y^* a_y + I_{\widetilde{\mathfrak{H}}}\right)$, $H^{\uparrow}_{\rm int} = -i(a_x a_y^* - a_y a_x^*)$. $H^{\uparrow}$ does not appear to be diagonal even in this form, so that another *change of variables* is required.

Using the operators $Q_{\pm}, P_{\pm}$, given in (41) and (44), let us define

$$\begin{aligned} A_+ &= \frac{1}{\sqrt{2}}(Q_+ + iP_+) = \frac{3}{4}(a_x - ia_y) - \frac{1}{4}(a_x^* + ia_y^*)\,, \\ A_+^* &= \frac{1}{\sqrt{2}}(Q_+ - iP_+) = \frac{3}{4}(a_x^* + ia_y^*) - \frac{1}{4}(a_x - ia_y)\,, \\ A_- &= \frac{1}{\sqrt{2}}(iQ_- - P_-) = \frac{3}{4}(a_x + ia_y) - \frac{1}{4}(a_x^* + ia_y^*)\,, \\ A_-^* &= \frac{1}{\sqrt{2}}(-iQ_- - P_-) = \frac{3}{4}(a_x^* - ia_y^*) - \frac{1}{4}(a_x - ia_y)\,. \end{aligned} \tag{52}$$

These satisfy the commutation relations,

$$[A_{\pm}, A_{\pm}^*] = 1\,, \tag{53}$$

with all other commutators being zero. In terms of these, we may write the two Hamiltonians as (see (42) and (43)),

$$H^{\uparrow} = N_- + \frac{1}{2}I_{\widetilde{\mathfrak{H}}}\,, \quad H^{\downarrow} = N_+ + \frac{1}{2}I_{\widetilde{\mathfrak{H}}}\,, \quad \text{with} \quad N_{\pm} = A_{\pm}^* A_{\pm}\,. \tag{54}$$

Furthermore,

$$H_0 = \frac{1}{2}(N_+ + N_- + 1) \quad \text{and} \quad H^{\uparrow}_{\text{int}} = -\frac{1}{2}(N_+ - N_-)\,, \quad H^{\downarrow}_{\text{int}} = \frac{1}{2}(N_+ - N_-)\,. \tag{55}$$

The eigenstates of $H^{\uparrow}$ are now easily written down. Let Ψ_{00} be such that $A_-\Psi_{00} = A_+\Psi_{00} = 0$. Then we define

$$\Psi_{n\ell} := \frac{1}{\sqrt{n!\ell!}} \left(A_+^*\right)^n \left(A_-^*\right)^{\ell} \Psi_{00}, \tag{56}$$

where $n, \ell = 0, 1, 2, \ldots$. All the relevant operators are now diagonal in this basis: $N_+\Psi_{n\ell} = n\Psi_{n\ell}$, $N_-\Psi_{n\ell} = \ell\Psi_{n\ell}$, $H_0\Psi_{n\ell} = \frac{1}{2}(n + \ell + 1)\Psi_{n\ell}$ and $H^{\uparrow}_{\text{int}}\Psi_{n\ell} = \frac{1}{2}(n - \ell)\Psi_{n\ell}$. Hence

$$H^{\uparrow}\Psi_{n\ell} = \left(\ell + \frac{1}{2}\right)\Psi_{n\ell}. \tag{57}$$

This means that, as already stated, each level ℓ is infinitely degenerate, with n being the degeneracy index. Again, this degeneracy can be lifted in a physically interesting way namely, by considering the *reflected* magnetic field with vector potential $\vec{A}^{\downarrow} = \frac{1}{2}(y, -x, 0)$, with the magnetic field directed along the negative z-direction. The same electron considered above is now described by the other Hamiltonian, $H^{\downarrow}$, which can be written as

$$\begin{cases} H^{\downarrow} = \frac{1}{2}\left(\vec{p} - \vec{A}^{\downarrow}\right)^2 = H_0 + H^{\downarrow}_{\text{int}}, \\ H^{\downarrow}_{\text{int}} = -H^{\uparrow}_{\text{int}} \end{cases} \tag{58}$$

Thus, since $H^{\downarrow}$ can also be written as in (54), its eigenstates are again the same $\Psi_{n\ell}$ given in (56). (Recall that $[H^{\uparrow}, H^{\downarrow}] = 0$, so that they can be simultaneously diagonalized.) This also means that, as in (48), $\mathcal{W}X_{n\ell} = \Psi_{n\ell}$ and the closure of the linear span of the $\Psi_{n\ell}$'s is the Hilbert space $\tilde{\mathfrak{H}} = L^2(\mathbb{R}^2, dx\, dy)$.

It should be mentioned that in[1] yet another representation has been considered, relating the two different directions of the magnetic field to holomorphic and anti-holomorphic functions. We refer to[1] for more details on this subject, and in particular to the interesting possibility of introducing a map $\mathcal{J}$ which is, at the same time, (i) the modular map of the Tomita-Takesaki theory; (ii) the complex conjugation map; (iii) the map which reverses the uniform magnetic field, from $\vec{B}$ to $-\vec{B}$, thus transforming $\mathcal{H}^{\uparrow}$ to $\mathcal{H}^{\downarrow}$; (iv) the operator interchanging the two mutually commuting von Neumann algebras $\mathfrak{U}_{\pm}$; (v) an interwining operator in the sense of.[9] Hence the structure is quite rich both from the mathematical and from the physical sides.

We also refer to[1] for the analysis of the coherent and bi-coherent states associated to the structure considered here.

3. What for unbounded operators?

It is well known that in many physical problems, mainly involving quantum systems with infinite degrees of freedom, unbounded operators play a crucial role.[10] This makes the above construction impossible to be used, as it is. The reason is that, in this case, algebras of operators cannot be introduced easily since unbounded operators cannot be multiplied without *danger*. Several partial structures have been introduced along the years, see[10] and references therein, which are essentially set of operators for which the multiplication is only partially defined. So, in order to extend what we have described above to this case, we have first to replace the Hilbert-Schmidt operators with a different set with similar properties and which, moreover, contains unbounded operators. This can be done with a little effort[11] . We begin with some notation:

let $\mathfrak{A}$ be a linear space, $\mathfrak{A}_0 \subset \mathfrak{A}$ a *-algebra with unit $1\!\!1$: $\mathfrak{A}$ is a *quasi* **-algebra over* $\mathfrak{A}_0$ if

[i] the right and left multiplications of an element of $\mathfrak{A}$ and an element of $\mathfrak{A}_0$ are always defined and linear;

[ii] $x_1(x_2 a) = (x_1 x_2)a, (ax_1)x_2 = a(x_1 x_2)$ and $x_1(ax_2) = (x_1 a)x_2$, for each $x_1, x_2 \in \mathfrak{A}_0$ and $a \in \mathfrak{A}$;

[iii] an involution * (which extends the involution of $\mathfrak{A}_0$) is defined in $\mathfrak{A}$ with the property $(ab)^* = b^* a^*$ whenever the multiplication is defined.

A quasi *-algebra $(\mathfrak{A}, \mathfrak{A}_0)$ is *locally convex* (or *topological*) if in $\mathfrak{A}$ a locally convex topology τ is defined such that (a) the involution is continuous and the multiplications are separately continuous; and (b) $\mathfrak{A}_0$ is dense in $\mathfrak{A}[\tau]$.

Let $\{p_\alpha\}$ be a directed set of seminorms which defines τ. The existence of such a directed set can always be assumed. We can further also assume that $\mathfrak{A}[\tau]$ is *complete.*

An explicit realization of a quasi *-algebra is constructed as follows: let $\mathcal{H}$ be a separable Hilbert space and N an unbounded, densely defined, self-adjoint operator. Let $D(N^k)$ be the domain of the operator N^k, $k \in \mathbb{N}$, and $\mathcal{D}$ the domain of all the powers of N: $\mathcal{D} \equiv D^\infty(N) = \cap_{k\geq 0} D(N^k)$. This set is dense in $\mathcal{H}$. Let us now introduce $\mathcal{L}^\dagger(\mathcal{D})$, the *-algebra of all the closable operators defined on $\mathcal{D}$ which, together with their adjoints, map $\mathcal{D}$ into itself. Here the adjoint of $X \in \mathcal{L}^\dagger(\mathcal{D})$ is $X^\dagger = X^*_{\restriction\mathcal{D}}$.

In $\mathcal{D}$ the topology is defined by the following N-depending seminorms: $\phi \in \mathcal{D} \to \|\phi\|_n \equiv \|N^n\phi\|$, $n \in \mathbb{N}_0$, while the topology τ_0 in $\mathcal{L}^\dagger(\mathcal{D})$ is introduced by the seminorms

$$X \in \mathcal{L}^\dagger(\mathcal{D}) \to \|X\|^{f,k} \equiv \max\left\{\|f(N)XN^k\|, \|N^kXf(N)\|\right\},$$

where $k \in \mathbb{N}_0$ and $f \in \mathcal{C}$, the set of all the positive, bounded and continuous functions on $\mathbb{R}_+$, which are decreasing faster than any inverse power of x: $\mathcal{L}^\dagger(\mathcal{D})[\tau_0]$ is a complete *-algebra.

Let further $\mathcal{L}(\mathcal{D}, \mathcal{D}')$ be the set of all continuous maps from $\mathcal{D}$ into $\mathcal{D}'$, with their topologies (in $\mathcal{D}'$ this is the strong dual topology), and let τ denotes the topology defined by the seminorms

$$X \in \mathcal{L}(\mathcal{D}, \mathcal{D}') \to \|X\|^f = \|f(N)Xf(N)\|,$$

$f \in \mathcal{C}$. Then $\mathcal{L}(\mathcal{D}, \mathcal{D}')[\tau]$ is a complete vector space.

In this case $\mathcal{L}^\dagger(\mathcal{D}) \subset \mathcal{L}(\mathcal{D}, \mathcal{D}')$ and the pair

$$(\mathcal{L}(\mathcal{D},\mathcal{D}')[\tau], \mathcal{L}^\dagger(\mathcal{D})[\tau_0])$$

is a *concrete realization* of a locally convex quasi *-algebra.

We will discuss how to define a sort of *quantum dynamics* and a state over a *left* and a *right* version of $\mathcal{L}^\dagger(\mathcal{D})$ which is a KMS state with respect to this dynamics. For that, as already stated, we first need to identify an ideal of $\mathcal{L}^\dagger(\mathcal{D})$ which should play the same role that the Hilbert-Schmidt operators play in our previous construction. There was a lot of interest in ideals in algebras of unbounded operators, and many possible sets were proposed,[12] . In this section we will assume that $\mathcal{D}$ is a dense subset of the Hilbert space $\mathcal{H}$, and that an orthonormal basis of $\mathcal{H}$, $\mathcal{F} = \{\varphi_j,\ j \in I\}$, exists whose vectors all belong to $\mathcal{D}$: $\varphi_j \in \mathcal{D}$, $\forall j \in I$, where I is a given set of indexes. We also assume that $\mathcal{L}^\dagger(\mathcal{D})$ possesses the identity $\mathbb{1}$. Here we consider the following set

$$\mathcal{N} = \left\{X \in \mathcal{L}^\dagger(\mathcal{D}) : |tr(AXB)| < \infty, \quad \forall A, B \in \mathcal{L}^\dagger(\mathcal{D})\right\}, \tag{59}$$

where tr is the trace in $\mathcal{H}$. Of course $\mathcal{N}$ is a linear vector space. It is closed under the adjoint and it is an ideal for $\mathcal{L}^\dagger(\mathcal{D})$: if $X \in \mathcal{N}$ then $AXB \in \mathcal{N}$, for all A and B in $\mathcal{L}^\dagger(\mathcal{D})$. Hence, for each $X \in \mathcal{N}$, both $\left|tr(AXX^\dagger B)\right|$ and $\left|tr(AX^\dagger XB)\right|$ exist finite, for all A and B in $\mathcal{L}^\dagger(\mathcal{D})$.

Let $\Phi, \Psi \in \mathcal{D}$. It is possible to check that the operator X defined on $\mathcal{H}$ as $Xf = \langle\Psi, f\rangle\,\Phi$ is an operator in $\mathcal{N}$, which using the Dirac's notation we will indicate as $X = |\Phi\rangle\langle\Psi|$. Then the set $\mathcal{N}$ is rather rich. In particular it contains all the operators $X_{i,j} := |\varphi_i\rangle\langle\varphi_j|$, $i, j \in I$. We call $\mathcal{G}$ the set of all these operators. As for the Hilbert-Schmidt operators we can introduce on $\mathcal{N}$ a scalar product defined as

$$\langle X, Y\rangle_2 = tr(X^\dagger Y), \tag{60}$$

for all $X, Y \in \mathcal{N}$. This map is well defined on $\mathcal{N}$ and, moreover, it is a scalar product. The set $\mathcal{G}$ is orthonormal with respect to this product. Indeed we find

$$\langle X_{i,j}, X_{k,l}\rangle_2 = \delta_{i,k}\delta_{j,l}, \tag{61}$$

for all i, j, k, and l in I. Moreover, $\mathcal{G}$ is complete in $\mathcal{N}$ with respect to this scalar product: indeed, the only element $Z \in \mathcal{N}$ which is orthogonal to all the $X_{i,j}$'s is $Z = 0$. We can also check that the only element of $\mathcal{L}^\dagger(\mathcal{D})$ which is orthogonal to all the $X_{i,j}$'s is again the 0 operator. In this sense, therefore, $\mathcal{G}$ is also complete in $\mathcal{L}^\dagger(\mathcal{D})$.

The operators in $\mathcal{N}$ are very good ones. Indeed it is possible to check that, if $X \in \mathcal{N}$, then X is bounded and $\|X\| \leq \|X\|_2$, where $\|.\|$ is the usual norm in $\mathcal{L}(\mathcal{H})$. This means that $\mathcal{N}$ is a subset of the Hilbert space of the Hilbert-Schmidt operators, $\mathcal{B}_2(\mathcal{H})$: $\mathcal{N} \subseteq \mathcal{B}_2(\mathcal{H})$. However, it is still not clear wether N is equal to $\mathcal{B}_2(\mathcal{H})$ or not. This is part of the work in progress.[13]

Let us now define $\sigma_2 = \overline{\mathcal{N}}^{\|.\|_2}$. This is an Hilbert space with scalar product $\langle ., .\rangle_2$ which can, at most, coincide with $\mathcal{B}_2(\mathcal{H})$. $\mathcal{G}$ is an orthonormal basis of σ_2.

As before we now introduce a map $\vee$ which, for all $A, B \in \mathcal{L}^\dagger(\mathcal{D})$ and $X \in \mathcal{N}$, associates an element of $\mathcal{N}$:

$$(A \vee B)(X) := AXB^\dagger \tag{62}$$

Of course, if $\mathcal{N} \subset \sigma_2$ then the problem of extending $\vee$ to σ_2 arises.[13]

This map, analogously to the one introduced in Section 1, satisfies the following properties:

$$(A \vee B)^\dagger = A^\dagger \vee B^\dagger, \quad (A_1 \vee B_1)(A_2 \vee B_2) = (A_1A_2) \vee (B_1B_2)$$

for all $A, B, A_1, B_1, A_2, B_2 \in \mathcal{L}^\dagger(\mathcal{D})$. Next, in analogy with what done before, we associate to $\mathcal{L}^\dagger(\mathcal{D})$ two different algebras, $\mathcal{L}_l^\dagger(\mathcal{D})$ and $\mathcal{L}_r^\dagger(\mathcal{D})$ defined as follows:

$$\mathcal{L}_l^\dagger(\mathcal{D}) := \{A_l := A \vee 1\!\!1, \quad A \in \mathcal{L}^\dagger(\mathcal{D})\}, \mathcal{L}_r^\dagger(\mathcal{D}) := \{A_l := 1\!\!1 \vee A, \quad A \in \mathcal{L}^\dagger(\mathcal{D})\}$$

We also define a map J acting on $\mathcal{G}$ as follows: $J(X_{i,j}) = X_{j,i}$, $i, j \in I$. Properties similar to those discussed in Section 1 can be recovered for $\mathcal{L}_l^\dagger(\mathcal{D})$, $\mathcal{L}_r^\dagger(\mathcal{D})$ and J,[13].

Let now λ_i, $i \in I$ be a sequence of strictly positive numbers such that $\sum_{i\in I} \lambda_i = 1$. We define a vector $\Phi \in \sigma_2$ as

$$\Phi = \sum_{i\in I} \sqrt{\lambda_i} X_{i,i}. \tag{63}$$

We will assume here that, in particular, $\Phi \in \mathcal{N}$.

Using Φ we can now define two states over $\mathcal{L}_l^\dagger(\mathcal{D})$ and $\mathcal{L}_r^\dagger(\mathcal{D})$ as follows:

$$\varphi(A_l) := \langle \Phi, A_l \Phi \rangle_2 = \langle \Phi, A\Phi \rangle_2, \qquad A_l \in \mathcal{L}_l^\dagger(\mathcal{D}), \tag{64}$$

and

$$\tilde{\varphi}(A_r) := \overline{\langle \Phi, A_r \Phi \rangle_2} = \langle \Phi, A\Phi \rangle_2, \qquad A_r \in \mathcal{L}_r^\dagger(\mathcal{D}). \tag{65}$$

We deduce that, if $A_r = 1\!\!1 \vee A$ and $A_l = A \vee 1\!\!1$, $\tilde{\varphi}(A_r) = \varphi(A_l)$. Hence the two states are "very close" to each other. Therefore we will concentrate only on the properties of φ, since those of $\tilde{\varphi}$ are completely analogous and can be proved in the same way. First we observe that φ can be written in a trace form by introducing the operator $\rho_\varphi := \sum_{n\in I} \lambda_n X_{n,n}$. Indeed we find that

$$\varphi(A_l) = tr(\rho_\varphi A), \qquad A_l \in \mathcal{L}_l^\dagger(\mathcal{D}). \tag{66}$$

This suggests to call φ, with a little abuse of notation, a *normal state* over $\mathcal{L}_l^\dagger(\mathcal{D})$. We remark that ρ_φ is bounded, $\|\rho_\varphi\| \leq \sum_{n\in I} \lambda_n = 1$, and, more than this, it is trace class, so that φ is normalized: $\varphi(1\!\!1_l) = tr(\rho_\varphi) = \sum_{n\in I} \lambda_n = 1$. We can also also check that $\rho_\varphi \in \sigma_2$, since $\langle \rho_\varphi, \rho_\varphi \rangle_2 = \sum_{n\in I} \lambda_n^2 < \infty$. Moreover, ρ_φ is self-adjoint and satisfies the eigenvalue equation $\rho_\varphi \varphi_j = \lambda_j \varphi_j$, $j \in I$.

The state φ is positive and faithful on $\mathcal{L}_l^\dagger(\mathcal{D})$. Positivity is clear. As for faithfulness, since for all $A_l \in \mathcal{L}_l^\dagger(\mathcal{D})$ $\varphi(A_l^\dagger A_l) = tr(\rho_\varphi A^\dagger A) = \sum_{n\in I} \lambda_n \|A\varphi_n\|^2$, then if $\varphi(A_l^\dagger A_l) = 0$, the positivity of the λ_n's implies that $A\varphi_n = 0$ for all $n \in I$, so that $A = 0$ and, as a consequence, $A_l = 0$.

The vector Φ is cyclic for $\mathcal{L}_l^\dagger(\mathcal{D})$: let $X \in \mathcal{N}$ be orthogonal to $A_l\Phi$, for all $A_l \in \mathcal{L}_l^\dagger(\mathcal{D})$. Then we can show that $X = 0$. Indeed we have, for all $A_l \in \mathcal{L}_l^\dagger(\mathcal{D})$

$$0 = \langle X, A_l \Phi \rangle_2 = \sum_{n\in I} \lambda_n \left\langle \varphi_n, X^\dagger A \varphi_j \right\rangle.$$

Hence, in particular, $\langle X, A_l \Phi \rangle_2 = 0$ if $A_l = A \vee 1\!\!1$ with $A = X_{k,l}$, for $k, l \in I$. But, since $\lambda_n > 0$ for all $n \in I$, $\langle X\varphi_l, \varphi_k \rangle = 0$ for all $k, l \in I$. Then $X = 0$.

The vector Φ is also separating for $\mathcal{L}_l^\dagger(\mathcal{D})$. Suppose in fact that $A_l\Phi = 0$. Then we can prove that $A_l = 0$. This follows from the fact that $0 =$

$\langle A_l\Phi, A_l\Phi\rangle_2 = \sum_{n\in I} \lambda_n \|A\varphi_n\|^2$, which implies again that $A = 0$ and $A_l = 0$.

We can now define an operator $P_{i,j}$ mapping $\mathcal{N}$ into $\mathcal{N}$ in the following way:

$$P_{i,j} := X_{i,i} \vee X_{j,j} \tag{67}$$

It is possible to check that $P_{i,j}^\dagger = P_{i,j}$ and that $P_{i,j}P_{k,l} = \delta_{i,k}\delta_{j,l}P_{i,j}$. Let us further introduce an operator H_φ acting on $\mathcal{F}$ as follows:

$$H_\varphi\, \varphi_j = -\frac{1}{\beta}\log(\lambda_j)\,\varphi_j, \tag{68}$$

$j \in I$. Here β is a positive constant which, in the physical literature, is usually called *the inverse temperature.* Then H_φ can be written as $H_\varphi = -\frac{1}{\beta}\sum_{j\in I}\log(\lambda_j)\, X_{j,j}$, where, of course, the sum converges on the dense domain of all the finite linear combinations of the vectors of $\mathcal{F}$.

Via the spectral theorem we deduce that ρ_φ and H_φ are related to each other. Indeed we have $\rho_\varphi = e^{-\beta H_\varphi}$. Extending the map $\vee$ we can also define the following operators:

$$H_\varphi^l = H_\varphi \vee 1\!\!1; \quad H_\varphi^r = 1\!\!1 \vee H_\varphi; \quad h_\varphi = H_\varphi^l - H_\varphi^r \tag{69}$$

which can be written as

$$\begin{array}{l} H_\varphi^l = -\frac{1}{\beta}\sum_{i,j\in I}\log(\lambda_i)\, P_{i,j}; \quad H_\varphi^r = -\frac{1}{\beta}\sum_{i,j\in I}\log(\lambda_j)\, P_{i,j}; \\ h_\varphi = -\frac{1}{\beta}\sum_{i,j\in I}\log\left(\frac{\lambda_i}{\lambda_j}\right) P_{i,j}. \end{array} \tag{70}$$

In particular h_φ can be used to define the operator Δ_φ as follows:

$$\Delta_\varphi = e^{-\beta h_\varphi} = \sum_{i,j\in I}\frac{\lambda_i}{\lambda_j}\, P_{i,j}, \tag{71}$$

and Δ_φ can be used to define a *Schrödinger dynamics*:

$$e^{ith_\varphi} = (\Delta_\varphi)^{-it/\beta} = \sum_{i,j\in I}\left(\frac{\lambda_i}{\lambda_j}\right)^{-it/\beta} P_{i,j}. \tag{72}$$

It is possible to check that $e^{ith_\varphi} = e^{itH_\varphi} \vee e^{itH_\varphi}$ and that $e^{ith_\varphi}(\Phi) = \Phi$, which is therefore invariant under our *time evolution*. Using e^{ith_φ} we can also define an Heisemberg dynamics on $\mathcal{L}_l^\dagger(\mathcal{D})$:

$$\alpha_\varphi^t(A_l) = e^{ith_\varphi}\, A_l\, e^{-ith_\varphi} = \left(e^{itH_\varphi}\, A\, e^{-itH_\varphi}\right) \vee 1\!\!1 \tag{73}$$

If $e^{itH_\varphi} \in \mathcal{L}^\dagger(\mathcal{D})$, $\alpha_\varphi^t(A_l)$ belongs to $\mathcal{L}_l^\dagger(\mathcal{D})$ for all $A_l \in \mathcal{L}_l^\dagger(\mathcal{D})$. The invariance of Φ under the action of e^{ith_φ} implies that $\varphi(\alpha_\varphi^t(A_l)) = \varphi(A_l)$ for all $A_l \in \mathcal{L}_l^\dagger(\mathcal{D})$. Moreover, we can also check that φ is a KMS state with respect to the time evolution α_φ^t.

Of course, now the next step is to apply this procedure to some concrete example, like the one we have considered in Section 2. Particularly interesting for us will be consider some concrete application of the procedure to QM_∞. This is work in progress,[13] .

Acknowledgements

I want to express my gratitude to the local organizers of the conference, and in particular to Prof. Rolando Rebolledo, for their patience.

References

1. F. Bagarello, S.T. Ali, G. Honnouvo, *Modular Structures on Trace Class Operators and Applications to Landau Levels*, J. Phys. A, doi:10.1088/1751-8113/43/10/105202, **43**, 105202 (2010) (17pp)
2. S.T. Ali and G.G. Emch, *Geometric quantization: Modular reduction theory and coherent states*, J. Math. Phys. **27**, 2936-2943 (1986).
3. G.G. Emch, *Prequantization and KMS structures*, Intern. J. Theoret. Phys. **20**, 891 - 904 (1981).
4. M. Takesaki, *Tomita's Theory of Modular Hilbert Algebras and its Applications*, (Springer-Verlag, New York, 1970).
5. G.L. Sewell, *Quantum Mechanics and its Emergent Macrophysics*, (Princeton University Press, Princeton, 2002).
6. S. Stratila, *Modular Theory in Operator Algebras*, (Abacus Press, Tunbridge Wells, 1981).
7. S.T. Ali, J-P. Antoine and J-P. Gazeau, *Coherent States, Wavelets and Their Generalizations*, (Springer-Verlag, New York, 2000).
8. S.T. Ali, and F. Bagarello, *Some physical appearances of vector coherent states and coherent states related to degenerate Hamiltonians*, J. Math Phys. **46**, 053518 1 - 053518 18 (online version) (2005).
9. S. Kuru, A. Tegmen and A. Vercin, *Intertwined isospectral potentials in an arbitrary dimension*, J. Math. Phys, **42**, 3344-3360, (2001); S. Kuru, B. Demircioglu, M. Onder and A. Vercin, *Two families of superintegrable and isospectral potentials in two dimensions*, J. Math. Phys, **43**, 2133-2150, (2002); K. A. Samani and M. Zarei, *Intertwined hamiltonians in two-dimensional curved spaces*, Ann. of Phys., **316**, 466-482, (2005).
10. F. Bagarello *Algebras of unbounded operators and physical applications: a survey*, Reviews in Math. Phys, **19**, No. 3, 231-272 (2007)
11. J.-P. Antoine, A. Inoue and C. Trapani *Partial *-algebras and Their Operator Realizations*, Kluwer, Dordrecht, 2002
12. W. Timmermann, *Ideals in algebras of unbounded operators*, Math. Nachr.,

92, 99-110 (1979); W. Timmermann, *Ideals in algebras of unbounded operators.II*, Math. Nachr., **93**, 313-318 (1979)

13. F. Bagarello, A. Inoue and C. Trapani *Modular structure in $\mathcal{L}^\dagger(\mathcal{D})$*, in preparation

STOCHASTIC SCHRÖDINGER EQUATIONS AND MEMORY

A. BARCHIELLI

Department of Mathematics, Politecnico di Milano,
Piazza Leonardo da Vinci, I-20133 Milano, Italy.
Also: Istituto Nazionale di Fisica Nucleare, Sezione di Milano.

P. DI TELLA

Institut für Stochastik Friedrich-Schiller-Universität Jena, Fakultät für Mathematik und Informatik, Ernst-Abbe-Platz 1-4, 07743 Jena, Germany.

C. PELLEGRINI

Laboratoire de Statistique et Probabilités, Université Paul Sabatier,
118, Route de Narbonne, 31062 Toulouse Cedex 4, France.

F. PETRUCCIONE

School of Physics and National Institute for Theoretical Physics, University of KwaZulu-Natal, Private Bag X54001, Durban 4000, South Africa

By starting from the stochastic Schrödinger equation and quantum trajectory theory, we introduce memory effects by considering stochastic adapted coefficients. As an example of a natural non-Markovian extension of the theory of white noise quantum trajectories we use an Ornstein-Uhlenbeck coloured noise as the output driving process. Under certain conditions a random Hamiltonian evolution is recovered. Moreover, we show that our non-Markovian stochastic Schrödinger equations unravel some master equations with memory kernels.

Keywords: Stochastic Schrödinger equation; Non Markovian quantum master equation; Unravelling; Quantum trajectories.

1. Introduction

The problem of how to describe the reduced dynamics of a quantum open system interacting with an environment is very important.[1–3] More and more applications demand to treat dissipative effects, tendency to equilibrium, decoherence,... or how to have more equilibrium states, survival of coherences and entanglement,... in spite of the interaction with the external environment. The open system dynamics is often described in terms of *quantum master equations* which give the time evolution of the density matrix of the small system. When the Markov approximation is good (no memory effects) the situation is well understood: if the generator of the dynamics has the "Lindblad structure", then the dynamics sends statistical operators into statistical operators and it is completely positive.[4,5]

However, for many new applications the Markovian approximation is not applicable. Such a situation appears in several concrete physical models: strong coupled systems, entanglement and correlation in the initial state, finite reservoirs... This gives rise to the theory of *non-Markovian quantum dynamics*, for which does not exist a general theory, but many different approaches.[6–23]

Non Markovian reduced dynamics are usually obtained from the total dynamics of system plus bath by projection operator techniques such as Nakajima-Zwanzig operator technique, time-convolutionless operator technique,[2,10] correlated projection operator or Lindblad rate equations.[16,19] These techniques yield in principle exact master equations for the evolution of the subsystem. For example Nakajima-Zwanzig technique gives rise to an integro-differential equation with a memory kernel involving a retarded time integration over the history of the small system. However, in most of the cases the exact evolutions remain of formal interest: no analytic expression of the solution, impossible to simulate... Usually, some approximations have to be used to obtain a manageable description. But as soon as an approximation is done, the resulting equation can violate the complete positivity property; let us stress that the complete positivity (and even positivity) is a major question in non-Markovian systems.[18,22]

The easiest way to preserve complete positivity is to introduce approximate or phenomenological equations at the Hilbert space level, an approach which is useful also for numerical simulations. We can say that in this way one is developing a non-Markovian theory of *unravelling* and of "Quantum Monte Carlo methods".[2,6–9,12,24] In the Markovian case such an approach is related to the so called stochastic Schrödinger equation, quantum trajectory theory, measurements in continuous time. It provides wide applications for optical quantum systems and description of experiments such as *photodetection* or *heterodyne/homodyne detection*.[24–31] In the non-Markovian case, an active line of research concentrates on a similar interpretation of non-Markovian unravelling. In this context, the question is more involved (for example, when complete positivity is violated the answer is hopeless) and remains an open problem. For the usual scheme of indirect quantum measurement it has been shown that in general such an interpretation is not accessible.[21,23] Actually, only few positive answers for very special cases have been found and this question is still highly debated.[12,13,20,21]

Our aim is to introduce memory at Hilbert space level, in order to guarantee at the end a completely positive dynamics, and to maintain the possibility of the measurement interpretation. Our approach is based on the introduction of stochastic coefficients depending on the past history and on the use of coloured noises.[32–34]

In Sect. 2 we introduce a special case of stochastic Schrödinger equation with memory. The starting point is a generalisation of the usual theory of the linear stochastic Schrödinger equation,[31–33] based upon the introduction of random coefficients. This approach introduces memory effects in the underlying dynamics. The main interest is that the complete positivity is preserved and a measurement interpretation can be developed. We present this theory only in the context of the diffusive stochastic Schrödinger equation.

In Sect. 3 we attach the problem of memory by introducing an example of *coloured bath* and we show that we obtain a model of random Hamiltonian evolution,[34] while we remain in the general framework of Sect. 2.

Finally, in Sect. 4, by using Nakajima-Zwanzig projection techniques, we show that the mean states satisfy closed master equations with memory kernels, which automatically preserve complete positivity. Moreover, we can say that the stochastic Schrödinger equations of the previous sections are unravellings of these memory master equations.

2. A non Markovian stochastic Schrödinger equation

The linear stochastic Schrödinger equation (lSSE) is the starting point to construct unravelling of master equations and models of measurements in continuous time.[27,31] By introducing random coefficients in such equation, but maintaining its structure, we get memory in the dynamical equations, while complete positivity of the dynamical maps and the continuous measurement interpretation are preserved.[32,33] To simplify the theory we consider only diffusive contributions and bounded operators.

Assumption 2.1 (The linear stochastic Schrödinger equation). *Let $\mathcal{H}$ be a complex, separable Hilbert space, the space of the quantum system, and $(\Omega, \mathcal{F}, (\mathcal{F}_t), \mathbb{Q})$ be a stochastic basis satisfying the usual hypotheses, where a d-dimensional continuous Wiener process is defined; $\mathbb{Q}$ will play the* *role of a reference probability measure. The lSSE we consider is*

$$\mathrm{d}\psi(t) = K(t)\psi(t)\mathrm{d}t + \sum_{j=1}^{d} R_j(t)\psi(t)\mathrm{d}W_j(t), \tag{1}$$

$$\psi(0) = \psi_0 \in L^2(\Omega, \mathcal{F}_0, \mathbb{Q}; \mathcal{H}).$$

Let us denote by $\mathcal{T}(\mathcal{H})$ the trace class on $\mathcal{H}$, by $\mathcal{S}(\mathcal{H})$ the subset of the statistical operators and by $\mathcal{L}(\mathcal{H})$ the space of the linear bounded operators.

Assumption 2.2 (The random coefficients). *The coefficient in the drift has the structure*

$$K(t) = -\mathrm{i}H(t) - \frac{1}{2}\sum_{j=1}^{d} R_j(t)^* R_j(t). \tag{2}$$

The coefficients $H(t)$, $R_j(t)$ are random bounded operators with $H(t) = H(t)^*$, say predictable càglàd processes in $\big(\Omega, \mathcal{F}, (\mathcal{F}_t), \mathbb{Q}\big)$.

Moreover, $\forall T > 0$, we have

$$\int_0^T \mathbb{E}_{\mathbb{Q}}\left[\|H(t)\|\right] \mathrm{d}t < +\infty, \tag{3a}$$

$$\mathbb{E}_{\mathbb{Q}}\left[\exp\left\{2\sum_{j=1}^{d}\int_0^T \|R_j(t)\|^2 \,\mathrm{d}t\right\}\right] < +\infty. \tag{3b}$$

Theorem 2.1. *Under Assumptions 2.1, 2.2, the lSSE (1) has a pathwise unique solution. The square norm $\|\psi(t)\|^2$ is a continuous positive martingale given by*

$$\|\psi(t)\|^2 = \|\psi_0\|^2 \exp\left\{\sum_j \left[\int_0^t m_j(s)\mathrm{d}W_j(s) - \frac{1}{2}\int_0^t m_j(s)^2 \,\mathrm{d}s\right]\right\}, \tag{4}$$

$$m_j(t) := 2\,\mathrm{Re}\left\langle \hat{\psi}(t) \middle| R_j(t)\hat{\psi}(t)\right\rangle, \tag{5}$$

$$\hat{\psi}(t) := \begin{cases} \psi(t)/\left\|\psi(t)\right\|, & \text{if } \|\psi(t)\| \neq 0, \\ v \text{ (fixed unit vector)}, & \text{if } \|\psi(t)\| = 0. \end{cases} \tag{6}$$

Proof. Assumptions 2.1, 2.2 imply the Hypotheses of [32, Proposition 2.1 and Theorem 2.4], but Hypothesis 2.3.A of page 295. According to the discussion at the end of p. 297, this last hypothesis can be substituted by (3b), which implies Novikov condition, a sufficient condition for an exponential supermartingale to be a martingale. Then, all the statements hold. □

Remark 2.1. By expression (4) we get that on the set $\{\|\psi_0\| > 0\}$ we have $\|\psi(t)\| > 0$ $\mathbb{Q}$-a.s. This means that, if $\psi_0 \neq 0$ $\mathbb{Q}$-a.s., then the process $\hat{\psi}(t)$ (6) is almost surely defined by the normalisation of $\psi(t)$ and the arbitrary vector v does not play any role with probability one.

Remark 2.2. Let us define the positive, $\mathcal{T}(\mathcal{H})$-valued process

$$\sigma(t) := |\psi(t)\rangle\langle\psi(t)|. \tag{7}$$

By applying the Itô formula to $\langle\psi(t)|a\psi(t)\rangle$, $a \in \mathcal{L}(\mathcal{H})$, we get the weak-sense linear stochastic master equation (lSME)

$$\mathrm{d}\sigma(t) = \mathcal{L}(t)[\sigma(t)]\mathrm{d}t + \sum_{j=1}^{d} \mathcal{R}_j(t)[\sigma(t)]\mathrm{d}W_j(t), \tag{8}$$

$$\mathcal{R}_j(t)[\rho] := R_j(t)\rho + \rho R_j(t)^*, \tag{9}$$

$$\mathcal{L}(t)[\rho] = -\mathrm{i}[H(t),\rho] + \sum_{j=1}^{d}\left(R_j(t)\rho R_j(t)^* - \frac{1}{2}\{R_j(t)^*R_j(t),\rho\}\right); \tag{10}$$

$\mathcal{L}(t)$ is the random Liouville operator [32, Proposition 3.4].

Assumption 2.3 (The initial condition). *Let us assume that the initial condition ψ_0 is normalised, in the sense that $\mathbb{E}_{\mathbb{Q}}\left[\|\psi_0\|^2\right] = 1$. Then, $\varrho_0 := \mathbb{E}_{\mathbb{Q}}\left[|\psi_0\rangle\langle\psi_0|\right] \in \mathcal{S}(\mathcal{H})$ represents the initial statistical operator.*

Remark 2.3. Under the previous assumptions $p(t) := \|\psi(t)\|^2$ is a positive, mean-one martingale and, $\forall T > 0$, we can define the new probability law on $(\Omega, \mathcal{F}_T)$

$$\forall F \in \mathcal{F}_T \qquad \mathbb{P}^T(F) := \mathbb{E}_{\mathbb{Q}}[p(T)1_F]. \tag{11}$$

By the martingale property these new probabilities are consistent in the sense that, for $0 \leq s < t$ and $F \in \mathcal{F}_s$, we have $\mathbb{P}^t(F) = \mathbb{P}^s(F)$.

The new probabilities are interpreted as the physical ones, the law of the output of the time continuous measurement. Let us stress that it is possible to express the physical probabilities in agreement with the axiomatic formulation of quantum mechanics by introducing positive operator valued measures and completely positive instruments.[32,33]

Remark 2.4. By Girsanov theorem, the d-dimensional process

$$\widehat{W}_j(t) := W_j(t) - \int_0^t m_j(s)\,\mathrm{d}s, \qquad j = 1,\dots,d, \quad t \in [0,T], \tag{12}$$

is a standard Wiener process under the physical probability $\mathbb{P}^T$ [32, Proposition 2.5, Remark 2.6].

By adding further sufficient conditions two more important equations can be obtained.

Assumption 2.4 ([32, Hypotheses 2.3.A]). *Let us assume that we have*

$$\sup_{\omega\in\Omega}\int_0^t\Big\|\sum_{j=1}^d R_j(s,\omega)^*R_j(s,\omega)\Big\|\mathrm{d}s<+\infty. \tag{13}$$

Theorem 2.2. *Let Assumptions 2.1–2.4 hold. Under the physical probability the normalized state $\hat{\psi}(t)$, introduced in Eq.* (6), *satisfies the non-linear* stochastic Schrödinger equation *(SSE)*

$$\mathrm{d}\hat{\psi}(t)=\sum_j\left[R_j(t)-\mathrm{Re}\,n_j\left(t,\hat{\psi}(t)\right)\right]\hat{\psi}(t)\,\mathrm{d}\widehat{W}_j(t)+K(t)\hat{\psi}(t)\,\mathrm{d}t$$
$$+\sum_j\left[\left(\mathrm{Re}\,n_j\left(t,\hat{\psi}(t)\right)\right)R_j(t)-\frac{1}{2}\left(\mathrm{Re}\,n_j\left(t,\hat{\psi}(t)\right)\right)^2\right]\hat{\psi}(t)\,\mathrm{d}t\,, \tag{14}$$

where $n_j(t,x):=\langle x|R_j(t)x\rangle$, $\forall t\in[0,+\infty)$, $j=1,\dots,d$, $x\in\mathcal{H}$.

*Moreover, the process (*a priori or average states*) defined by*

$$\eta(t):=\mathbb{E}_{\mathbb{Q}}[\sigma(t)],\quad \text{or}\quad \mathrm{Tr}\,\{a\eta(t)\}=\mathbb{E}_{\mathbb{Q}}\left[\langle\psi(t)|a\psi(t)\rangle\right]\quad \forall a\in\mathcal{L}(\mathcal{H}) \tag{15}$$

satisfies the master equation

$$\eta(t)=\varrho_0+\int_0^t\mathbb{E}_{\mathbb{Q}}\left[\mathcal{L}(s)[\sigma(s)]\right]\mathrm{d}s. \tag{16}$$

Proof. One can check that all the hypotheses of [32, Theorem 2.7] hold. Then, the SSE for $\hat{\psi}(t)$ follows.

As in the proof of [32, Propositions 3.2], one can prove that the stochastic integral in the lSME (8) has zero mean value. Then, Eq. (16) follows. □

Note that, by the definition of the physical probabilities, we have also

$$\mathrm{Tr}\,\{a\eta(t)\}=\mathbb{E}_{\mathbb{P}^T}\left[\langle\hat{\psi}(t)|a\hat{\psi}(t)\rangle\right],\quad \forall a\in\mathcal{L}(\mathcal{H}),\quad \forall t,T:0\le t\le T. \tag{17}$$

The SSE (14) is the starting point for numerical simulations; the key point is that norm of its solution $\hat{\psi}(t)$ is constantly equal to one. We underline that Eq. (16) is not a closed equation for the mean state of the system. In the last section we shall see how to obtain, in principle, a closed equation for the *a priori* states of the quantum system.

The finite dimensional case

If we assume a finite dimensional Hilbert space and we strengthen the conditions on the coefficients, we obtain a more rich theory. We discuss here below the situation.[33]

Assumption 2.5.

(1) The Hilbert space of the quantum system is finite dimensional, say $\mathcal{H} := \mathbb{C}^n$. We write $M_n(\mathbb{C})$ for the space of the linear operators on $\mathcal{H}$ into itself ($n \times n$ complex matrices).
(2) The coefficient processes R_j and H are $M_n(\mathbb{C})$-valued and progressive with respect to the reference filtration.
(3) The coefficients satisfy the following conditions: for every $T > 0$ there exist two positive constants $M(T)$ and $L(T)$ such that

$$\begin{aligned} &\sup_{\omega\in\Omega}\sup_{t\in[0,T]}\left\|\sum_{j=1}^{d} R_j(t,\omega)^* R_j(t,\omega)\right\| \leq L(T) < \infty, \\ &\sup_{\omega\in\Omega}\sup_{t\in[0,T]} \|H(t,\omega)\| \leq M(T) < \infty. \end{aligned} \tag{18}$$

Under Assumption 2.5 it is possible to prove existence and pathwise uniqueness of the solution of the lSSE just modifying classical results for existence and uniqueness of the solution for stochastic differential equation with deterministic coefficients.

Moreover, it is possible to prove that the solution of lSSE fulfil some L^p estimate: in this point the finite dimension of the Hilbert space plays a fundamental role because the bounds we obtain for the process $\psi(t)$ involve constants depending on n.

Obviously, in this context the martingale property of the norm of the solution is still valid and so one can define the consistent family of physical probabilities. It is also possible to introduce the propagator of the lSSE, that is the two times $M_n(\mathbb{C})$-valued stochastic process $A(t,s)$ such that $A(t,s)\psi(s) = \psi(t)$, for all $t, s \geq 0$ s.t. $s \leq t$. We are able to obtain a stochastic differential equation (with pathwise unique solution) for the propagator and, by means of it, to prove that the propagator takes almost surely values in the space of the invertible matrices. The L^p estimates for $\psi(t)$ are useful to obtain L^p estimates on $A(t,s)$. Furthermore, the propagator satisfies the typical composition law of an evolution: $A(t,s) = A(t,r)A(r,s)$ for all $t, r, s \geq 0$ s.t. $s \leq r \leq t$.

The almost sure invertibility of the propagator guarantees that the process $\hat{\psi}(t)$ can be almost surely defined and that this process satisfies, under the physical probabilities, a non linear SSE, similar to Eq. (14). It is possible to prove that in this case the SSE has a pathwise unique solution.

When we go on extending the theory to the space of the statistical operators, we can take as initial condition a random statistical operator or a deterministic one. We define the process $\sigma(t)$ as in Eq. (7) and, by using the Itô formula, we obtain an equation formally similar to the lSME, but in this case we are able to prove the uniqueness of its solution given the initial statistical operator ϱ_0 (the existence comes out by construction). In this way we can say that the lSME is the evolution equation of the quantum system, when the initial condition is a deterministic (or even random) statistical operator ϱ_0. We can introduce the propagator of the lSME, which is a two times-linear map valued stochastic process, say $\Lambda(t,s)$, such that $\Lambda(t,0)[\varrho_0] = \sigma(t)$, $\Lambda(t,s) = \Lambda(t,r) \circ \Lambda(r,s)$, for all $t,r,s \geq 0$ s.t. $s \leq r \leq t$ and $\Lambda(t,s)[\tau] = A(t,s)\tau A(t,s)^*$. From the last expression of the propagator of the lSME, it comes out that this is a completely positive map valued process.

Also in this case we can introduce a consistent family of physical probabilities. Indeed, the process $\mathrm{Tr}\{\sigma(t)\}$ is an exponential mean-one martingale that can be used to define the new probability laws, as we did in the Hilbert space.

It is then possible to define the normalisation of $\sigma(t)$ with respect to its trace,

$$\varrho(t) = \frac{\sigma(t)}{\mathrm{Tr}\{\sigma(t)\}}$$

and, under the physical probabilities, we have the following non linear equation for $\varrho(t)$, with pathwise unique solution

$$\begin{cases} \mathrm{d}\varrho(t) = \mathcal{L}(t)[\varrho(t)]\mathrm{d}t + \sum_{j=1}^{d} \{\mathcal{R}_j(t)[\varrho(t)] - v_j(t)\varrho(t)\}\, \mathrm{d}\widehat{W}_j(t)\,, & t \geq 0 \\ \varrho(0) = \varrho_0\,, \end{cases} \tag{19}$$

where $v_j(t) := \mathrm{Tr}\{(R_j(t) + R_j(t)^*)\varrho(t)\}$, and $\widehat{W}(t)$ is a Wiener process under the physical probabilities defined by

$$\widehat{W}_j(t) := W_j(t) - \int_0^t v_j(s)\mathrm{d}s\,, \quad \forall j = 1, \ldots, d\,.$$

3. Random Hamiltonian

In the previous section, we have presented a non Markovian generalisation of the usual diffusive lSSE by using random coefficients to introduce memory. In this section we adopt an alternative strategy and we start with a usual lSSE with non random coefficients, but driven by a coloured noise; in this way the memory is encoded in the driving noise of the lSSE, not in the coefficients. As we shall see, this model too turns out to be a particular case of the general theory presented in Section 2. Moreover, the new lSSE will be norm-preserving and will represent a quantum system evolving under a random Hamiltonian dynamics, while the Hamiltonian is very singular and produces dissipation.

As our aim is just to explore some possibility, we keep things simple and we consider a one-dimensional driving noise and two non-random, bounded operators A and B on $\mathcal{H}$ in the drift and in the diffusive terms. The starting point is then the basic linear stochastic Schrödinger equation

$$\mathrm{d}\psi(t) = A\psi(t)\mathrm{d}t + B\psi(t)\mathrm{d}X(t). \tag{20}$$

The simplest choice of a coloured noise is the stationary Ornstein-Uhlenbeck process defined by

$$X(t) = \mathrm{e}^{-\gamma t} Z + \int_0^t \mathrm{e}^{-\gamma(t-s)}\mathrm{d}W(s), \qquad \gamma > 0.$$

where $(W(t))$ is a one dimensional Wiener process, defined on the stochastic basis $(\Omega, \mathcal{F}, \mathcal{F}_t, \mathbb{Q})$ and Z is an $\mathcal{F}_0$-measurable, normal random variable with mean 0 and variance $1/(2\gamma)$. The Ornstein-Uhlenbeck process $(X(t))$ is a Gaussian process with zero mean and correlation function

$$\mathbb{E}_{\mathbb{Q}}[X(t)X(s)] = \frac{\mathrm{e}^{-\gamma|t-s|}}{2\gamma}; \tag{21}$$

it satisfies the stochastic differential equation

$$\mathrm{d}X(t) = -\gamma X(t)\mathrm{d}t + \mathrm{d}W(t), \qquad X(0) = Z. \tag{22}$$

Formally, Eq. (20) is driven by the derivative of the Ornstein-Uhlenbeck process (heuristically, $\mathrm{d}X(t) = \dot{X}(t)\mathrm{d}t$), whose two-time correlation is no more a delta, as in the case of white noise, but it is formally given by $\mathbb{E}_{\mathbb{Q}}[\dot{X}(t)\dot{X}(s)] = \delta(t-s) - \frac{\gamma}{2}\,\mathrm{e}^{-\gamma|t-s|}$. Note that the Markovian regime is recovered in the limit $\gamma \downarrow 0$.

It is then straightforward that Eq. (20) can be rewritten in the form

$$\mathrm{d}\psi(t) = \big(A - \gamma X(t)B\big)\psi(t)\mathrm{d}t + B\psi(t)\mathrm{d}W(t), \tag{23}$$

on $(\Omega, \mathcal{F}, \mathcal{F}_t, \mathbb{Q})$. The initial condition is assumed to satisfy Assumption 2.3. Assumption 2.1 is satisfied with $d = 1$, $K(t) = A - \gamma X(t)B$, $R(t) = B$.

The key point of the construction of Section 2 and of its interpretation is the fact that $\|\psi(t)\|^2$ is a martingale. To this end we compute its stochastic differential by using the Itô rules and we get

$$\begin{aligned} \mathrm{d}\langle\psi(t)|\psi(t)\rangle &= \langle \mathrm{d}\psi(t)|\psi(t)\rangle + \langle\psi(t)|\mathrm{d}\psi(t)\rangle + \langle \mathrm{d}\psi(t)|\mathrm{d}\psi(t)\rangle \\ &= \langle\psi(t)|\left[A^* + A - \gamma X(t)\left(B^* + B\right) + B^*B\right]\psi(t)\rangle \mathrm{d}t \\ &\qquad + \langle\psi(t)|\left(B^* + B\right)\psi(t)\rangle \mathrm{d}W(t). \qquad (24) \end{aligned}$$

Then, the process $(\|\psi(t)\|^2)$ can be a martingale only if the term in front of $\mathrm{d}t$ (the drift term) is equal to zero. This imposes that

$$A^* + A - \gamma X(t)\left(B^* + B\right) + B^*B = 0, \qquad \forall t. \qquad (25)$$

By taking the mean of this equation we get $A^* + A + B^*B = 0$; then, we need also $B^* + B$. These conditions impose that there are two self-adjoint operators L and H_0 such that $B = -\mathrm{i}L$ and $A = -\mathrm{i}H_0 - \frac{1}{2}L^2$. As a consequence the initial equation (20) becomes

$$\mathrm{d}\psi(t) = \left[-\mathrm{i}\left(H_0 - \gamma X(t)L\right) - \frac{1}{2}L^2\right]\psi(t)\mathrm{d}t - \mathrm{i}L\psi(t)\mathrm{d}W(t). \qquad (26)$$

Now, being $X(t)$ a continuous adapted process and $H_0 = H_0{}^* \in \mathcal{L}(\mathcal{H})$, $L = L^* \in \mathcal{L}(\mathcal{H})$, also Assumption 2.2 holds with

$$H(t) = H_0 - \gamma X(t)L, \quad R(t) = -\mathrm{i}L, \quad K(t) = -\mathrm{i}(H_0 - \gamma X(t)L) - \frac{1}{2}L^2.$$

Moreover, we have $\mathbb{E}_\mathbb{Q}[|X(t)|] \leq \sqrt{\mathbb{E}_\mathbb{Q}[X(t)^2]} \leq 1/\sqrt{2\gamma}$,

$$\mathbb{E}_\mathbb{Q}[\|H(t)\|] \leq \|H_0\| + \gamma\,\|L\|\,\mathbb{E}_\mathbb{Q}[|X(t)|] \leq \|H_0\| + \sqrt{\frac{\gamma}{2}}\,\|L\|\,,$$

which implies condition (3a). Condition (3b) and Assumption 2.4 are trivially satisfied because $R(t)$ is non random, time independent and bounded.

As all Assumptions 2.1–2.4 hold, also all statements of Theorems 2.1, 2.2 and Remarks 2.1–2.4 hold. In particular the lSSE (26) has a pathwise unique solution.

What is peculiar of the present model is that Eqs. (24)–(26) give $\|\psi(t)\|^2 = \|\psi(0)\|^2$ or that the probability densities are independent of time, $p(t) = p(0)$, cf. Eqs. (5) and (12), which give $m(t) = 0$ and $\widehat{W}(t) = W(t)$. We have also, from Eq. (6), $\hat{\psi}(t) = \psi(t)/\,\|\psi(0)\|$, if $\|\psi(0)\| \neq 0$, and, from Remark 2.3, $\mathbb{P}^t(F) = \mathbb{Q}(F)$, for all events $F \in \mathcal{F}_t$, independent of $\mathcal{F}_0$. As a consequence the change of probability has no effect (the new probability is equal to the initial for events independent of $\mathcal{F}_0$). In other terms, no information has been extracted from the measurement interpretation.

Moreover, the property $\|\psi(t)\| = \|\psi_0\|$ is in agreement with a purely Hamiltonian evolution. More precisely, let $\overleftarrow{\mathrm{T}}\exp\{\cdots\}$ denotes the time ordered exponential; then, the formal solution of Eq. (26) is given by

$$\psi(t) = \overleftarrow{\mathrm{T}}\exp\left\{-\mathrm{i}\int_0^t (H_0 - \gamma X(s)L)\,\mathrm{d}s - \mathrm{i}\int_0^t L\,\mathrm{d}W(s)\right\}\psi_0.$$

The evolution of the quantum system is then completely determined by the time-dependent, random Hamiltonian

$$\hat{H}_t = H_0 + \left(\dot{W}(t) - \gamma X(t)\right)L.$$

Let us stress that it is a formal expression, due to the presence of $\dot{W}(t)$.

This shows that the usual measurement interpretation of (20) *coloured* with an Ornstein-Uhlenbeck process gives raise to a random Hamiltonian evolution. As announced, we recover the framework of the evolution of a closed system incorporating a random environment characterised in terms of an Ornstein-Uhlenbeck noise.

One can investigate the evolution of the corresponding density matrices. To this end, we consider the pure state process $(\sigma(t))$ defined by $\sigma(t) = |\psi(t)\rangle\langle\psi(t)|$, Eq. (7). By using Itô rules, the process $(\sigma(t))$ satisfies the stochastic differential equation (SDE)

$$\mathrm{d}\sigma(t) = -\mathrm{i}[H_0 - \gamma X(t)L, \sigma(t)]\mathrm{d}t - \mathrm{i}[L, \sigma(t)]\mathrm{d}W(t) - \frac{1}{2}[L, [L, \sigma(t)]]\mathrm{d}t, \quad (27)$$

which is, of course, equivalent to (26) with random Liouville operator (10) given by

$$\mathcal{L}(t) = -\mathrm{i}[H_0 - \gamma X(t)L, \cdot]\mathrm{d}t - \frac{1}{2}[L, [L, \cdot]]\mathrm{d}t.$$

Let us stress that the presence of the Ornstein-Uhlenbeck process implies that the solution $(\sigma(t))$ of Eq. (27) is not a Markov process.

Taking the expectation, we get the evolution of the mean $\eta(t) = \mathbb{E}_{\mathbb{Q}}[\sigma(t)]$, Eq. (15), which turns out to be

$$\frac{\mathrm{d}}{\mathrm{d}t}\eta(t) = -\mathrm{i}[H_0, \eta(t)] - \frac{1}{2}[L, [L, \eta(t)]] + \mathrm{i}\gamma\big[L, \mathbb{E}_{\mathbb{Q}}[X(t)\sigma(t)]\big]. \quad (28)$$

Note that it is not a closed master equation for the mean state $\eta(t)$. Actually, we have derived a model with memory for the mean state. Indeed, the term $\mathrm{i}\gamma\big[L, \mathbb{E}_{\mathbb{Q}}[X(t)\sigma(t)]\big]$ introduces non-Markovian memory effects in the dynamics. Moreover, Eq. (26) is an unravelling of the master equation (28).

4. Projection techniques and closed master equations with memory

As we have seen in Eq. (16), the *a priori* states or average states $\eta(t) = \mathbb{E}_{\mathbb{Q}}[\sigma(t)] = \mathbb{E}_{\mathbb{P}^t}[\rho(t)]$ satisfy the equation $\dot{\eta}(t) = \mathbb{E}_{\mathbb{Q}}\left[\mathcal{L}(t)[\sigma(t)]\right]$, which is not closed because both $\mathcal{L}(t)$ and $\sigma(t)$ are random. However, at least heuristically, some kind of generalised master equations can be obtained by using the Nakajima-Zwanzig projection technique [2, Section 9.1.2].

Let us introduce the projection operators on the relevant part (the mean) and on the non relevant one:

$$\mathcal{P}[\cdots] := \mathbb{E}_{\mathbb{Q}}[\cdots], \qquad \mathcal{Q} := \mathbb{1} - \mathcal{P}.$$

Then, we have $\eta(t) = \mathcal{P}[\sigma(t)]$ and we define the non relevant part of the state, the mean Liouville operator and the difference from the mean of the Liouville operator

$$\sigma_{\perp}(t) := \mathcal{Q}[\sigma(t)] = \sigma(t) - \eta(t),$$

$$\mathcal{L}_{\mathrm{M}}(t) := \mathbb{E}_{\mathbb{Q}}[\mathcal{L}(t)], \qquad \Delta\mathcal{L}(t) := \mathcal{L}(t) - \mathcal{L}_{\mathrm{M}}(t).$$

By using the projection operators and the fact that the stochastic integrals have zero mean, which means $\mathcal{P}\int_0^t \mathrm{d}W_j(s)\cdots = 0$, from (8) we get the system of equations

$$\dot{\eta}(t) = \mathcal{L}_{\mathrm{M}}(t)[\eta(t)] + \mathcal{P}\circ\Delta\mathcal{L}(t)[\sigma_{\perp}(t)], \tag{29a}$$

$$\begin{aligned} \mathrm{d}\sigma_{\perp}(t) = \mathcal{Q}\circ\mathcal{L}(t)[\sigma_{\perp}(t)]\mathrm{d}t + \sum_{j=1}^{d}\mathcal{R}_j(t)[\sigma_{\perp}(t)]\mathrm{d}W_j(t) \\ + \mathcal{Q}\circ\mathcal{L}(t)[\eta(t)]\mathrm{d}t + \sum_{j=1}^{d}\mathcal{R}_j(t)[\eta(t)]\mathrm{d}W_j(t). \end{aligned} \tag{29b}$$

As one can check by using Itô formula, the formal solution of Eq. (29b) can be written as

$$\begin{aligned} \sigma_{\perp}(t) = \mathcal{Q}\circ\mathcal{V}(t,0)[\sigma_{\perp}(0)] + \int_0^t \mathcal{Q}\circ\mathcal{V}(t,s)\circ\Big(\mathcal{L}(s) - \sum_j \mathcal{R}_j(s)^2\Big)[\eta(s)]\mathrm{d}s \\ + \mathcal{Q}\circ\mathcal{V}(t,0)\left[\sum_{j=1}^{d}\int_0^t \mathcal{V}(s,0)^{-1}\circ\mathcal{R}_j(s)[\eta(s)]\mathrm{d}W_j(s)\right], \end{aligned} \tag{30}$$

where $\mathcal{V}(t,r)$ is the fundamental solution (or propagator) of the lSME (8) and satisfies the SDE

$$\mathcal{V}(t,r) = \mathbb{1} + \int_r^t \mathrm{d}s\, \mathcal{L}(s) \circ \mathcal{V}(s,r) + \sum_{j=1}^d \int_r^t \mathrm{d}W_j(s)\, \mathcal{R}_j(s) \circ \mathcal{V}(s,r).$$

Let us stress that if one includes $\mathcal{V}(t,0)$ into the stochastic integrals, from one side one gets the simpler expression $\mathcal{V}(t,0) \circ \mathcal{V}(s,0)^{-1} = \mathcal{V}(t,s)$. But the propagator is a stochastic process and could be non adapted. To overcome this difficulty one should use some definition of anticipating stochastic integral. So, we prefer the formulation with $\mathcal{V}(t,0)$ outside the stochastic integral, in order to have only adapted integrands.

By introducing the quantity (30) into Eq. (29a), we get the generalised master equation for the *a priori* states

$$\begin{aligned} \dot{\eta}(t) &= J(t) + \mathcal{L}_{\mathrm{M}}(t)[\eta(t)] + \int_0^t \mathcal{K}(t,s)[\eta(s)]\mathrm{d}s \\ &+ \mathbb{E}_{\mathbb{Q}}\left[\Delta\mathcal{L}(t) \circ \mathcal{Q} \circ \mathcal{V}(t,0)\left[\sum_{j=1}^d \int_0^t \mathcal{V}(s,0)^{-1} \circ \mathcal{R}_j(s)[\eta(s)]\mathrm{d}W_j(s)\right]\right], \end{aligned} \tag{31}$$

where

$$J(t) := \mathbb{E}_{\mathbb{Q}}\left[\Delta\mathcal{L}(t) \circ \mathcal{Q} \circ \mathcal{V}(t,0)[\sigma_\perp(0)]\right]$$

is an inhomogeneous term which disappears if the initial state is non random, i.e. $\sigma_\perp(0) = 0$, and

$$\mathcal{K}(t,s) := \mathbb{E}_{\mathbb{Q}}\left[\Delta\mathcal{L}(t) \circ \mathcal{Q} \circ \mathcal{V}(t,s) \circ \left(\mathcal{L}(s) - \sum_j \mathcal{R}_j(s)^2\right)\right]$$

is an integral memory kernel. Also the last term in (31) is a memory contribution, linear in η and depending on its whole trajectory up to t.

Let us stress that to compute the terms appearing in Eq. (31) and to solve it is not simpler than to solve Eq. (8) and to compute the mean of the solution. The meaning of Eq. (31) is theoretical: it is a quantum master equation with memory and Eq. (14) gives an unravelling of it.

While the best way to study a concrete model is to simulate the stochastic Schrödinger equation, Eq. (31) could be the starting point for some approximation. A possibility is to approximate $\mathcal{V}(t,r)$ by the deterministic evolution generated by the mean Liouville operator:

$$\mathcal{V}_{\mathrm{M}}(t,r) = \mathbb{1} + \int_r^t \mathrm{d}s\, \mathcal{L}_{\mathrm{M}}(s) \circ \mathcal{V}_{\mathrm{M}}(s,r).$$

If we take also $\sigma_\perp(0)=0$, we get

$$\dot\eta(t) \simeq \mathcal{L}_{\rm M}(t)[\eta(t)] + \int_0^t \mathcal{K}_1(t,s)[\eta(s)]{\rm d}s \\ + \mathbb{E}_{\mathbb{Q}}\left[\Delta\mathcal{L}(t)\left[\sum_{j=1}^d \int_0^t \mathcal{V}_{\rm M}(t,s)\circ\mathcal{R}_j(s)[\eta(s)]{\rm d}W_j(s)\right]\right], \qquad (32)$$

$$\mathcal{K}_1(t,s) := \mathbb{E}_{\mathbb{Q}}\left[\Delta\mathcal{L}(t)\circ\mathcal{V}_{\rm M}(t,s)\circ\left(\Delta\mathcal{L}(s)-\sum_j \Delta\mathcal{R}_j^2(s)\right)\right],$$

where $\Delta\mathcal{R}_j^2(s)=\mathcal{R}_j(s)^2-\mathbb{E}_{\mathbb{Q}}[\mathcal{R}_j(s)^2]$.

For the model of the previous section we have: $\mathcal{R}_j(s)=\mathcal{R}=-{\rm i}[L,\cdot]$,

$$\mathcal{L}_{\rm M}(t)=\mathcal{L}_{\rm M}=-{\rm i}[H_0,\cdot]-\frac{1}{2}[L,[L,\cdot]], \qquad \mathcal{V}_{\rm M}(t,s)={\rm e}^{\mathcal{L}_{\rm M}(t-s)},$$

$$\mathbb{E}_{\mathbb{Q}}\left[\Delta\mathcal{L}(t)\left[\sum_{j=1}^d\int_0^t \mathcal{V}_{\rm M}(t,s)\circ\mathcal{R}_j(s)[\eta(s)]{\rm d}W_j(s)\right]\right] \\ = -\gamma\,\mathbb{E}_{\mathbb{Q}}\left[X(t)\int_0^t {\rm d}W(s)\,\mathcal{R}\circ{\rm e}^{\mathcal{L}_{\rm M}(t-s)}\circ\mathcal{R}[\eta(s)]\right] \\ = -\gamma\int_0^t {\rm d}s\,\mathcal{R}\circ{\rm e}^{(\mathcal{L}_{\rm M}-\gamma)(t-s)}\circ\mathcal{R}[\eta(s)],$$

$$\mathcal{K}_1(t,s)=\gamma^2\,\mathbb{E}_{\mathbb{Q}}[X(t)X(s)]\mathcal{R}\circ{\rm e}^{\mathcal{L}_{\rm M}(t-s)}\circ\mathcal{R}=\frac{\gamma}{2}\,\mathcal{R}\circ{\rm e}^{(\mathcal{L}_{\rm M}-\gamma)(t-s)}\circ\mathcal{R}.$$

Finally, the approximation of the non Markovian master equation turns out to be

$$\dot\eta(t)\simeq -{\rm i}[H_0,\eta(t)]-\frac{1}{2}[L,[L,\eta(t)]] \\ +\frac{\gamma}{2}\int_0^t {\rm d}s\left[L,{\rm e}^{(\mathcal{L}_{\rm M}-\gamma)(t-s)}\big[[L,\eta(s)]\big]\right]. \qquad (33)$$

However, we have no results on the positivity preserving properties of such an approximate evolution equation, while the unravelling of the complete equation guarantees complete positivity and feasibility of numerical simulations.

Acknowledgments

PDT thanks his PhD advisor, Prof. Dr. Hans-Jurgen Engelbert, and acknowledges the financial support of the Marie Curie Initial Training Network (ITN), FP7-PEOPLE-2007-1-1-ITN, no.213841-2, Deterministic and Stochastic Controlled Systems and Applications.

CP acknowledges the financial support of the ANR "Hamiltonian and Markovian Approach of Statistical Quantum Physics" (A.N.R. BLANC no ANR-09-BLAN-0098-01).

References

1. R. Alicki, M. Fannes, *Quantum dynamical systems* (Oxford University Press, Oxford, 2001).
2. H.-P. Breuer, F. Petruccione, *The Theory of Open Quantum Systems* (Oxford University Press, Oxford, 2002).
3. C.W. Gardiner, P. Zoller, *Quantum Noise: A Handbook of Markovian and Non-Markovian Quantum Stochastic Methods with Applications to Quantum Optics*, Springer Series in Synergetics (Springer, Berlin, 2004).
4. V. Gorini, A. Kossakowski, E.C.G. Sudarshan, *Completely Positive Dynamical Semigroups of N-Level Systems*, J. Math. Phys. **17** (1976) 821–825.
5. G. Lindblad, *On the generators of quantum dynamical semigroups*, Commun. Math. Phys **48** (1976) 119–130.
6. L. Diósi, W.T. Strunz, *The non-Markovian stochastic Schrödinger equation for open systems*, Phys. Lett. A **235** (1997) 569–573.
7. L. Diósi, N. Gisin, W.T. Strunz, *Non-Markovian quantum state diffusion*, Phys. Rev. A **58** (1998) 1699–1712.
8. L. Diósi, N. Gisin, W. T. Strunz, *Open system dynamics with non-Markovian quantum trajectories*, Phys. Rev. Lett. **82** (1999) 1801-1805.
9. H.-P. Breuer, B. Kappler, F. Petruccione, *Stochastic wave function method for non-Markovian quantum master equations*, Phys. Rev. A **59** (1999) 1633–1643.
10. H.-P. Breuer, B. Kappler, F. Petruccione, *The time-convolutionless projection operator technique in the quantum theory of dissipation and decoherence* Ann. Physics **291** (2001) 36–70.
11. J. Gambetta, H.M. Wiseman, *A perturbative approach to non-Markovian stochastic Schrödinger equations*, Phys. Rev. A **66** (2002) 052105.
12. J. Gambetta, H.M. Wiseman, *Non-Markovian stochastic Schrödinger equations: Generalization to real-valued noise using quantum measurement theory*, Phys. Rev. A **66** (2002) 012108.
13. J. Gambetta, H.M. Wiseman, *Interpretation of non-Markovian stochastic Schrödinger equation as a hidden-variable theory*, Phys. Rev. A **68** (2003) 062104.
14. A. Budini, *Random Lindblad equations from complex environments*, Phys. Rev. E **72** (2005) 056106 .
15. A. Budini, H. Schomerus, *Non-Markovian master equations from entangle-*

ment with stationary unobserved degrees of freedom, J. Phys. A: Math. Gen. **38** (2005) 9251.
16. A. Budini, *Lindblad rate equations*, Phys. Rev. A **74** (2006) 053815.
17. S. Maniscalco, F. Petruccione, *Non-Markovian dynamics of a qubit*, Phys. Rev. A **73** (2006) 012111.
18. A. Kossakowski, R. Rebolledo, *On non-Markovian Time Evolution in Open Quantum Systems*, Open Syst. Inf. Dyn. **15** (2007) 135–142.
19. H.-P. Breuer, *Non-Markovian generalization of the Lindblad theory of open quantum systems* Phys. Rev. A **75** (2007) 022103.
20. L. Diósi, *Non-Markovian continuous quantum measurement of retarded observables*, Phys. Rev. Lett. **100** (2008) 080401.
21. J. Gambetta, H.M. Wiseman, *Pure-state quantum trajectories for general non-Markovian systems do not exist*, Phys. Rev. Lett. **101** (2008) 140401.
22. H.-P. Breuer, B. Vacchini, *Structure of completely positive quantum master equations with memory*, Phys. Rev. E **79** (2009) 041147.
23. C. Pellegrini, F. Petruccione, *Non Markovian Quantum Repeated Interactions and Measurements*, J. Phys. A: Math. Theor. **42** (2009) 425304.
24. H.J. Carmichael, *Statistical Methods in Quantum Optics*, Vol 2 (Springer, Berlin, 2008).
25. A. Barchielli, *Measurement theory and stochastic differential equations in quantum mechanics*, Phys. Rev. A **34** (1986) 1642–1649.
26. V.P. Belavkin, *Nondemolition measurements, nonlinear filtering and dynamic programming of quantum stochastic processes.* In A. Blaquière (ed.), *Modelling and Control of Systems*, Lecture Notes in Control and Information Sciences **121** (Springer, Berlin, 1988) pp. 245–265.
27. A. Barchielli, V.P. Belavkin, *Measurements continuous in time and a posteriori states in quantum mechanics*, J. Phys. A: Math. Gen. **24** (1991) 1495–1514.
28. H.M. Wiseman, *Quantum trajectories and quantum measurement theory*, Quantum Semiclass. Opt. **8** (1996) 205–222.
29. G.J. Milburn, *Classical and quantum conditional statistical dynamics*, Quantum Semiclass. Opt. **8** (1996) 269–276.
30. A.S. Holevo, *Statistical Structure of Quantum Theory*, Lecture Notes in Physics **m 67** (Springer, Berlin, 2001).
31. A. Barchielli, M. Gregoratti, *Quantum Trajectories and Measurements in Continuous Time — The diffusive case*, Lecture Notes in Physics **782** (Springer, Berlin, 2009).
32. A. Barchielli, A.S. Holevo, *Constructing quantum measurement processes via classical stochastic calculus*, Stoch. Proc. Appl. **58** (1995) 293–317.
33. P. Di Tella, *Quantum trajectories: Memory and feedback*, master thesis (2009), `http://www1.mate.polimi.it/biblioteca/tesiview.php?id=302&L=i`
34. A. Barchielli, C. Pellegrini, F. Petruccione, *Stochastic Schrödinger equations with coloured noise*; arXiv:0911.2554v1 [quant-ph].

ON SPECTRAL APPROACH TO PASCAL WHITE NOISE FUNCTIONALS

ABDESSATAR BARHOUMI
Department of Mathematics
Higher School of Sci. and Tech. of Hammam-Sousse
University of Sousse, Sousse, Tunisia
E-mail: abdessatar.barhoumi@ipein.rnu

HABIB OUERDIANE
Department of Mathematics
Faculty of Sciences of Tunis
University of Tunis El-Manar, Tunis, Tunisia
E-mail: habib.ouerdiane@fst.rnu.tn

ANIS RIAHI
Department of Mathematics
Higher Institute of Appl. Sci. and Tech. of Gabes
University of Gabes, Gabes, Tunisia
E-mail: a1riahi@yahoo.fr

We introduce a one-mode type interacting Fock space $\mathcal{F}_{NB}(\mathcal{H})$ naturally associated to the negative binomial distribution $\mu_{r,\alpha}$. The Fourier transform in generalized joint eigenvectors of a family $\{\mathbf{J}_\phi \,;\, \phi \in \mathcal{E}\}$ of Pascal Jacobi fields provides a way to explicit a unitary isomorphism $\mathfrak{U}_{r,\alpha}$ between $\mathcal{F}_{NB}(\mathcal{H})$ and the so-called Pascal white noise space $L^2(\mathcal{E}', \Lambda_{r,\alpha})$. Then, we derive a chaotic decomposition property of the quadratic integrable functionals of the Pascal white noise process in terms of an appropriate wick tensor product.

Keywords: Chaos decomposition, Generating function, Interacting Fock space, Jacobi field, Pascal white noise process, Spectral measure.

1. Introduction

Over the past few decades a considerable number of studies have been made on calculus on Fock space with application to operator formalism of quantum physics. White noise analysis initiated by Hida[14] has become a subject of much current interest and has been expected to offer one possible framework for these purposes. Investigations in this line of research are based on the concept of Gaussian measure and the associated expansion into Hermite polynomials. Later on an extension of white noise theory to non-Gaussian analysis was established in[1] and developed further in.[11,16,17] The main tool in this theory is the biorthogonal decomposition which extends the Wiener-Itô chaos expansion.

The main purpose of this paper is to propound a white noise framework to study Pascal white noise functionals. In,[5,6,20] the authors provide an expansion of the Jacobi fields in generalized joint vectors. The corresponding fourier transform appears to be a unitary operator between an interacting Fock space and an appropriate L^2-space. The notion of a Jacobi field in the Fock space, first appeared in the paper,[7] is devoted to the axiomatic quantum field theory, and then was further developed in.[9]

During last years, one witnesses the growth of interest to the study of compound (or generally marked Poisson) processes and applications in stochastic analysis. The Pascal process is an important example of a compound Poisson process with bounded intensity and logarithmic distributed jumps in the Lévy formula. Despite the close connection to Poisson processes new interesting features are revealed. Especially in contrast to the well-known Gaussian and Poisson cases, the multiple stochastic integrals no longer are capable of generating the Hilbert space of quadratic integrable functionals with respect to the so-called Pascal white noise measure.

The present paper is organized as follows. In Section 2, the basic facts about Jacobi matrix, spectral measure and Fock space are presented. In particular, an explicit form of a one-mode type interacting Fock space $\mathcal{F}_{NB}(\mathcal{H})$, naturally associated to the negative binomial distribution, is introduced. In Section 3, we introduce a family $\{\mathbf{J}_\phi;\ \phi \in \mathcal{E}\}$ of commuting essentially self-adjoint operators on $\mathcal{F}_{NB}(\mathcal{H})$. For the family $\left\{\widetilde{\mathbf{J}}_\phi;\ \phi \in \mathcal{E}\right\}$ ($\widetilde{\mathbf{J}}_\phi$ being the closure of $\mathbf{J}_\phi$), we construct the Fourier transform, denoted by $\mathfrak{U}_{r,\alpha}$, in generalized joint vectors. Then we show that $\mathfrak{U}_{r,\alpha}$ is a unitary isomorphism between $\mathcal{F}_{NB}(\mathcal{H})$ and the space $L^2(\mathcal{E}', \Lambda_{r,\alpha})$, where $\Lambda_{r,\alpha}$ is the so-called Pascal white noise measure obtain by using the spectral approach to commutative Jacobi field in the interacting Fock space. Therefore, we prove that, under $\mathfrak{U}_{r,\alpha}$, $\widetilde{\mathbf{J}}_\phi$ goes over into the operator of multiplication by the monomial $\langle \cdot, \phi \rangle$, so that, $\widetilde{\mathbf{J}}_\phi$'s are called Pascal field operators. Furthermore, by using $\mathfrak{U}_{r,\alpha}$, we derive a chaos decomposition of the quadratic integrable functionals of the Pascal process in terms of an appropriate wick tensor product. This is expected to offer, in the Pascal case, a possible way to overcome the main difficulty concerning the fact that compound Poisson processes do not possess the chaos representation property. Finally in Section 4, we give an essence to our approach for defining the Pascal white noise functionals by expanding more features of the Pascal white noise process.

2. Frameworks

2.1. *Jacobi matrix and its spectral measure*

In this subsection we review from[4] the basic concepts of Jacobi matrix and its spectral measures. By ℓ_2 we denote the Hilbert space spanned by the orthonormal basis $(e_n)_{n=0}^{\infty}$ with $e_n = (0, \cdots, 0, 1, 0, \cdots)$. An infinite matrix $J = (J_{n,m})_{n,m=0}^{\infty}$ is called a Jacobi matrix if $J_{n,n} = a_n \in \mathbb{R}, J_{n,n-1} = J_{n-1,n} = b_n > 0$ for $n \in \mathbb{N}$, and $J_{n,m} = 0$ for $|n-m| > 1$. We denote by $\ell_{2,fin}$ the dense subset of ℓ_2 consisting of all finite vectors, i.e.,

$$\ell_{2,fin} := \left\{ (f^{(n)})_{n\in\mathbb{N}},\ \exists m \in \mathbb{N} \quad \text{such that} \quad f^{(n)} = 0 \quad \text{for all} \quad n \geq m \right\}.$$

Each Jacobi matrix J determines a linear symmetric operator in ℓ_2 with domain $\ell_{2,fin}$ by the following formula

$$Je_n = b_{n+1}e_{n+1} + a_n e_n + b_n e_{n-1}, \quad n \in \mathbb{N}, \quad e_{-1} = 0. \tag{1}$$

We denote by $\widetilde{J}$ the closure of J, which evidently exists since the operator J is symmetric. Under some appropriate conditions on the behavior of the coefficients a_n, b_n at infinity, the operator $\widetilde{J}$ can be shown to be self-adjoint.

Theorem 2.1. *Assume the operator $\widetilde{J}$ is self-adjoint. Then, there exists a unique probability measure μ on $(\mathbb{R}, \mathcal{B}(\mathbb{R}))$ ($\mathcal{B}(\mathbb{R})$ denoting the Borel σ-algebra on $\mathbb{R}$) and a unique unitary operator*

$$U : \ell_2 \longrightarrow L^2(\mathbb{R}, \mu)$$

such that $Ue_0 = 1$ and, under U, the operator $\widetilde{J}$ corresponds to the multiplication operator by x on $L^2(\mathbb{R}, \mu)$, i.e.,

$$\left(U\widetilde{J}U^{-1}f\right)(x) = xf(x),$$

$$f \in U(Dom(\widetilde{J})) = \left\{ g \in L^2(\mathbb{R}, \mu)\,;\ \int_{\mathbb{R}} x^2 g^2(x)\mu(dx) < \infty \right\}.$$

The measure μ in Theorem 2.1 is called the spectral measure of the Jacobi matrix J. Since $e_0 \in Dom(J^n)$, for each $n \in \mathbb{N}$, the measure μ has finite moments of all orders, i.e.,

$$\int_{\mathbb{R}} |x|^n \mu(dx) < \infty, \quad n = 0, 1, 2, \cdots.$$

Thus, we can implement the procedure of orthogonalization of monomials x^n, $n \in \mathbb{N}$, to obtain a sequence $\{P_n\}_{n=0}^{\infty}$ of normalized orthogonal polynomials with leading coefficient equals to 1, satisfying the recursion formula

$$\begin{cases} xP_n(x) = P_{n+1}(x) + a_n P_n(x) + b_n^2 P_{n-1}(x), & n \in \mathbb{N} \\ \\ P_{-1}(x) := 0, \ \ P_0(x) := 1 \end{cases} \tag{2}$$

In our setting, we are interested by the converse of the above problem. Let μ be an arbitrary probability measure on $(\mathbb{R}, \mathcal{B}(\mathbb{R}))$ having finite moments of all orders and such that the set of all corresponding polynomials is dense in $L^2(\mathbb{R}, \mu)$. Additionally, we suppose that the support of μ has an infinite number points. Via the orthogonalization procedure, one may construct a sequence of orthogonal polynomials $\{P_n\}_{n=0}^{\infty}$, satisfying the recursion relation (2) with some $a_n \in \mathbb{R}$ and $b_n > 0$.

Theorem 2.2. *Let μ be an arbitrary probability measure on $(\mathbb{R}, \mathcal{B}(\mathbb{R}))$ as described above. Then, μ is the spectral measure of a Jacobi matrix J having the elements a_n on the main diagonal and the elements b_n on the off-diagonals.*

2.2. *The Pascal one-mode type interacting Fock space*

According to the background mentioned above, the main tool in the present paper is the negative binomial distribution, $\mu_{r,\alpha}$, with parameters $r > 0$ and $0 < \alpha < 1$, given by

$$\mu_{r,\alpha}(dx) = \sum_{k=0}^{\infty} \alpha^r \binom{-r}{k} (-\beta)^k \, \delta_k(dx) \,, \quad \beta = 1 - \alpha,$$

where, for $\gamma \in \mathbb{R}$, $p \in \mathbb{N} \backslash \{0\}$,

$$\binom{\gamma}{p} = \frac{\gamma(\gamma-1)\cdots(\gamma-p+1)}{p!}.$$

Apply the Gram-Schmidt orthogonalization process to the sequence $\{1, x, x^2, \cdots, x^n, \cdots\}$ to get a sequence $\{P_{r,\alpha,n}; \, n = 0, 1, 2, \cdots\}$ of orthogonal polynomials in $L^2(\mathbb{R}, \mu_{r,\alpha})$ given by $P_{r,\alpha,0}(x) = 1$ and

$$P_{r,\alpha,n}(x) = \frac{n!}{\alpha^n} \sum_{k=0}^{n} (-1)^k \binom{x}{n-k} \binom{x+r+k-1}{k} \beta^k \,, \; n \geq 1.$$

This family satisfies the recurrence formula

$$x \, P_{r,\alpha,n}(x) = P_{r,\alpha,n+1}(x) + \upsilon_{r,\alpha,n} P_{r,\alpha,n}(x) + \omega_{r,\alpha,n} P_{r,\alpha,n-1}(x), \tag{3}$$

with

$$v_{r,\alpha,n} = \frac{n(\beta+1)}{\alpha} + r\,\frac{\beta}{\alpha}, \quad \omega_{r,\alpha,n} = n(r+n-1)\,\frac{\beta}{\alpha^2}. \tag{4}$$

For future use, we recall that the generating function of the measure $\mu_{r,\alpha}$ is given by

$$\psi_{\mu_{r,\alpha}}(t;x) = (1+t)^x(1+\beta t)^{-x-r} = \sum_{n=0}^{\infty} \frac{\alpha^n}{n!}\,t^n P_{r,\alpha,n}(x). \tag{5}$$

From the Favard theorem,[10] one can easily obtain

$$\|P_{r,\alpha,n}\|^2_{L^2(\mathbb{R},\mu_{r,\alpha})} = \frac{n!\,\beta^n(r)_n}{\alpha^{2n}},$$

where the shifted factorials are given by

$$(a)_0 = 1\ ;\ (a)_k = a(a+1)\cdots(a+k-1) = \frac{\Gamma(a+k)}{\Gamma(a)}, \quad k \in \mathbb{N}\backslash\{0\},$$

and $\Gamma(x)$ is the Γ-function at x. Set

$$Q_{r,\alpha,n}(x) = \frac{\alpha^n}{\left(n!\,\beta^n(r)_n\right)^{1/2}}\,P_{r,\alpha,n}(x), \tag{6}$$

we get an orthonormal basis $\{Q_{r,\alpha,n};\, n = 0,1,2,\cdots\}$ for $H := L^2(\mathbb{R},\mu_{r,\alpha})$.

It is well-known (see[15]) that the family of polynomials $\{P_{r,\alpha,n}\}_{n=0}^{\infty}$ satisfies the difference equation

$$\beta(x+r)P_{r,\alpha,n}(x+1) + xP_{r,\alpha,n}(x-1)$$
$$-(1+\beta)(x+\frac{r}{2})P_{r,\alpha,n}(x) + \alpha\left(n+\frac{r}{2}\right)P_{r,\alpha,n}(x) = 0.$$

This motivate the introduction of the operator $A_{r,\alpha}$ defined by

$$A_{r,\alpha} = 1 - \beta(x+r)T_{-1} - xT_1 + (1+\beta)(x+\frac{r}{2}),$$

where, for $y \in \mathbb{R}$, T_y is the translation operator on H, i.e.,

$$T_y f(x) := f(x-y), \quad x \in \mathbb{R}.$$

Notice that $A_{r,\alpha}$ is diagonalized by $\{Q_{r,\alpha,n}\}$:

$$A_{r,\alpha}Q_{r,\alpha,n} = \lambda_{r,\alpha,n}Q_{r,\alpha,n}$$

with

$$\lambda_{r,\alpha,n} = 1 + \alpha\left(n+\frac{r}{2}\right), \quad n = 0,1,2,\cdots,$$

moreover, $A_{r,\alpha}^{-p}$ is a Hilbert-Schmidt operator satisfying

$$\|A_{r,\alpha}^{-p}\|_{HS}^2 = \sum_{n=0}^{\infty} \lambda_{r,\alpha,n}^{-2p} < \infty, \quad p > \frac{1}{2}.$$

Now, for each $p \in \mathbb{R}$, define a norm $|\cdot|_p$ on H by

$$|f|_p = |A_{r,\alpha}^p f|_0 = \left(\sum_{n=0}^{\infty} \lambda_{r,\alpha,n}^{2p} \langle f, Q_{r,\alpha,n} \rangle^2 \right)^{1/2},$$

where $|\cdot|_0$ and $\langle \cdot, \cdot \rangle$ are , respectively, the norm and the inner product of H. For $p \geq 0$, let $\mathcal{E}_p$ be the Hilbert space consisting of all $f \in H$ with $|f|_p < \infty$, and let $\mathcal{E}_{-p}$ be the completion of H with respect to $|\cdot|_{-p}$. Since $A_{r,\alpha}^{-1}$ is of Hilbert-Schmidt type, identifying H with its dual space we come to the real standard nuclear triple[12]

$$\mathcal{E} := \bigcap_{p \geq 0} \mathcal{E}_p \subset H \subset \bigcup_{p \geq 0} \mathcal{E}_{-p} =: \mathcal{E}'.$$

Finally, by taking complexification we have a complex Gel'fand triple :

$$\mathcal{E}_{\mathbb{C}} \subset \mathcal{H} \subset \mathcal{E}'_{\mathbb{C}},$$

where $\mathcal{E}_{\mathbb{C}}$, $\mathcal{H}$ and $\mathcal{E}'_{\mathbb{C}}$ are the complexifications of $\mathcal{E}$, H and $\mathcal{E}'$, respectively. The canonical $\mathbb{C}$-bilinear form on $\mathcal{E}'_{\mathbb{C}} \times \mathcal{E}_{\mathbb{C}}$ which is compatible with the inner product of $\mathcal{H}$ is denoted by $\langle \cdot, \cdot \rangle$ again.

Denote by $\mathcal{F}_{NB}^{(n)}(\mathcal{H})$ the n-th symmetric tensor power of $\mathcal{H}$, i.e.,

$$\mathcal{F}_{NB}^{(0)}(\mathcal{H}) := \mathbb{C} \quad \text{and} \quad \mathcal{F}_{NB}^{(n)}(\mathcal{H}) := \mathcal{H}^{\widehat{\otimes} n}, \quad n \geq 1.$$

By a simple embedding argument the sesquilinear form

$$\langle \phi^{\otimes n}, \psi^{\otimes n} \rangle_{\mathcal{F}_{NB}^{(n)}(\mathcal{H})} := \sum_{i_1+2i_2+\cdots+ki_k=n} \frac{n! \langle \phi, \psi \rangle^{i_1} \langle \phi^2, \psi^2 \rangle^{i_2} \cdots \langle \phi^k, \psi^k \rangle^{i_k}}{(i_1!)(i_2! 2^{i_2}) \cdots (i_k! 2^{i_k})} \tag{7}$$

defines a scalar product on $\mathcal{F}_{NB}^{(n)}(\mathcal{H})$.

Theorem 2.3. *The scalar product* (7) *satisfies the identity :*

$$\langle \phi^{\otimes n}, \psi^{\otimes n} \rangle_{\mathcal{F}_{NB}^{(n)}(\mathcal{H})} = \sum_{k=0}^{n-1} \binom{n-1}{k} \langle \phi^{\otimes k}, \psi^{\otimes k} \rangle_{\mathcal{F}_{NB}^{(k)}(\mathcal{H})} \langle \phi^{n-k}, \psi^{n-k} \rangle \tag{8}$$

Proof. From[8] (see also[13]), we have

$$\frac{d^n}{dt^n} e^{f(t)} = \sum_{i_1+2i_2+\cdots+ki_k=n} \frac{n!}{i_1! \cdots i_k!} \left(\frac{f^{(1)}(t)}{1!} \right)^{i_1} \cdots \left(\frac{f^{(k)}(t)}{k!} \right)^{i_k} e^{f(t)}.$$

By using this identity for $f(t) = -\langle \log(1 - t\phi\psi)\rangle_{r,\alpha}$, the scalar product (7) can be rewritten as

$$\left\langle \phi^{\otimes n}, \psi^{\otimes n} \right\rangle_{\mathcal{F}^{(n)}_{NB}(\mathcal{H})} = \frac{d^n}{dt^n}\bigg|_{t=0} \left\{ \exp\left(-\langle \log(1 - t\phi\psi)\rangle_{r,\alpha} \right) \right\}, \quad \phi,\, \psi \in \mathcal{H}, \tag{9}$$

where $\langle g\rangle_{r,\alpha} := \int_{\mathbb{R}} g(x)\mu_{r,\alpha}(dx)$. From (9) we have

$$F(t) := \exp(f(t)) = \sum_{n=0}^{\infty} \frac{t^n}{n!} \langle \phi^{\otimes n}, \psi^{\otimes n} \rangle_{\mathcal{F}^{(n)}_{NB}(\mathcal{H})}. \tag{10}$$

On the other hand

$$\begin{aligned}
F'(t) = f'(t)F(t) &= \sum_{n=0}^{\infty} \frac{t^{n-1}}{(n-1)!} \langle \phi^{\otimes n}, \psi^{\otimes n} \rangle_{\mathcal{F}^{(n)}_{NB}(\mathcal{H})} \\
&= \left(\sum_{n=0}^{\infty} \frac{t^n}{n!} \langle \phi^{\otimes n}, \psi^{\otimes n} \rangle_{\mathcal{F}^{(n)}_{NB}(\mathcal{H})} \right) \left(\sum_{n=0}^{\infty} \frac{t^n}{n!} f^{(n+1)}(0) \right) \\
&= \sum_{n=0}^{\infty} \left(\sum_{k=0}^{n} \binom{n}{k} f^{(n-k+1)}(0) \langle \phi^{\otimes k}, \psi^{\otimes k} \rangle_{\mathcal{F}^{(k)}_{NB}(\mathcal{H})} \right) \frac{t^n}{n!} \\
&= \sum_{n=1}^{\infty} \left(\sum_{k=0}^{n-1} \binom{n-1}{k} f^{(n-k)}(0) \langle \phi^{\otimes k}, \psi^{\otimes k} \rangle_{\mathcal{F}^{(k)}_{NB}(\mathcal{H})} \right) \frac{t^{n-1}}{(n-1)!}.
\end{aligned}$$

Hence (8) follows from (10) and the fact $f^{(n-k)}(0) = \langle \phi^{n-k}, \psi^{n-k} \rangle$. □

Let be given $w \in \mathcal{E}'$. The k-linear function

$$(\phi_1, \phi_2, \cdots, \phi_k) \longmapsto \langle w, \phi_1\phi_2\cdots\phi_k \rangle$$

is a symmetric continuous on $\mathcal{E}^k$. Then, by the kernel theorem there exists a unique element $\tau_k(w) \in \mathcal{E}'^{\widehat{\otimes} k}$ such that

$$\langle \tau_k(w), \phi_1 \widehat{\otimes} \phi_2 \widehat{\otimes} \cdots \widehat{\otimes} \phi_k \rangle = \langle w, \phi_1\phi_2\cdots\phi_k \rangle.$$

$\tau_k(w)$ is called the trace of order k with kernel w. In the scalar case $w = 1$, we write $\tau_k = \tau_k(1)$ for short.

Definition 2.1. The generalized trace of order n, $\Upsilon_n \in \mathcal{E}'^{\widehat{\otimes} n}$, is defined recursively by

$$\Upsilon_0 = 1, \quad \Upsilon_n = \sum_{k=0}^{n-1} \binom{n-1}{k} \Upsilon_k \widehat{\otimes} \tau_{n-k}, \quad n = 1, 2, \cdots \tag{11}$$

In view of (8) and (11), we have

$$\langle \phi^{\otimes n}, \psi^{\otimes n} \rangle_{\mathcal{F}_{NB}^{(n)}(\mathcal{H})} = \langle \Upsilon_n, (\phi\psi)^{\otimes n} \rangle.$$

Then, by a density imbedding arguments, for arbitrary $\varphi^{(n)}, \psi^{(n)} \in \mathcal{F}_{NB}^{(n)}(\mathcal{H})$, the inner product $\langle \cdot, \cdot \rangle_{\mathcal{F}_{NB}^{(n)}(\mathcal{H})}$ is given by

$$\langle \varphi^{(n)}, \psi^{(n)} \rangle_{\mathcal{F}_{NB}^{(n)}(\mathcal{H})} = \langle \Upsilon_n, \varphi^{(n)} \psi^{(n)} \rangle. \tag{12}$$

Definition 2.2. The Pascal one-mode type interacting Fock space $\mathcal{F}_{NB}(\mathcal{H})$ is defined as the following direct orthogonal Hilbert sum :

$$\mathcal{F}_{NB}(\mathcal{H}) := \bigoplus_{n=0}^{+\infty} n! \left(\frac{\beta}{\alpha^2} \right)^n \mathcal{F}_{NB}^{(n)}(\mathcal{H}). \tag{13}$$

Let $\mathcal{F}_{fin}(\mathcal{E})$ be the algebraic direct sum of the spaces $\mathcal{E}_{\mathbb{C}}^{\widehat{\otimes} n}$. Observe that $\mathcal{F}_{fin}(\mathcal{E})$ is dense in $\mathcal{F}_{NB}(\mathcal{H})$. Moreover, if $\mathcal{F}_{fin}^*(\mathcal{E})$ denote the dual of $\mathcal{F}_{fin}(\mathcal{E})$ with respect to the zero space $\mathcal{F}_{NB}(\mathcal{H})$, we have the Gel'fand triple

$$\mathcal{F}_{fin}(\mathcal{E}) \subset \mathcal{F}_{NB}(\mathcal{H}) \subset \mathcal{F}_{fin}^*(\mathcal{E}).$$

The space $\mathcal{F}_{fin}^*(\mathcal{E})$ consists of infinite sequences $\Phi = (\Phi^{(n)})_{n=0}^{\infty}$, where $\Phi^{(n)} \in \mathcal{E}'^{\widehat{\otimes} n}$ and its action on $\varphi = (\varphi^{(n)})_{n=0}^{\infty} \in \mathcal{F}_{fin}(\mathcal{E})$ is given by

$$\langle\langle \Phi, \varphi \rangle\rangle = \sum_{n=0}^{\infty} n! \left(\frac{\beta}{\alpha^2} \right)^n \left\langle \Phi^{(n)}, \varphi^{(n)} \right\rangle,$$

where $\langle \cdot, \cdot \rangle$ is the canonical $\mathbb{C}$-bilinear form on $\mathcal{E}_{\mathbb{C}}'^{\widehat{\otimes} n} \times \mathcal{E}_{\mathbb{C}}^{\widehat{\otimes} n}$.

3. Pascal white noise functionals

In this Section, we will define a family of Pascal field operators and we construct an associated Fourier transform in generalized joint eigenvectors. For completeness, we start with a summary on the "projection spectral theorem" as a main building bloc for our results. Fore more details see.[6,20]

3.1. *Jacobi fields and spectral theorem*

In the interacting Fock space $\mathcal{F}_{NB}(\mathcal{H})$, as an operator version of the background given in subsection 2.1 we consider a family $J = \{J_\phi\}_{\phi \in \mathcal{E}}$ of operator-valued Jacobi matrices

$$J_\phi = \begin{pmatrix} b_0(\phi) & a_0(\phi) & 0 & 0 & 0 \cdots \\ a_0^\dagger(\phi) & b_1(\phi) & a_1(\phi) & 0 & 0 \cdots \\ 0 & a_1^\dagger(\phi) & b_2(\phi) & a_2(\phi) & 0 \cdots \\ \vdots & \vdots & \vdots & \vdots & \vdots \ddots \end{pmatrix}$$

with the following entries

$$a_n^\dagger(\phi) : \mathcal{F}_{NB}^{(n)}(\mathcal{H}) \longrightarrow \mathcal{F}_{NB}^{(n+1)}(\mathcal{H}),$$

$$a_n(\phi) = (a_n^\dagger(\phi))^* : \mathcal{F}_{NB}^{(n)}(\mathcal{H}) \longrightarrow \mathcal{F}_{NB}^{(n-1)}(\mathcal{H}),$$

$$b_n(\phi) = (b_n(\phi))^* : \mathcal{F}_{NB}^{(n)}(\mathcal{H}) \longrightarrow \mathcal{F}_{NB}^{(n)}(\mathcal{H}).$$

Every matrix J_ϕ determines a densely defined Hermitian operator $\mathbf{J}_\phi$ on $\mathcal{F}_{NB}(\mathcal{H})$ defined by

$$\mathbf{J}_\phi \varphi := \left\{ \mathbf{J}_\phi \varphi^{(n)} \right\}_{n=0}^{\infty} \in \mathcal{F}_{NB}(\mathcal{H}), \quad \varphi = \{\varphi^{(n)}\}_{n=0}^{\infty} \in \mathcal{F}_{fin}(\mathcal{H})$$

where

$$\mathbf{J}_\phi \varphi^{(n)} := a_{n-1}^\dagger(\phi)\varphi^{(n-1)} + b_n(\phi)\varphi^{(n)} + a_n(\phi)\varphi^{(n+1)}, \quad n \in \mathbb{N},$$

with the convention $a_{-1}(\phi) = 0$ and $\varphi^{(-1)} = 0$.

The family $\mathbf{J} = \{\mathbf{J}_\phi\}_{\phi\in\mathcal{E}}$ is called a (commutative) Jacobi field if the following three assumptions are true :

(a) The operators $a_n(\phi)$ and $b_n(\phi)$, $\phi \in \mathcal{E}$, $n \in \mathbb{N}$, are bounded and real, i.e., they transform real vectors into real vectors.

(b) The operators $\mathbf{J}_\phi$, $\phi \in \mathcal{E}$, are essentially self-adjoint and their closures $\widetilde{\mathbf{J}}_\phi$ are strongly commuting.

(c) For $\varphi^{(n)} \in \mathcal{E}_{\mathbb{C}}^{\widehat{\otimes} n}$, the maps

$$\phi \longmapsto a_n(\phi)\varphi^{(n)}, \quad \phi \longmapsto b_n(\phi)\varphi^{(n)}, \quad \phi \longmapsto a_n^\dagger(\phi)\varphi^{(n)}$$

are continuous linear operators.

Theorem 3.1. *(Projection spectral theorem)*
Let $\mathbf{J}$ be a commutative Jacobi field as described above. Then there exist a vector-valued function

$$\Phi : \mathcal{E}' \longrightarrow \mathcal{F}_{fin}^*(\mathcal{E})$$

and a Borel probability measure $\Lambda_{r,\alpha}$ on $\mathcal{E}'$ called spectral measure of the Jacobi field $\{\mathbf{J}_\phi;\ \phi \in \mathcal{E}\}$ such that the following statements holds

(1) For every $z \in \mathcal{E}'$, the vector $\Phi_z \equiv \Phi(z)$ is a generalized joint eigenvector of $\boldsymbol{J}$ with eigenvalue z, i.e.,

$$\langle\!\langle \Phi_z, \widetilde{\boldsymbol{J}}_\phi \varphi \rangle\!\rangle = \langle z, \phi \rangle \langle\!\langle \Phi_z, \varphi \rangle\!\rangle, \quad \phi \in \mathcal{E}, \ \varphi \in \mathcal{F}_{fin}(\mathcal{E}).$$

(2) The Fourier transform

$$\mathfrak{U}_{r,\alpha} : \mathcal{F}_{fin}(\mathcal{E}) \longrightarrow L^2(\mathcal{E}', \Lambda_{r,\alpha})$$

$$\varphi \longmapsto \mathfrak{U}_{r,\alpha}\varphi = \langle\!\langle \Phi(\cdot), \varphi \rangle\!\rangle$$

can be extended to a unitary operator acting from $\mathcal{F}_{NB}(\mathcal{H})$ to $L^2(\mathcal{E}', \Lambda_{r,\alpha})$. Moreover, we have

$$\mathfrak{U}_{r,\alpha} \widetilde{\boldsymbol{J}}_\phi \mathfrak{U}_{r,\alpha}^{-1} = M_{\langle \cdot , \phi \rangle},$$

where $M_{\langle \cdot , \phi \rangle}$ is the multiplication operator by $\langle \cdot , \phi \rangle$.

For the sake of simplicity, in all the remainder of this paper, for $\phi \in \mathcal{E}$ we use the common symbol T_ϕ for an operator $T_n(\phi)$ acting on $\mathcal{F}^{(n)}_{NB}(\mathcal{H})$ and its natural implementation to $\mathcal{F}_{NB}(\mathcal{H})$.

3.2. *Pascal Jacobi field and its spectral measure*

For each $\phi \in \mathcal{E}$, let $\mathbf{a}^\dagger_\phi$ denote the standard creation operator acting on $\mathcal{F}_{fin}(\mathcal{E})$ by

$$\mathbf{a}^\dagger_\phi \varphi^{(n)} := \phi \widehat{\otimes} \varphi^{(n)}, \quad \varphi^{(n)} \in \mathcal{E}^{\widehat{\otimes} n}, \quad n \in \mathbb{N}. \tag{14}$$

From the obvious inequality

$$\left| \phi \widehat{\otimes} \varphi^{(n)} \right|_p \le |\phi|_p \left| \varphi^{(n)} \right|_p ,$$

$\mathbf{a}^\dagger_\phi$ is a densely defined continuous linear operator on $\mathcal{F}_{NB}(\mathcal{H})$.

Lemma 3.1. *The adjoint operator, $\mathbf{a}_\phi$, of $\mathbf{a}^\dagger_\phi$ in $\mathcal{F}_{NB}(\mathcal{H})$ is given by*

$$\mathbf{a}_\phi \varphi^{\otimes n} = \frac{\beta}{\alpha^2} n \langle \phi, \varphi \rangle \varphi^{\otimes(n-1)} + \frac{\beta}{\alpha^2} n(n-1)(\phi\varphi^2) \widehat{\otimes} \varphi^{\otimes(n-2)}.$$

Proof. First, we shall prove the following much simpler recursion formulas

$$\Upsilon_{n+1} = \Upsilon_n \widehat{\otimes} \tau_1 + n \Upsilon_n \widehat{\otimes}_1 \tau_2, \quad n = 1, 2, \cdots, \tag{15}$$

where $\Upsilon_n \widehat{\otimes}_1 \tau_2 \in \mathcal{E}'^{\widehat{\otimes}(n+1)}$ is defined by

$$\langle \Upsilon_n \widehat{\otimes}_1 \tau_2, \phi^{\otimes(n+1)} \rangle = \langle \Upsilon_n, \phi^2 \widehat{\otimes} \phi^{\otimes(n-1)} \rangle, \quad \phi \in \mathcal{E}.$$

In fact, (11) yields

$$\begin{aligned}\Upsilon_{n+1} &= \sum_{k=0}^{n} \binom{n}{k} \Upsilon_k \widehat{\otimes} \tau_{n+1-k} \\ &= \Upsilon_n \widehat{\otimes} \tau_1 + \sum_{k=0}^{n-1} \binom{n}{k} \Upsilon_k \widehat{\otimes} \tau_{n+1-k} \\ &= \Upsilon_n \widehat{\otimes} \tau_1 + n \sum_{k=0}^{n-1} \binom{n-1}{k} \Upsilon_k \widehat{\otimes} \tau_{n-k} \widehat{\otimes}_1 \tau_2 \\ &= \Upsilon_n \widehat{\otimes} \tau_1 + n \Upsilon_n \widehat{\otimes}_1 \tau_2.\end{aligned}$$

By virtues of (13), (14) and (15), we have for arbitrary ϕ, φ, $\psi \in \mathcal{E}$,

$$\begin{aligned}&\left\langle \mathbf{a}_\phi^\dagger \varphi^{\otimes n}, \psi^{\otimes (n+1)} \right\rangle_{\mathcal{F}_{NB}(\mathcal{H})} = (n+1)! \Big(\frac{\beta}{\alpha^2}\Big)^{n+1} \langle \Upsilon_{n+1}, (\phi\psi) \widehat{\otimes} (\varphi\psi)^{\otimes n} \rangle \\ &= (n+1)! \Big(\frac{\beta}{\alpha^2}\Big)^{n+1} \langle \Upsilon_n \widehat{\otimes} \tau_1 + n \Upsilon_n \widehat{\otimes}_1 \tau_2, (\phi\psi) \widehat{\otimes} (\varphi\psi)^{\otimes n} \rangle \\ &= (n+1)! \left(\frac{\beta}{\alpha^2}\right)^{n+1} \Big[\langle \phi, \psi \rangle \langle \Upsilon_n, (\varphi\psi)^{\otimes n} \rangle + n \langle \Upsilon_n, (\phi\varphi\psi^2) \widehat{\otimes} (\varphi\psi)^{\otimes (n-1)} \rangle \Big] \\ &= \left\langle \varphi^{\otimes n}, \frac{\beta}{\alpha^2}(n+1) \langle \phi, \psi \rangle \psi^{\otimes n} + \frac{\beta}{\alpha^2} n(n+1)(\phi\psi^2) \widehat{\otimes} \psi^{\otimes (n-1)} \right\rangle_{\mathcal{F}_{NB}(\mathcal{H})}.\end{aligned}$$

Thus $\mathbf{a}_\phi$ is defined on $\mathcal{F}_{fin}(\mathcal{E})$ by

$$\mathbf{a}_\phi \varphi^{\otimes n} = \frac{\beta}{\alpha^2} n \langle \phi, \varphi \rangle \varphi^{\otimes (n-1)} + \frac{\beta}{\alpha^2} n(n-1)(\phi\varphi^2) \widehat{\otimes} \varphi^{\otimes (n-2)}. \qquad \square$$

Finally, we define on $\mathcal{F}_{fin}(\mathcal{E})$ the preservation operator $\mathbf{n}_\phi$ at $\phi \in \mathcal{E}$ by

$$\mathbf{n}_\phi \varphi^{\otimes n} = n(\phi\varphi) \widehat{\otimes} \varphi^{\otimes (n-1)}, \quad \varphi \in \mathcal{E}.$$

Now, we are ready to define the *Pascal field operator*, $\mathbf{J}_\phi$, on $\mathcal{F}_{fin}(\mathcal{E})$:

$$\mathbf{J}_\phi := \mathbf{a}_\phi^\dagger + \mathbf{a}_\phi + \frac{\beta+1}{\alpha} \mathbf{n}_\phi + \frac{\beta}{\alpha} \langle \phi \rangle_{r,\alpha} \mathbf{I},$$

where I denote the identity operator.

Note that $\mathbf{J}_\phi$, $\phi \in \mathcal{E}$, are symmetric operators. Furthermore, if we denote $\widetilde{\mathbf{J}}_\phi$ the closure of $\mathbf{J}_\phi$, one can show that $\widetilde{\mathbf{J}}_\phi$, $\phi \in \mathcal{E}$ are self-adjoint operators. Thus, by construction, the family of operators $\{\mathbf{J}_\phi;\ \phi \in \mathcal{E}\}$ has a Jacobi field structure in the interacting Fock space $\mathcal{F}_{NB}(\mathcal{H})$. This family will be called *Pascal-Jacobi field.* For more details on the notion of Jacobi field, see.[5–7,9]

In the following crucial lemma, we show that $\left\{\widetilde{\mathbf{J}}_\phi;\ \phi \in \mathcal{E}\right\}$ is commutative in the sense of the resolution of the identity.

Lemma 3.2. *$\{\widetilde{\mathbf{J}}_\phi;\ \phi \in \mathcal{E}\}$ is a family of commuting operators in $\mathcal{F}_{NB}(\mathcal{H})$.*

Proof. Any operators $\mathbf{a}^\dagger_{\phi_1}$ and $\mathbf{a}^\dagger_{\phi_2}$ evidently commute. Hence, their adjoints $\mathbf{a}_{\phi_1}$ and $\mathbf{a}_{\phi_2}$ also commute. Next, the preservation operators $\mathbf{n}_{\phi_1}$ and $\mathbf{n}_{\phi_2}$ commute. On the other hand we have

$$\mathbf{a}_\phi\left(\varphi_1\widehat{\otimes}\varphi_2^{\otimes n}\right) = \frac{\beta}{\alpha^2}2n(\phi\varphi_1\varphi_2)\widehat{\otimes}\varphi_2^{\otimes(n-1)} + \frac{\beta}{\alpha^2}n(n-1)\varphi_1\widehat{\otimes}(\phi\varphi_2^2)\widehat{\otimes}\varphi_2^{\otimes(n-2)}$$
$$+\frac{\beta}{\alpha^2}\langle\phi,\varphi_1\rangle\varphi_2^{\otimes n} + \frac{\beta}{\alpha^2}(n-1)\langle\phi,\varphi_2\rangle\varphi_1\widehat{\otimes}\varphi_2^{\otimes(n-1)}$$

and

$$\mathbf{a}^\dagger_\phi\left(\varphi_1\widehat{\otimes}\varphi_2^{\otimes n}\right) = \phi\widehat{\otimes}\varphi_1\widehat{\otimes}\varphi_2^{\otimes n}.$$

Then we obtain

$$(\mathbf{a}^\dagger_{\phi_1}\mathbf{a}_{\phi_2} + \mathbf{a}_{\phi_1}\mathbf{a}^\dagger_{\phi_2})\varphi^{\otimes n}$$
$$= \frac{\beta}{\alpha^2}\Big[n\langle\phi_2,\varphi\rangle\phi_1\widehat{\otimes}\varphi^{\otimes(n-1)} + n(n-1)\phi_1\widehat{\otimes}(\phi_2\varphi^2)\widehat{\otimes}\varphi^{\otimes(n-2)}$$
$$+\langle\phi_1,\phi_2\rangle\varphi^{\otimes n} + n\langle\phi_1,\varphi\rangle\phi_2\widehat{\otimes}\varphi^{\otimes(n-1)}$$
$$+2n(\phi_1\phi_2\varphi)\widehat{\otimes}\varphi^{\otimes(n-1)} + n(n-1)\phi_2\widehat{\otimes}(\phi_1\varphi^2)\widehat{\otimes}\varphi^{\otimes(n-2)}\Big] \quad (16)$$

which is symmetric in ϕ_1 and ϕ_2. Thus, arbitrary $\mathbf{J}_{\phi_1}$ and $\mathbf{J}_{\phi_2}$ commute as desired. □

Theorem 3.2. *There exist a unique probability measure $\Lambda_{r,\alpha}$ on $(\mathcal{E}',\mathcal{B}(\mathcal{E}'))$ and a unique unitary isomorphism*

$$\mathfrak{U}_{r,\alpha} : \mathcal{F}_{NB}(\mathcal{H}) \longrightarrow L^2(\mathcal{E}',\mathcal{B}(\mathcal{E}'),\Lambda_{r,\alpha})$$

defined on the dense set $\mathcal{F}_{fin}(\mathcal{E})$ as follows

$$\mathcal{F}_{fin}(\mathcal{E}) \ni \varphi = \left(\varphi^{(n)}\right)_{n=0}^{\infty} \longmapsto \mathfrak{U}_{r,\alpha}\,\varphi \equiv (\mathfrak{U}_{r,\alpha}\,\varphi)(z) = \sum_{n=0}^{+\infty}\left\langle :z^{\otimes n}:_{NB}, \varphi^{(n)}\right\rangle, \quad (17)$$

where for $z \in \mathcal{E}'$, $: z^{\otimes n} :_{NB}$ is the continuous linear functional on $\mathcal{E}'^{\widehat{\otimes} n}$ given recursively by

$$\begin{aligned}
\left\langle : z^{\otimes(n+1)} :_{NB}, \phi^{\otimes(n+1)} \right\rangle &:= \langle z, \phi\rangle \left\langle : z^{\otimes n} :_{NB}, \phi^{\otimes n} \right\rangle \\
&- \left\langle : z^{\otimes n} :_{NB}, \frac{\beta+1}{\alpha} n(\phi^2)\widehat{\otimes}\phi^{\otimes(n-1)} + \frac{\beta}{\alpha}\langle\phi\rangle_{r,\alpha}\phi^{\otimes n} \right\rangle \\
&- \left\langle : z^{\otimes(n-1)} :_{NB}, \frac{\beta}{\alpha^2} n\left\langle\phi^2\right\rangle_{r,\alpha}\phi^{\otimes(n-1)} + \frac{\beta}{\alpha^2} n(n-1)(\phi^3)\widehat{\otimes}\phi^{\otimes(n-2)} \right\rangle
\end{aligned} \tag{18}$$

with the conventions $: z^{\otimes(-1)} :_{NB} = 0$ and $: z^{\otimes 0} :_{NB} = 1$. Under the isomorphism $\mathfrak{U}_{r,\alpha}$, each operator $\widetilde{\mathbf{J}}_\phi$, $\phi \in \mathcal{E}$, goes over into the operator of multiplication by the monomial $\langle \cdot, \phi\rangle$ in $L^2(\mathcal{E}', \Lambda_{r,\alpha})$, i.e.,

$$\mathfrak{U}_{r,\alpha} \widetilde{\mathbf{J}}_\phi \mathfrak{U}_{r,\alpha}^{-1} = M_{\langle \cdot, \phi\rangle}.$$

Proof. As easily seen, for any $\phi \in \mathcal{E}$ and $n \in \mathbb{N}$, the operators $\mathbf{a}^\dagger_\phi$, $\mathbf{n}_\phi$ and $\mathbf{a}_\phi$ act continuously from $\mathcal{E}_{\mathbb{C}}^{\widehat{\otimes} n}$ into $\mathcal{E}_{\mathbb{C}}^{\widehat{\otimes}(n+1)}$, $\mathcal{E}_{\mathbb{C}}^{\widehat{\otimes} n}$ and $\mathcal{E}_{\mathbb{C}}^{\widehat{\otimes}(n-1)}$, respectively. Therefore, for any $\phi \in \mathcal{E}$, $\mathbf{J}_\phi$ acts continuously on $\mathcal{F}_{fin}(\mathcal{E})$. Furthermore, for any fixed $\varphi \in \mathcal{F}_{fin}(\mathcal{E})$, the mapping $\mathcal{E} \ni \phi \longmapsto \mathbf{J}_\phi\varphi \in \mathcal{F}_{fin}(\mathcal{E})$ is linear and continuous. Finally, the vacuum Ω is evidently a cyclic vector for the operators $\mathbf{a}^\dagger_\phi$, $\phi \in \mathcal{E}$ in $\mathcal{F}_{fin}(\mathcal{E})$, and hence in $\mathcal{F}_{NB}(\mathcal{H})$. Then, by using the Jacobi field structure of $\mathbf{J}_\phi$, we can easily show that Ω is a cyclic vector for $\mathbf{J}_\phi$, $\phi \in \mathcal{E}$, in $\mathcal{F}_{NB}(\mathcal{H})$. Thus, by using Theorem 3.1, we deduce the existence of a unique probability measure $\Lambda_{r,\alpha}$ on $(\mathcal{E}', \mathcal{B}(\mathcal{E}'))$ and a unique unitary operator $\mathfrak{U}_{r,\alpha} : \mathcal{F}_{NB}(\mathcal{H}) \longrightarrow L^2(\mathcal{E}', \Lambda_{r,\alpha})$ such that, for each $\phi \in \mathcal{E}$, under $\mathfrak{U}_{r,\alpha}$, $\widetilde{\mathbf{J}}_\phi$ goes over into the operator of multiplication by the function $\langle \cdot, \phi\rangle$ and $\mathfrak{U}_{r,\alpha}\Omega = 1$.

Now, for $\Lambda_{r,\alpha}$-a.e. $z \in \mathcal{E}'$, there exists a generalized joint vector $\Phi_z = (\Phi_z^{(n)})_{n\in\mathbb{N}} \in \mathcal{F}^*_{fin}(\mathcal{E})$ of the family $\{\mathbf{J}_\phi;\ \phi \in \mathcal{E}\}$:

$$\langle\!\langle \Phi_z, \mathbf{J}_\phi\varphi\rangle\!\rangle = \langle z, \phi\rangle \langle\!\langle \Phi_z, \varphi\rangle\!\rangle, \quad \forall\varphi \in \mathcal{F}_{fin}(\mathcal{E}), \tag{19}$$

and, for each $\varphi = (\varphi^{(n)})_{n\in\mathbb{N}} \in \mathcal{F}_{fin}(\mathcal{E})$, the action of $\mathfrak{U}_{r,\alpha}$ on φ is given by

$$(\mathfrak{U}_{r,\alpha}\varphi)(z) = \langle\!\langle \Phi_z, \varphi\rangle\!\rangle = \sum_{n=0}^{\infty} n! \left(\frac{\beta}{\alpha^2}\right)^n \left\langle \Phi_z^{(n)}, \varphi^{(n)} \right\rangle, \quad z \in \mathcal{E}'. \tag{20}$$

For $\Lambda_{r,\alpha}$-a.e. $z \in \mathcal{E}'$, put

$$\Xi_n(z) := n!\left(\frac{\beta}{\alpha^2}\right)^n \Phi_z^{(n)},$$

which is an element of $\mathcal{E}'^{\widehat{\otimes} n}$. By (19) and (20), for $\varphi = (0, \cdots, 0, \phi^{\otimes n}, 0, \cdots) \in \mathcal{F}_{fin}(\mathcal{E})$, $\phi \in \mathcal{E}$, we have

$$\begin{aligned}\langle z, \phi\rangle \langle \Xi_n(z), \phi^{\otimes n}\rangle &= \langle \Xi_{n+1}(z), \phi^{\otimes(n+1)}\rangle \\ &\quad - \left\langle \Xi_n(z), \frac{\beta+1}{\alpha} n(\phi^2)\widehat{\otimes}\phi^{\otimes(n-1)} + \frac{\beta}{\alpha}\langle\phi\rangle_{r,\alpha}\phi^{\otimes n}\right\rangle \\ &\quad - \left\langle \Xi_{n-1}, \frac{\beta}{\alpha^2} n\left\langle\phi^2\right\rangle_{r,\alpha}\phi^{\otimes(n-1)} + \frac{\beta}{\alpha^2} n(n-1)(\phi^3)\widehat{\otimes}\phi^{\otimes(n-2)}\right\rangle.\end{aligned}$$

Hence the $\Xi_n(z)$' s are given by the recurrence relation (18). Thus, we conclude that, for any $n \in \mathbb{N}$, $\Xi_n(z) =: z^{\otimes n} :_{NB}$ and (17) follows. This complete the proof. □

Definition 3.1. The probability measure $\Lambda_{r,\alpha}$ on $\mathcal{E}'$, highlighted in the above theorem is called *the Pascal white noise measure with parameters r and α*. The probability space $(\mathcal{E}', \mathcal{B}(\mathcal{E}'), \Lambda_{r,\alpha})$ is called *the Pascal white noise space.*

We denote by $\mathcal{P}(\mathcal{E}')$ the set of continuous polynomials on $\mathcal{E}'$, i.e., functions on $\mathcal{E}'$ of the form

$$\varphi(z) = \sum_{k=0}^{n} \langle z^{\otimes n}, \varphi^{(k)}\rangle, \quad \varphi^{(k)} \in \mathcal{E}_{\mathbb{C}}^{\widehat{\otimes} k}, \quad z^{\otimes 0} = 1, \quad n \in \mathbb{N}.$$

The greatest number k for which $\varphi^{(k)} \neq 0$ is called the power of the polynomials φ. We denote by $\mathcal{P}_n(\mathcal{E}')$ the set of continuous polynomials of order $\leq n$.

Corollary 3.1. *We have $\mathfrak{U}_{r,\alpha}(\mathcal{F}_{fin}(\mathcal{E})) = \mathcal{P}(\mathcal{E}')$. In particular, $\mathcal{P}(\mathcal{E}')$ is a dense subset of $L^2(\mathcal{E}', \Lambda_{r,\alpha})$. Furthermore, let $\mathcal{P}_{r,\alpha}^{(n)}(\mathcal{E}')$ denote the closure of $\mathcal{P}_n(\mathcal{E}')$ in $L^2(\mathcal{E}', \Lambda_{r,\alpha})$, and let $(L^2_{r,\alpha})^{(n)}$ denote the orthogonal difference $\mathcal{P}_{r,\alpha}^{(n)}(\mathcal{E}') \ominus \mathcal{P}_{r,\alpha}^{(n-1)}(\mathcal{E}')$ in $L^2(\mathcal{E}', \Lambda_{r,\alpha})$. Then, we have the chaos decomposition*

$$L^2(\mathcal{E}', \Lambda_{r,\alpha}) = \bigoplus_{n=0}^{\infty} (L^2_{r,\alpha})^{(n)}. \tag{21}$$

Proof. Using the recurrence relation (18), we obtain by induction the inclusion $\mathfrak{U}_{r,\alpha}(\mathcal{F}_{fin}(\mathcal{E})) \subset \mathcal{P}(\mathcal{E}')$, and moreover, the equality

$$\langle : z^{\otimes n} :_{NB}, \varphi^{(n)}\rangle = \langle z^{\otimes n}, \varphi^{(n)}\rangle + p_{n-1}(z), \quad \varphi^{(n)} \in \mathcal{E}_{\mathbb{C}}^{\widehat{\otimes} n}, \tag{22}$$

where $p_{n-1} \in \mathcal{P}_{n-1}(\mathcal{E}')$. Using (22), we then obtain by induction also the inverse inclusion $\mathcal{P}(\mathcal{E}') \subset \mathfrak{U}_{r,\alpha}(\mathcal{F}_{fin}(\mathcal{E}))$. That $\mathcal{P}(\mathcal{E}')$ is dense in $L^2(\mathcal{E}', \Lambda_{r,\alpha})$ follows from the fact that $\mathcal{F}_{fin}(\mathcal{E})$ is dense in $\mathcal{F}_{NB}(\mathcal{H})$. Decomposition (21) now becomes evident. □

4. The white noise Pascal process

In the present Section, our aim is to make the Pascal white noise measure $\Lambda_{r,\alpha}$ more useful by giving explicitly its Fourier transform and its generating function. The major step in the proof of these facts is the following lemma.

Lemma 4.1. *Let $\mathcal{B}_b(\mathbb{R})$ denote the set of all bounded Borel sets on $\mathbb{R}$, and for $D \in \mathcal{B}_b(\mathbb{R})$ let χ_D denote the indicator function of D. Then the distribution of the random variable $\langle \cdot, \chi_D \rangle$ is the negative binomial distribution $\mu_{|D|_r, \alpha}$ with parameters α and $|D|_r := \mu_{r,\alpha}(D) = \int_{\mathbb{R}} \chi_D(x)\mu_{r,\alpha}(dx)$.*

Proof. Let $\widetilde{\mathbf{J}}_{\chi_D}$ be the operator acting on $\mathcal{F}_{NB}(\mathcal{H})$ which, under the unitary $\mathfrak{U}_{r,\alpha}$, goes over into the operator of multiplication by the function

$$\langle z, \chi_D \rangle = \langle : z^{\otimes 1} :_{NB}, \chi_D \rangle + \frac{\beta}{\alpha}|D|_r = \mathfrak{U}_{r,\alpha}(\chi_D)(z) + \frac{\beta}{\alpha}|D|_r.$$

For each $n \in \mathbb{N}$, the vectors $\chi_D^{\otimes n}$, with $\chi_D^{\otimes 0} := \Omega$ (the vacuum vector), belongs to $\mathcal{F}_{NB}(\mathcal{H})$. Let $\mathfrak{X}_D$ stand for the subset of $\mathcal{F}_{NB}(\mathcal{H})$ spanned by these vectors. We consider an approximating sequence $\{\varphi_n \in \mathcal{E};\ n \in \mathbb{N}\}$ of χ_D, in such away $\varphi_n(x) \to \chi_D(x)$ as $n \to \infty$ in $L^2(\mathbb{R}, \mu_{r,\alpha})$. By the dominated convergence theorem, we get

$$\begin{aligned}\widetilde{\mathbf{J}}_{\chi_D} \chi_D^{\otimes n} = \chi_D^{\otimes(n+1)} &+ \left(\tfrac{n(\beta+1)}{\alpha} + \tfrac{\beta|D|_r}{\alpha}\right)\chi_D^{\otimes n} \\ &+ \frac{\beta}{\alpha^2} n\,(n-1+|D|_r)\,\chi_D^{\otimes(n-1)}. \end{aligned} \tag{23}$$

Therefore, $\mathfrak{X}_D$ is an invariant subspace for the operator $\widetilde{\mathbf{J}}_{\chi_D}$. Let $\mathbf{J}_D$ stand for the restriction of $\widetilde{\mathbf{J}}_{\chi_D}$ to $\mathfrak{X}_D$. Then $\mathbf{J}_{\chi_D}$ is a densely defined operator in $\mathfrak{X}_D$. Let $\lambda_{n,D} = \|\chi_D^{\otimes n}\|_{\mathcal{F}_{NB}(\mathcal{H})}$, then the vectors $\left(\varepsilon_D^{(n)}\right)_{n\in\mathbb{N}}$, with $\varepsilon_D^{(n)} = \lambda_{n,D}^{-1}\chi_D^{\otimes n}$, constitute an orthonormal basis for $\mathfrak{X}_D$. (23) yields

$$\begin{aligned}\mathbf{J}_D \varepsilon_D^{(n)} = \frac{\lambda_{n+1,D}}{\lambda_{n,D}}\,\varepsilon_D^{(n+1)} &+ \left(\frac{n(\beta+1)}{\alpha} + \frac{\beta|D|_r}{\alpha}\right)\varepsilon_D^{(n)} \\ &+ \frac{\beta}{\alpha^2} n\Big(n-1+|D|_r\Big)\,\frac{\lambda_{n-1,D}}{\lambda_{n,D}}\,\varepsilon_D^{(n-1)}. \end{aligned} \tag{24}$$

The matrix of the operator $\mathbf{J}_D$ in the orthonormal basis $(\varepsilon_D^{(n)})_{n\in\mathbb{N}}$ must be symmetric. Thus, together with the formula (24), this implies that this matrix is a Jacobi one with recurrence formula :

$$\mathbf{J}_D\,\varepsilon_D^{(n)} = v_{n+1,D}\,\varepsilon_D^{(n+1)} + w_{n,D}\,\varepsilon_D^{(n)} + w_{n,D}\,\varepsilon_D^{(n-1)}. \tag{25}$$

By comparing (24) and (25) we obtain the equality

$$\frac{\lambda_{n,D}}{\lambda_{n-1,D}} = \frac{\beta}{\alpha^2}n\Big(n-1+|D|_r\Big)\frac{\lambda_{n-1,D}}{\lambda_{n,D}},$$

from which we derive

$$\frac{\lambda_{n,D}}{\lambda_{n-1,D}} = \sqrt{\frac{\beta}{\alpha^2}n(n-1+|D|_r)}. \tag{26}$$

Substituting (26) into (24), we conclude that the coefficients $w_{n,D}$ and $v_{n,D}$ in (25) equal

$$w_{n,D} = \sqrt{\frac{\beta}{\alpha^2}n\big(n-1+|D|_r\big)}\,, \quad v_{n,D} = \frac{\beta+1}{\alpha}n + \frac{\beta}{\alpha}|D|_r.$$

By using (1) and (2), with $a_n = v_{n,D}$ and $b_n = w_{n,D}$, Theorem 2.1 guarantees the existence of a unique probability measure $\mu_{|D|_r,\alpha}$ on $\mathbb{R}$ such that the family of polynomials $\{P_{|D|_r,\alpha,n}\}_{n\in\mathbb{N}}$, defined by the recurrence formula

$$P_{|D|_r,\alpha,n+1}(x) = (x - v_{n,D})P_{|D|_r,\alpha,n}(x) - w_{n,D}^2 P_{|D|_r,\alpha,n-1}(x), \tag{27}$$

forms an orthonormal basis for $L^2(\mathbb{R},\mu_{|D|_r,\alpha})$. Denote $\widetilde{\mathbf{J}}_D$ the closure in $\mathfrak{X}_D$ of the essentially self-adjoint operator $\mathbf{J}_D$. Then, under the unitary operator $U_D : \mathfrak{X}_D \to L^2(\mathbb{R},\mu_{|D|_r,\alpha})$ given by $U_D\varepsilon_D^{(n)} := P_{|D|_r,\alpha,n}$, $\widetilde{\mathbf{J}}_D$ goes over into the operator of multiplication by x. Moreover, from (3), (4) and (27) we observe that $\mu_{|D|_r,\alpha}$ is nothing but the negative binomial distribution with parameters $|D|_r$ and α. On the other hand, we know from Theorem 3.2 that $\mathfrak{U}_{r,\alpha}\widetilde{\mathbf{J}}_{\chi_D}\mathfrak{U}_{r,\alpha}^{-1}$ coincides with the operator of multiplication by $\langle z,\chi_D\rangle$. Hence, (25) and (27) imply that

$$(\mathfrak{U}_{r,\alpha}\,\varepsilon_D^{(n)})(z) = P_{|D|_r,\alpha,n}\Big(\langle z,\chi_D\rangle\Big)\,.$$

Therefore, $\mu_{|D|_r,\alpha}$ coincides with the distribution of $\langle\cdot\,,\chi_D\rangle$ regarded as a random variable in the probability space $(\mathcal{E}',\mathcal{B}(\mathcal{E}'),\Lambda_{r,\alpha})$. □

Theorem 4.1. *The Fourier transform of $\Lambda_{r,\alpha}$ is given by the following formula*

$$\int_{\mathcal{E}'} e^{iu\langle z,\varphi\rangle}d\Lambda_{r,\alpha}(z) = \exp\left\{\int_{\mathbb{R}}\log\Big(\frac{\alpha}{1-\beta e^{iu\varphi(x)}}\Big)\mu_{r,\alpha}(dx)\right\}\,,\ \varphi\in\mathcal{E},\ u\in\mathbb{R}. \tag{28}$$

Proof. Let $D \in \mathcal{B}_b(\mathbb{R})$ be given. For $u \in \mathbb{R}$, by using Lemma 4.1, we get

$$\int_{\mathcal{E}'} e^{iu\langle z,\chi_D\rangle} d\Lambda_{r,\alpha}(z) = \exp\left\{|D|_r \log\left(\frac{\alpha}{1-\beta e^{iu}}\right)\right\}$$

$$= \exp\left\{\int_{\mathbb{R}} \log\left(\frac{\alpha}{1-\beta e^{iu\chi_D(x)}}\right)\mu_{r,\alpha}(dx)\right\}.$$

Let be given an n disjoint sets $D_1, \cdots, D_n \in \mathcal{B}_b(\mathbb{R})$. Then, the random variables $\langle\cdot, \chi_{D_1}\rangle, \cdots, \langle\cdot, \chi_{D_n}\rangle$ are independent. Hence, for any step function $\varphi = \sum_i u_i\chi_{D_i}$, formula (28) holds. Now, for any $\varphi \in \mathcal{E}$, choose any sequence of step functions $\{\varphi_n\}_{n\in\mathbb{N}}$ such that φ_n converges uniformly to φ as $n \to \infty$. Then, by the dominated convergence theorem, we deduce that the left-hand side of (28) converges to the right-hand side of (28). On the other hand, we can easily show that $\langle\cdot, \varphi_n\rangle$ converges to $\langle\cdot, \varphi\rangle$ in $L^2(\mathcal{E}', \Lambda_{r,\alpha})$, and therefore also in probability. Hence, again by the dominated convergence theorem, the left-hand side of (28) with $\varphi = \varphi_n$ converges to the right-hand side of (28). This completes the proof of the theorem. □

Corollary 4.1. *For any $D \in \mathcal{B}_b(\mathbb{R})$, we have*

$$\left\langle : z^{\otimes n} :_{NB}, \chi_D^{\otimes n}\right\rangle = P_{|D|_r,\alpha,n}\left(\langle z, \chi_D\rangle\right), \quad \Lambda_{r,\alpha} - a.e.\ z \in \mathcal{E}', \tag{29}$$

where $\{P_{|D|_r,\alpha,n}\}_{n=0}^{\infty}$ are the polynomials on $\mathbb{R}$ defined by the recurrence relation (27).

Proof. The statement follows from the proof of Theorem 3.2. □

Remark 4.1. For $t \in \mathbb{R}_+^*$, denote $|t|_r := \mu_{r,\alpha}([0,t])$. Then, by setting $\varphi = \lambda\,\chi_{[0,t]}$ in (28) we obtain

$$\int_{\mathcal{E}'} e^{i\langle z,\varphi\rangle} d\Lambda_{r,\alpha}(z) = \exp\left\{|t|_r \log\left(\frac{\alpha}{1-\beta e^{i\lambda}}\right)\right\}, \quad \lambda \in \mathbb{R}.$$

This function coincides with the Fourier transform of the measure $\mu_{|t|_r,\alpha}$. On the probability space $(\mathcal{E}', \mathcal{B}(\mathcal{E}'), \Lambda_{r,\alpha})$, the random variable $X_t := \langle\cdot, \chi_{[0,t]}\rangle$ has a negative binomial distribution with parameters $|t|_r$ and α. So the family of random variables

$$X = \{X_t;\ t \in \mathbb{R}\}, \quad X_0 = 0$$

is called *a white noise Pascal process.* The image of the Pascal white noise measure $\Lambda_{r,\alpha}$ under the random variable $X_t(\cdot)$ is the negative binomial distribution $\mu_{|t|_r,\alpha}$. Then, for $t > s > 0$, the characteristic function of $X_t - X_s$ has the following form :

$$\mathbb{E}\left[\exp\{i\lambda(X_t - X_s)\}\right] = \exp\left[|t-s|_r \log\left(\frac{\alpha}{1-\beta e^{i\lambda}}\right)\right]. \qquad (30)$$

We conclude that for $s, t \in \mathbb{R}$ the random variable $X_t - X_s$ has the negative binomial distribution with parameters $|t-s|_r$ and α. This gives an essence to our approach for defining the Pascal white noise functionals. Denote by $\langle \cdot , \phi \rangle$ the L^2-limit of a sequence $\{\langle \cdot , \phi_n \rangle\}$, $\phi_n \in \mathcal{E}$. When $\phi = \chi_{(s,t)}$, the characteristic function of $\langle \cdot , \phi \rangle$ is exactly the same as the one in (30). So the white noise Pascal process X on $(\mathcal{E}', \mathcal{B}(\mathcal{E}'), \Lambda_{r,\alpha})$ can be represented by

$$X(t; z) = \begin{cases} \langle z , \chi_{(0,t)} \rangle \text{ if } t \geq 0 \\ -\langle z , \chi_{(t,0)} \rangle \text{ if } t < 0 , \quad z \in \mathcal{E}' \end{cases}$$

Taking the time derivative formally, we get

$$\dot{X}(t; z) = z(t), \quad z \in \mathcal{E}'.$$

Thus, the elements of $\mathcal{E}'$ can be regarded as the sample paths of the Pascal white noise, and members of $L^2(\mathcal{E}', \Lambda_{r,\alpha})$ are called *Pascal white noise functionals.*

Theorem 4.2. *The Pascal white noise measure $\Lambda_{r,\alpha}$ is a compound Poisson measure on $(\mathcal{E}', \mathcal{B}(\mathcal{E}'))$ whose Lévy-Khintchine representation reads as follows:*

$$\int_{\mathcal{E}'} e^{i\langle z, \varphi\rangle} d\Lambda_{r,\alpha}(z) = \exp\left\{\int_{\mathbb{R}\times\mathbb{R}_+^*} \left(e^{is\varphi(x)} - 1\right) \mu_{r,\alpha}(dx)\nu_\alpha(ds)\right\}, \qquad (31)$$

where

$$\nu_\alpha(ds) = \sum_{k=1}^{\infty} \frac{\beta^k}{k} \delta_k(ds). \qquad (32)$$

Proof. It follows from the general theory of compound Poisson measures that there exists a compound Poisson measure $\widetilde{\Lambda}_{r,\alpha}$ whose Fourier transform is given by (31) with ν_α given by (32). It suffices to show that, for any $D \in \mathcal{B}_b(\mathbb{R})$, the distributions of the random variable $\langle z, \chi_D \rangle$ with respect to $\Lambda_{r,\alpha}$ and $\widetilde{\Lambda}_{r,\alpha}$ coincide. By direct computation we have

$$\int_{\mathcal{E}'} e^{iu\langle z, \chi_D\rangle} d\widetilde{\Lambda}_{r,\alpha}(z) = \exp\left\{|D|_r \int_{\mathbb{R}_+^*} (e^{ius} - 1)\nu_\alpha(ds)\right\}. \qquad (33)$$

On the other hand,

$$\int_{\mathbb{R}_+^*} (e^{ius} - 1)\nu_\beta(ds) = \sum_{k=1}^{\infty} \Big(\frac{(\beta e^{iu})^k}{k} - \frac{\beta^k}{k} \Big) = \log \Big(\frac{\alpha}{1 - \beta e^{iu}} \Big). \tag{34}$$

Hence, we conclude that

$$\int_{\mathcal{E}'} e^{iu\langle z, \chi_D \rangle} d\widetilde{\Lambda}_{r,\alpha}(z) = \exp \Big\{ |D|_r \log \Big(\frac{\alpha}{1 - \beta e^{iu}} \Big) \Big\}$$

$$= \int_{\mathcal{E}'} e^{iu\langle z, \chi_D \rangle} d\Lambda_{r,\alpha}(z).$$

This follows the proof. □

We recall that in one-dimensional case,[2] the generating function $\psi_{\mu_{|D|_r,\alpha}}$ of the sequence of polynomials $\{P_{|D|_r,\alpha,n}\}_{n=0}^{\infty}$ characterized by (27) is given by

$$\psi_{\mu_{|D|_r,\alpha}}(t; x) = \sum_{n=0}^{\infty} \frac{\alpha^n}{n!} t^n P_{|D|_r,\alpha,n}(x) = (1+t)^x (1+\beta t)^{-x-|D|_r}. \tag{35}$$

According to (29), as infinite dimensional analogue of (35), the generating function of the polynomials $\langle : z^{\otimes n} :_{NB}, \varphi^{\otimes n} \rangle$ is given by

$$\Psi_{\Lambda_{r,\alpha}}(z; \varphi) = \sum_{n=0}^{\infty} \frac{\alpha^n}{n!} \langle : z^{\otimes n} :_{NB}, \varphi^{\otimes n} \rangle, \quad z \in \mathcal{E}', \quad \varphi \in \mathcal{E}.$$

Theorem 4.3. *For each $\varphi \in \mathcal{E}$ satisfying $\sup_{x \in \mathbb{R}} |\varphi(x)| = \|\varphi\|_\infty < 1$, we have*

$$\Psi_{\Lambda_{r,\alpha}}(z; \varphi) = \exp \Big\{ \Big\langle z, \log \Big(\frac{1+\varphi}{1+\beta\varphi} \Big) \Big\rangle - \langle \log(1+\beta\varphi) \rangle_{r,\alpha} \Big\}. \tag{36}$$

Proof. The major step in the proof of this theorem is the following identity: for any $k_1, \cdots, k_m \in \mathbb{N}$ such that $k_1 + \cdots + k_m = n$, any $z \in \mathcal{E}'$ and any disjoint $D_1, \cdots, D_m \in \mathcal{B}_b(\mathbb{R})$, we have

$$\Big\langle : z^{\otimes n} :_{NB}, \chi_{D_1}^{\otimes k_1} \widehat{\otimes} \cdots \widehat{\otimes} \chi_{D_m}^{\otimes k_m} \Big\rangle = \prod_{i=1}^{m} P_{|D_i|_r,\alpha,k_i}(\langle z, \chi_{D_i} \rangle). \tag{37}$$

In fact, we prove (37) by induction in $n \in \mathbb{N}$. For $n = 1$, formula (37) trivially holds. Now, suppose that (37) holds for any $n \leq N$. Let $k_1, \cdots, k_m \in \mathbb{N}$ with $k_1 + \cdots + k_m = N + 1$. Applying the recurrence formula (18), for any $1 \leq j \leq m$, direct calculus gives

$$\left\langle :z^{\otimes(N+1)}:_{NB}, \chi_{D_1}^{\otimes k_1}\widehat{\otimes}\cdots\widehat{\otimes}\chi_{D_m}^{\otimes k_m}\right\rangle = \left\{\langle z, \chi_{D_j}\rangle - \left(\frac{\beta+1}{\alpha}N + |D_j|_r\frac{\beta}{\alpha}\right)\right\}$$
$$\times \left\langle :z^{\otimes N}:_{NB}, \chi_{D_1}^{\otimes k_1}\widehat{\otimes}\cdots\widehat{\otimes}\chi_{D_j}^{\otimes(k_j-1)}\widehat{\otimes}\cdots\widehat{\otimes}\chi_{D_m}^{\otimes k_m}\right\rangle$$

$$-\frac{\beta}{\alpha^2}N\Big(|D_j|_r + N - 1\Big)\left\langle :z^{\otimes(N-1)}:_{NB}, \chi_{D_1}^{\otimes k_1}\widehat{\otimes}\cdots\widehat{\otimes}\chi_{D_j}^{\otimes(k_j-2)}\widehat{\otimes}\cdots\widehat{\otimes}\chi_{D_m}^{\otimes k_m}\right\rangle.$$

Applying formula (37) with $n = N$ and $n = N-1$ and then using the recurrence relation (27) for the polynomials $P_{|D|_r,\alpha,n}$, we conclude the statement. Now, fix $D \in \mathcal{B}_b(\mathbb{R})$, by using the equality (29) we obtain

$$\sum_{n=0}^{\infty}\frac{\alpha^n}{n!}\langle :z^{\otimes n}:_{NB}, (u\,\chi_D)^{\otimes n}\rangle = \sum_{n=0}^{\infty}\frac{\alpha^n u^n}{n!}P_{|D|_r,\alpha,n}(\langle z, \chi_D\rangle).$$

Hence, it follows from (35) that (36) holds with $\varphi = u\,\chi_D$ and $|u| < 1$ from a neighborhood of zero in $\mathbb{R}$. Finally, for any disjoint $D_1, \cdots, D_m \in \mathcal{B}_b(\mathbb{R})$ and any $u_1, \cdots, u_m \in \mathbb{R}$ satisfying $|u_i| < 1$, $i = 1, \cdots, m$, put $f := \sum_{i=1}^{m} u_i\chi_{D_i}$. By (37), we get

$$\sum_{n=0}^{\infty}\frac{\alpha^n}{n!}\langle :z^{\otimes n}:_{NB}, f^{\otimes n}\rangle = \prod_{i=1}^{m}\left(\sum_{n=0}^{\infty}\frac{\alpha^n u_i^n}{n!}P_{|D_i|_r,\alpha,n}(\langle z, \chi_{D_i}\rangle)\right).$$

Hence, (36) holds with $\varphi = f$.

Finally, fix any $\varphi \in \mathcal{E}$. Choose any sequence of step functions $\{\varphi_n\}_{n\in\mathbb{N}}$ such that φ_n converges uniformly in the L^2-sense to φ as $n \to \infty$. Then, by the dominated convergence theorem, we conclude the statement. □

Remark 4.2. In the one-dimensional setting, the Laplace transform

$$\mathcal{L}_{\mu_{|D|_r,\alpha}}(u) := \int_{\mathbb{R}} e^{ux}\,d\mu_{|D|_r,\alpha}(x) = \exp\left\{|D|_r \log\left(\frac{\alpha}{1-\beta e^u}\right)\right\}$$

of the measure $\mu_{|D|_r,\alpha}$ is well defined and holomorphic on a neighborhood of zero in $\mathbb{R}$. Furthermore, for all $x \in \mathbb{R}$ and u from the neighborhood of zero, by using (35), we have

$$\psi_{\mu_{|D|_r;\alpha}}(x,u) = \frac{\exp(x\theta(u))}{\mathcal{L}_{\mu_{|D|_r,\alpha}}(\theta(u))} \quad \text{with} \quad \theta(u) = \log\left(\frac{1+u}{1+\beta u}\right). \tag{38}$$

Similarly, the Laplace transform of the Pascal white noise measure $\mathcal{L}_{\Lambda_{r,\alpha}}(\varphi) = \int_{\mathcal{E}'} e^{\langle z,\varphi\rangle}d\Lambda_{r,\alpha}(z)$ is well defined and holomorphic on some neighborhood of zero $\mathcal{O}$ in $\mathcal{E}$. If we consider the holomorphic $\mathcal{E}$-valued function Θ defined on $\mathcal{O}$ by $\Theta(\varphi) = \log\left(\frac{1+\varphi}{1+\beta\varphi}\right)$, we get the infinite dimensional counterpart of (38):

$$\Psi_{\Lambda_{r,\alpha}}(z;\varphi) = \frac{\exp\left(\langle z, \Theta(\varphi)\rangle\right)}{\mathcal{L}_{\Lambda_{r,\alpha}}(\Theta(\varphi))}.$$

References

1. S. Albeverio, Yu.G. Kondratiev and J.L. Silva, *How to generalize white noise analysis to non-Gaussian spaces.* In. Dynamics of complex and irregular systems (Ph. Blanchard, L. Streit, M. Sirugue-Collin and D. Testard, eds.), World Scientific, Singapore, (1993), 120-130.
2. N. Asai, I. Kubo and H.-H. Kuo, *Multiplicative Renormalization and Generating Functions II.*, Taiwanese Journal of Mathematics, Vol. **8**, No. 4 (2004), 583-628.
3. A. Barhoumi, H. Ouerdiane and A. Riahi, *Unitary Representations of the Witt and* $sl(2,\mathbb{R})$*-algebra through the Renormalized Powers of the Quantum Pascal White Noise*, Infinite Dimensional Analysis, Quantum Probability and Related Topics. Vol. **11**, No. 3 (2008), 323-350.
4. Yu.M. Berezansky, *Expansions in Eigenfunctions of Selfadjoint Operators*, Amer. Math. Soc., Providence, Rhode Island, 1968.
5. Yu.M. Berezansky, *Commutative Jacobi fields in Fock space*, Integral Equations Operator Theory Vol. **30** (1998), 163-190.
6. Yu.M. Berezansky and Yu.G. Kontratiev, *Spectral Methods in infinite Dimentional Analysis*, Kluwer Acad. Publ., Dordrecht, Boston, London 1994.
7. Yu.M. Berezansky and V.D. Koshmanenko, *An asymptotic field theory in terms of operator Jacobian matrices*, Soviet Physics Dokl. Vol. **14** (1969-1970), 1064-1066.
8. N. Bourbaki, *Functions d'une variable Réelle. Eléments de Mathématiques*, Livre IV. Hermann et Cie, Paris, 1950.
9. E. Bruning, *When is a field a Jacobi field. A characterization of states on tensor algebra*, Publ. Res. Inst. Math. Sci. Vol. **22** (1986), 209-246.
10. T.S. Chihara, *An Introduction to Orthogonal Polynomials*, Gordon and Breach, New York, 1978.
11. R. Gannoun, R. Hchaichi, H. Ouerdiane and A. Rezgui, *Un théorème de dualité entre espaces de fonctions holomorphes à croissance exponentielle*, J. Funct. Anal. Vol. **171**, No. 1 (2000), 1-14.
12. I.M. Gelfand and N.Ya. Vilenkin, *Generalized Functions*, Vol. **4**, Academic press New York and London 1964.
13. I. Gradstein and I. Ryshik, *Tables of Series, Products and Integrals*, Vol. **1** Verlag Harri Deutsch Thun, Frankfurt. M, 1981.
14. T. Hida, *Analysis of Brownian Functional*, Carleton Math. Lecture Notes No. **13**, Carleton University, Ottawa, 1975.
15. R. Koekoek and R. F. Swarttouw, *The askey-sheme of hypergeometric orthogonal polynomials and its q-analogue* , Delft Univ. of Tech., Report No. **98-17** (1998).

16. Y. Kondratiev, J.L. Silva and L. Streit, *Differential geometry on compound Poisson Space*, Meth. Funct. Anal. and Topol. Vol. **4**, No. 1 (1998), 32-58.
17. Y. Kondratiev, J.L. Silva, L. Streit and G.F. US, *Analysis on Poisson and Gamma spaces*, Infi. Dimen. Anal. Quant. Probab. Rel. Top. Vol. **1** (1998), 91-117.
18. H.-H. Kuo, *White noise distrubition theory*, CRC press, Boca Raton 1996.
19. N. Obata, *White noise calculus and Fock space*, Lecture Notes in Math. Vol. **1577**, Springer-Verlag, 1994.
20. A.D. Pulemyotov, *Support of a joint resolution of identity and the projection spectral theorem*, Infin. Dimen. Anal. Quant. Probab. Rel. Top. Vol. **6** (2003), 549-561.

QUANTUM WHITE NOISE ANALYSIS AND QUANTUM STOCHASTIC EQUATIONS

V. P. BELAVKIN

Mathematics Department
University of Nottingham
NG7 2RD, UK
vpb@maths.nott.ac.uk

JAESEONG HEO

Department of Mathematics
Research Institute for Natural Sciences
Hanyang University
Seoul 133-791, Korea
hjs@hanyang.ac.kr

UN CIG JI

Department of Mathematics
Research Institute of Mathematical Finance
Chungbuk National University
Cheongju 361-763, Korea
uncigji@chungbuk.ac.kr

We review the basic concepts of quantum stochastic analysis using the universal Itô B*-algebra approach. The thermal quantum white noise as a noncommutative analog of the Lévy-Itô process is characterized in terms of the modular B*-Itô algebra. The main notions and results of classical and quantum stochastics are reformulated in this unifying approach. The Belavkin master equation for a quasi-Markov quantum stochastic dynamics on the predual space of a W*-algebra is presented as a noncommutative version of the Zakai equation driven by the modular quantum white noise. This is done by the noncommutative analog of the Girsanov transformation which we introduce in full QS generality.

Keywords: Quantum Itô algebra, White Noise Analysis, Girsanov Transformation, Stochastic Master Equations.

1. Introduction

This review is based on the lecture notes on Quantum Stochastics given by Belavkin in Nottingham in February 2010.

Classical stochastic calculus developed by Itô, and its quantum stochastic analog, outlined by Hudson and Parthasarathy in Ref,[19] have been unified by Belavkin[1] in his $\star$-algebraic approach to the operator integration in Fock space,[8,10] in which the classical and quantum calculi became represented as two extreme commutative and noncommutative cases of an algebraically generalized Itô calculus.

Let us first remind the basic notions of noncommutative probability in the algebraic form unifying the classical and quantum cases. The standard classical approach based on the probability triple $(\Omega, \mathfrak{A}, \mathsf{P})$ is not appropriate for quantum world, it should be replaced by the triple $(H, \mathbb{M}, \epsilon)$, where H is a complex Hilbert space, $\mathbb{M}$ is a von Neumann algebra, a unital W*-algebra of not necessarily commuting random variables as adjointable operators on H, and ϵ is a linear functional of normal expectations satisfying the positivity property $\epsilon(\mathrm{X}) \geq 0$ on all Hermitian-positive operators $\mathrm{X} \geq 0$ in $\mathbb{M}$ and normalized on the identity operator I as $\epsilon(\mathrm{I}) = 1$. The space H has no objective meaning as Ω does but simply is a representational tool for $\mathbb{M}$ which can be chosen standard such that every normal ϵ is determined by a unit vector $h \in H$ as $\epsilon(\mathrm{X}) = \langle h|\mathrm{X}h\rangle$. It induces the weak (physical) topology on $\mathbb{M}$ in which every W*-algebra is characterized as closed $\star$-subalgebra of the maximal W*-algebra $\mathbb{B}(H)$. A physical state on $\mathbb{M}$ is defined as a weakly continuous normal functional ϵ given by a state vector h which is defined up to a unitary transformation $U : H \to H$ commuting with all $\mathrm{X} \in \mathbb{M}$, and in the standard representation every normal expectation is physical. The classical case $(\Omega, \mathfrak{A}, \mathsf{P})$ in the standard representation given by a choice of the reference probability measure Q corresponds to the Hilbert space $H = L^2_{\mathsf{Q}}(\Omega, \mathfrak{A})$ of measurable complex functions $h(\omega)$ on Ω defining Q-integrable densities $\varrho(\omega) = |h(\omega)|^2$ with $\mathrm{dP} = \varrho \mathrm{dQ}$ given by a normalized $h \in H$. It is defined by the commutative W*-algebra $\mathbb{M} = \breve{M}$ of the diagonal operators $\mathrm{X} = \breve{\chi}$ as the multiplications $\breve{\chi} : h \mapsto \chi h$ of the complex L^2_{Q}-functions $h(\omega)$ by L^∞_{Q}-functions $\chi \in L^\infty_{\mathsf{Q}}(\Omega, \mathfrak{A}) \equiv M$ with the normal expectation $\epsilon_h(\breve{\chi}) = \langle \chi, \varrho\rangle$ given by the integrals $\langle \chi, \varrho\rangle = \int \chi \varrho \mathrm{dQ}$ with the Radon-Nikodym derivative $\varrho = \mathrm{dP}/\mathrm{dQ}$ as the positive integrable functions $|h|^2 \in L^1_{\mathsf{Q}}(\Omega, \mathfrak{A})$ normalized as $\langle 1, \varrho\rangle = 1$.

In classical stochastics there are two types of random processes with independent increments which play a fundamental role: the diffusive processes and the jump processes. As it is suggested by celebrated Lévy-Khinchin theorem, every regular in a sense stochastics process x_t with independent increments is decomposed into a sum of smooth, diffusive and jumping parts given by independent Wiener and Poisson compound processes as standard processes respectively of diffusive and jump types. This theorem has very simple algebraic version[16] decomposing every classical (commutative) Itô B*-algebra, and even noncommutative Belavkin-Itô B*-algebra, into the orthogonal sum of three fundamentally different B*-algebras: the nilpotent B*-algebra, the Wiener-Itô B*-algebra as second order nilpotent, in general noncommutative B*-algebra and the Poisson-Itô B*-algebra as an idempotent, in general also noncommutative B*-algebra.

As part of the regularity in classical stochastic analysis the *corlol* convention (continuity on the right, limits on the left, or *càdlàg* in French) is usually adapted, but this condition does not show up and thus is redundant on the differential algebra level. Moreover, the corlol convention in general is not appropriate for the integrants $x(t) = x_{t-}$ of the stochastic integrals $x \centerdot y = \int x(t)\, \mathrm{d}y(t)$ and sometimes it is even inappropriate for applications in physics where the stochastic processes (fields) are naturally treated in the interaction picture as right discontinuous at the boundary (atoms) corresponding to the initial time $t = 0$, and the quantum jump processes like photo-countings on the increasing time intervals $[0, t)$ are obviously right discontinuous at any t. Instead, we have to assume the predictability of the integrants $x(t)$ which will be by default *corlol* regular (having limits on the right and continuous on the left, or *càglàd* in French), and we shall always take the corlol convention for the integrators $y(t)$ adapting the notation $y^t = \lim_{\varepsilon \searrow 0} y(t + \varepsilon)$ for their right limits $y(t^+)$. We may also denote by $\mathfrak{A}^t$ and $\mathfrak{A}^{t]}$ the corresponding σ-algebras generated respectively by $y(s)$ and $y^s = y(s^+)$, $s \leq t$ for each t. We shall exploit this convention for the notational simplicity also in the general (quantum) stochastics.

Moreover, unlike the induced dynamics on $\mathbb{A}^{t]}$, the predual dynamics is not closed but open, described by the QS forward maps from the smaller $\mathbb{B}_\star^{r]}$ into the larger $\mathbb{B}_\star^{t]}$ predual spaces $\forall r < t$. It is generated not by deterministic but quantum stochastic differential equation driven by the general output quantum Lévy process defining the quantum state on $\mathbb{A}^{t]}$ as a QS process with values in the predual reduced space $\mathbb{B}_\star^{t]}$. It defines the QS Radon-Nikodym derivative of the output state with respect to the reference "quantum white noise" product state as the reference temperature state corresponding to the modular B*-Itô state (quantum Lévy measure) on $\mathfrak{b}$. This equation in the classical case corresponds to the Zakai equation, and its semi-quantum (quantum object – classical output) derived in

Refs[4,6] has been already widely accepted as the QS master equation for "quantum state trajectories" in physics. Here we present the full noncommutative version of the QS master equation, the diffusive version of which, corresponding to a Wiener modular B*-Itô algebra, was obtained in Ref.[7]

2. Matrix and Noncommutative Itô Algebras

Here we give a brief introduction into the algebraic approach to universal Itô calculus introduced by Belavkin in Ref.[1] It is based on the notion of associative but nonunital and noncommutative algebra with involution, which is defined even in infinite dimensions as a Banach space called B*-Itô algebra.[17] The universality means that it is equally applicable for both, classical and quantum stochastic calculi based respectively on the commutative and noncommutative Itô B*-algebras corresponding to a classical or quantum Lévy process. Like a finite or infinite-dimensional C*-algebra it has matrix or operator representation constructed in Refs,[10,16] however, not in a Hilbert but pseudo-Hilbert space. The most effective use of Itô algebra is the algebraic proofs of the universal Lévy-Khinchin theorem given by Belavkin in Refs[16,17] and the Belavkin's universal Itô formula.[8,10]

2.1. *Matrix implementation of classical Itô algebras*

2.1.1. *Newton, Wiener and Poisson Itô algebras*

The deterministic Newton-Leibniz differential calculus based on the nilpotent multiplication rule $(\mathrm{d}t)^2 = 0$ can be easily extended from smooth to continuous trajectories $x(t)$ having right, say $\mathbb{C}$-valued corlol derivatives $\kappa(t)$. The differentials $\mathrm{d}x = x(t+\mathrm{d}t) - x(t) = \kappa\mathrm{d}t$ form obviously one-dimensional $\star$-algebra with the trivial products $\kappa \cdot \kappa^* = 0$ representing $|\mathrm{d}x|^2 = 0$. This formal algebra of differential calculus was generalized by Itô[20] to non-smooth continuous diffusions having conditionally independent increments $\mathrm{d}x \propto \sqrt{\mathrm{d}t}$ with no derivative at any t or having the right derivative for almost all t with forward conditionally independent jumps $\mathrm{d}x \propto \{0,1\}$ of corlol trajectories at some random t with right limits defined as $x(t^+) = x(t) + \mathrm{d}x(t)$ at every t. The first can be achieved by adding the increment of a real Wiener process $w(t) = w(t)^*$ in $\mathrm{d}x = \kappa\mathrm{d}t + \zeta\mathrm{d}w$ with the expectation $\mathbb{E}[\mathrm{d}w] = 0$ and the standard multiplication table

$$(\mathrm{d}w)^2 = \mathrm{d}t,\ \mathrm{d}w\mathrm{d}t = 0 = \mathrm{d}t\mathrm{d}w,\ (\mathrm{d}t)^2 = 0.$$

The second can be achieved by adding the increment of a compensated Poisson process $m(t) = \varepsilon n(t) - t/\varepsilon$ in $\mathrm{d}x = \kappa\mathrm{d}t + \zeta\mathrm{d}m$ with the multiplication table

$$(\mathrm{d}m)^2 = \mathrm{d}t + \varepsilon \mathrm{d}m, \quad \mathrm{d}m\mathrm{d}t = 0 = \mathrm{d}t\mathrm{d}m$$

following from $(\mathrm{d}n)^2 = \mathrm{d}n$ for a counting Poisson process $n(t) = n(t)^*$ of the intensity $\lambda(t) > 0$ defining its expectation by $\mathbb{E}[\mathrm{d}n] = \lambda \mathrm{d}t$ and $\varepsilon = \lambda^{1/2}$. These rules are known respectively as the differential multiplications for the standard Wiener process $w(t)$ and for the forward differentials of the standard Poisson process $n(t)$ compensated by its mean value t.

2.1.2. *Newton-Leibniz differentials as matrices*

The Newton-Leibniz calculus of the increments $\mathrm{d}x = x(t + \mathrm{d}t) - x(t)$ for smooth trajectories $t \mapsto x(t)$ is based on the nilpotent algebraic rule $(\mathrm{d}t)^2 = 0$. The linear space of formal differentials $\mathrm{d}x = \kappa \mathrm{d}t$ with $\kappa \in \mathbb{C}$ say, becomes one-dimensional complex nonunital associative $\star$-algebra $\mathfrak{d} = \mathbb{C}d$ of $c = \kappa d$ with respect to the trivial product $c \cdot c^\star = 0$ and involution $c^\star = \kappa^* d$ given by the complex conjugation $\kappa \mapsto \kappa^*$ of $\kappa = \mathrm{d}x/\mathrm{d}t \in \mathbb{C}$ in "real" basis of a nilpotent element $d = d^\star$, $d^2 = 0$ as the abstract notation for $\mathrm{d}t$. Such d has no realization in complex numbers, and no operator representation by a Hermitian nilpotent D in any C*-algebra. But as was noticed in Refs,[1,16] it can be implemented in the associative $\star$-algebra of complex quaternions, say by a complex pseudo-Hermitian symmetric nilpotent $\hat{d} = \frac{1}{2}(\hat{\sigma}_3 + \mathrm{i}\hat{\sigma}_1) = \hat{d}^\star$ in Stocks basis of standard Pauli matrices $\{\hat{\sigma}_i\}$. Here $\hat{a}^\star = \hat{\sigma}_3 \hat{a}^\dagger \hat{\sigma}_3$ denotes the pseudo-adjoint of any 2×2-matrix $\hat{a} = \sum_i \alpha^i \hat{\sigma}_i \equiv \hat{\sigma}_i \alpha^i$ with respect to the Minkowski metric $\langle \mathbf{z} | \mathbf{z} \rangle := |z^+|^2 - |z^-|^2 = \mathbf{z}^\dagger \hat{\sigma}_3 \mathbf{z}$ in the standard basis $\mathbf{z} = z^+ \mathbf{v}_+ + z^- \mathbf{v}_-$.

The nilpotent $\hat{d}$ has also the real pseudo-symmetric canonical form

$$\mathbf{d} = \begin{bmatrix} 0 & 1 \\ 0 & 0 \end{bmatrix} = \widetilde{\mathbf{d}} := \hat{\sigma}_1 \mathbf{d}^\intercal \hat{\sigma}_1, \quad \mathbf{d}^\intercal = \begin{bmatrix} 0 & 0 \\ 1 & 0 \end{bmatrix}$$

in another basis $\mathbf{f}_\pm = \frac{1}{\sqrt{2}}(\mathbf{v}_+ \pm \mathbf{v}_-)$ in which it is also pseudo-Hermitian, $\mathbf{d}^\ddagger := \widetilde{\mathbf{d}}^\dagger = \mathbf{d}$ as

$$(\mathbf{k} | \mathbf{dk}) = \mathbf{k}^\ddagger \mathbf{dk} = (\mathbf{dk} | \mathbf{k}) \quad \forall \mathbf{k}^\ddagger = (k_-, k_+) = \mathbf{k}^\dagger \hat{\sigma}_1$$

with respect to the new pseudo-Euclidean metric $(\mathbf{k} | \mathbf{k}) := k_- k^- + k_+ k^+ = \mathbf{k}^\dagger \hat{\sigma}_1 \mathbf{k}$ in terms of $k^\pm = (z^+ \pm z^-)/\sqrt{2} = k^*_\mp$. Note that the nilpotent $\hat{d}$ can also be represented by the Hermitian idempotent $\eth := \mathbf{d}\hat{\sigma}_1 = \eth^\dagger$ with the usual matrix product $\eth^2 = \eth\eth = \eth$ replaced by the new product

$\mathfrak{d} \cdot \mathfrak{d} = \mathfrak{d}\hat{\sigma}_1\mathfrak{d} = 0$, and such $\mathfrak{d}$ is "real" also in the sense of the pseudo-Hermitian conjugation $\mathfrak{d}^\star := \hat{\sigma}_3\mathfrak{d}^\dagger\hat{\sigma}_3 = \mathfrak{d}$ with respect to

$$\langle \mathbf{k}|\mathbf{k}\rangle = \mathbf{k}^\star\mathbf{k} = \mathbf{z}^\ddagger\mathbf{z} = (\mathbf{z}|\mathbf{z})$$

in terms $\mathbf{k} = k^+\mathbf{v}_+ + k^-\mathbf{v}_-$, $\mathbf{k}^\star = \mathbf{k}^\dagger\hat{\sigma}_3$ and $\mathbf{z} = k^-\mathbf{f}_- + k^+\mathbf{f}_+$.

2.1.3. *Matrix implementation of classical Itô algebras*

The linear complex spans $b_0 = \kappa d + \zeta e_w$ of formal increments $\{d, e_w\}$, having the only nonzero product $e_w^2 := e_w \cdot e_w = d$ given by a nilpotent in second order element $e_w = e_w^\star$, is two-dimensional commutative $\star$-algebra called the Wiener-Itô algebra $\mathfrak{b}_0$ with the state $l(b) = \kappa$ defining the drift $\mathbb{E}[\mathrm{d}x_0] = \kappa \mathrm{d}t$. Another basic Itô $\star$-algebra is the direct sum $\mathfrak{b}_1 = \mathfrak{d} \oplus \mathfrak{a}$ of the nilpotent $\mathfrak{d}$ and the unital $\star$-algebra $\mathfrak{a} = \mathbb{C}e_n$ with usual multiplication of complex numbers rescaling the unit jumps $e_n = e_n \cdot e_n \equiv e_n^2 \sim \mathrm{d}n$ of the standard Poisson process $n(t) = (m(t) + t/\varepsilon)/\varepsilon$ corresponding to $\varepsilon = 1$. It can be also written in the form of spans $b_1 = \kappa d + \zeta e_m$, where κ is defined by the drift $\mathbb{E}[\mathrm{d}x_1] = \kappa \mathrm{d}t$ of $\mathrm{d}x_1 = \kappa \mathrm{d}t + \zeta \mathrm{d}m$, with the formal increments $e_m = \varepsilon e_n - d/\varepsilon \sim \mathrm{d}m$ of the compensated Poisson process $m(t)$ having the products

$$e_m^2 = d + \varepsilon e_m = \varepsilon^2 p, \; e_m \cdot d = 0 = d \cdot e_m,$$

where $p = p \cdot p^\star \equiv p \star p$ is the Poisson idempotent $e_n = p = e_n^\star$. It is also associative and commutative algebra called the Itô-Poisson algebra with the state $l(b) = \kappa$. As in the case of Newton algebra $\mathfrak{d} = \mathbb{C}d$, the Itô $\star$-algebras $\mathfrak{b}_0$ and $\mathfrak{b}_1$ have no Euclidean operator realization, but they can be represented by nonunital $\star$-algebras of triangular 3×3-matrices $\mathbf{b}_0 = \kappa\mathbf{e}_\emptyset + \zeta\mathbf{e}_w$, $\mathbf{b}_1 = \kappa\mathbf{e}_\emptyset + \zeta\mathbf{e}_m$ in the matrix pseudo-Hermitian basis

$$\mathbf{e}_\emptyset = \begin{bmatrix} 0\,0\,1 \\ 0\,0\,0 \\ 0\,0\,0 \end{bmatrix}, \mathbf{e}_w = \begin{bmatrix} 0\,1\,0 \\ 0\,0\,1 \\ 0\,0\,0 \end{bmatrix}, \mathbf{e}_m = \begin{bmatrix} 0\,1\,0 \\ 0\,\varepsilon\,1 \\ 0\,0\,0 \end{bmatrix},$$

where $\mathbf{e}_\emptyset \equiv \mathbf{d}$ represents the death element d such that

$$\mathbf{e}_\emptyset^\ddagger := \boldsymbol{I}\mathbf{e}_\emptyset^\dagger\boldsymbol{I} = \mathbf{e}_\emptyset, \quad \mathbf{e}_w^\ddagger := \boldsymbol{I}\mathbf{e}_w^\dagger\boldsymbol{I} = \mathbf{e}_w, \quad \mathbf{e}_m^\ddagger := \boldsymbol{I}\mathbf{e}_m^\dagger\boldsymbol{I} = \mathbf{e}_m \;,$$

with respect to the "falling identity" (Itô) metric $\boldsymbol{I} = [\delta^\mu_{-\nu}]^{\mu=-,\circ,+}_{\nu=-,\circ,+}$ instead of $\hat{\sigma}_1$, not to be mixed with the identity $\mathbf{I} = [\delta^\mu_\nu]$. These commutative nilpotent and idempotent $\star$-algebras $\mathfrak{b} = \mathfrak{b}_0, \mathfrak{b}_1$ are the only possible two-dimensional extensions of the Newton $\star$-algebra $\mathfrak{d}$ such that $\mathfrak{d}\mathfrak{b} = 0 = \mathfrak{b}\mathfrak{d}$.

The algebraic Lévy-Khinchin theorem[16] suggests that every classical Itô $\star$-algebra $\mathfrak{b}$ as a commutative extension of $\mathfrak{d}$ can be decomposed into the orthogonal sum $\mathfrak{b}_0 + \mathfrak{b}_1$, $\mathfrak{b}_0\mathfrak{b}_1 = 0 = \mathfrak{b}_1\mathfrak{b}_0$ uniquely up to the $\star$-ideal $\mathfrak{d} = \mathfrak{b}_0 \cap \mathfrak{b}_1$. In particular, every nilpotent $\mathbf{e}_w$ must be orthogonal to every idempotent $\mathbf{e}_n$ in any classical Itô $\star$-algebra $\mathfrak{b} = \mathfrak{b}_0 + \mathfrak{b}_1$, $\mathbf{e}_w\mathbf{e}_n = 0 = \mathbf{e}_n\mathbf{e}_w$, given simply as $\mathbf{e}_n = \mathbf{e}_m + \mathbf{e}_\emptyset$ with $\varepsilon = 1$. This suggests that the minimal triangular matrix implementation of any commutative Itô $d+1$-dimensional algebra $\mathfrak{b}$ in the pseudo-Euclidean space $\mathbb{k} = \mathbb{C} \oplus \mathbb{C}^d \oplus \mathbb{C}$ with $d = d_w + d_n$ given by $\dim \mathfrak{b}_0 = d_w + 1$ and $\dim \mathfrak{b}_1 = d_n + 1$.

2.1.4. *Classical separable Lévy-Itô algebras*

The increments $\mathrm{d}m_i$ of a classical standard d-dimensional Lévy process with Wiener or Poisson components corresponding respectively to $\varepsilon_i = 0$ or $\varepsilon_i = 1$ form together with $\mathrm{d}t$ a basis of $d+1$-dimensional, respectively nilpotent $\mathfrak{b}_0$ and idempotent $\mathfrak{b}_1$, or mixed commutative Itô $\star$-algebra with usual complex conjugation of its coefficients if $m_i^* := \overline{m_i} = m_i$ for all $i \in J$. We shall call such, in general infinite-dimensional, Lévy-Itô algebra separable if it has discrete basis satisfying a closed multiplication table

$$\mathrm{d}m_i \mathrm{d}m_k = \sigma_{ik}\mathrm{d}t + \varepsilon_{ik}^j \mathrm{d}m_j, \quad \mathrm{d}m_j \mathrm{d}t = 0 = \mathrm{d}t \mathrm{d}m_j$$

indexed by a countable set J of the cardinality $|J| \geq d$ (Here and below the tensor summation rule is assumed for the convoluting indices $j \in J$). As it was proved in Ref,[17] the Itô product in such $\mathfrak{b}$ can also be matrix-implemented by the triangular block-matrices $\mathbf{e}_i$ for $\mathrm{d}m_i$ and $\mathbf{e}_\emptyset$ for $\mathrm{d}t$ such that $\mathbf{e}_i\mathbf{e}_k = \sigma_{ik}\mathbf{e}_\emptyset + \varepsilon_{ik}^j \mathbf{e}_j$ and $\mathbf{e}_j\mathbf{e}_\emptyset = 0 = \mathbf{e}_\emptyset\mathbf{e}_j$ in the $d+1$-dimensional pseudo-Hilbert space $\mathbb{k} = \mathbb{C} \oplus \mathfrak{k} \oplus \mathbb{C}$ with the Hilbert component $\mathfrak{k} \subseteq \mathbb{C}^{|J|}$.

In the case of pseudo-Hermitian basis $\mathbf{e}_i^\ddagger = \mathbf{e}_i$ corresponding to real $m_i = m_i^*$, the standard assumption is $\sigma_{ik} = \delta_{ik}$ and $\varepsilon_{ik}^j = \varepsilon\delta_i^j\delta_k^j$, where $\varepsilon = 0$ in the Wiener case and $\varepsilon = 1$ in the Poisson case. However, it is often convenient to consider a reflect-Hermitian basis $\mathbf{e}_i = \mathbf{e}_{-i}^\ddagger$, say, implementing the increments $\mathrm{d}m_i$, $i \in \mathbb{Z}$ of the reflect-real Fourier components $\bar{m}_i := m_{-i}^* = m_i$ for a real Lévy field $M(t,\theta) = \sum_{j=-\infty}^{\infty} m_j(t)\, \mathrm{e}^{2\pi \mathrm{i}\theta j}$ periodic in $\theta \in \mathbb{R}$. It is defined by the covariances

$$\mathbb{E}\left[M(\mathrm{d}t,\theta)\, M(\mathrm{d}t,\theta')\right] = \sum_{j=-\infty}^{\infty} \nu_j(t)\, \mathrm{e}^{2\pi\mathrm{i}(\theta'-\theta)j}\mathrm{d}t$$

corresponding to the diagonal $\sigma_{ik} = \nu_k\delta_k^{-i}$ and $\varepsilon_{ik}^j = \varepsilon\delta_{i+k}^j$ indexed by $J = \mathbb{Z}$. To be concrete we shall always assume that the separable Lévy-Itô algebra is given in such discrete symmetric basis corresponding to $m_{-i} = \overline{m_i}$ with respect to a general reflection as involutive bijection $j \mapsto -j$, $-(-j) = j$ on J which can be chosen trivial, $i = -i$ if $\overline{m_i} = m_i$ for all $i \in J$.

Every matrix-implemented classical Lévy-Itô algebra $\mathfrak{b}$ has faithful, or thermal state $l(\mathbf{b}) = (\mathbf{v}_+|\mathbf{b}\mathbf{v}_+) = b_+^-$ induced by the pseudo-vacuum vector $\mathbf{v}_+^\ddagger = (1,0,0)$ such that $l(\mathbf{b}^\ddagger\mathbf{b}) = 0 \Rightarrow \mathbf{b} = 0$ and $l(\mathbf{e}) = 1$. It defines the unique orthogonal decomposition $\mathbf{b} = \mathbf{b}^\circ + l(\mathbf{b})\mathbf{e}_\emptyset$ of the commutative $\star$-algebra $\mathfrak{b}$ into the $\star$-ideal $\mathbb{C}\mathbf{e}_\emptyset$ and the quotient $\star$-algebra $\mathfrak{b}/\mathfrak{d} = \{\mathbf{b} \in \mathfrak{b} : l(\mathbf{b}) = 0\}$ and is described by covariations σ_{ik} as

$$\mathbb{E}\left[\mathrm{d}m_i^*\mathrm{d}m_i\right] = l\left(\mathbf{e}_i^\ddagger\mathbf{e}_k\right)\mathrm{d}t = \sigma_k^i\mathrm{d}t, \quad \sigma_k^{-i} = \sigma_{ik},$$

with $\sigma_k^i = \delta_k^i$ for the standard basis m_i with the diagonal matrix $\boldsymbol{\sigma} = \left[\sigma_k^i\right]$ given by its eigenvalues $\nu_i = 1$. The $\star$-algebra $\mathfrak{b}^\circ = \mathfrak{b}/\mathfrak{d}$ identified with the vector subspace $\mathfrak{b}^\circ\mathbf{v}_+$ and equipped with the quotient product $\left(\mathbf{b}^\ddagger\mathbf{b}\right)^\circ = \mathbf{b}^\ddagger\mathbf{b} - l\left(\mathbf{b}^\ddagger\mathbf{b}\right)\mathbf{e}_\emptyset$ is a commutative pre-Hilbert algebra with respect to the scalar product

$$\langle\zeta|\zeta\rangle = l\left(\mathbf{b}^\ddagger\mathbf{b}\right) = l\left(\mathbf{b}^\ddagger\mathbf{b}\right) = \langle\zeta^*|\zeta^*\rangle$$

and the isometric involution $J\zeta := \mathbf{b}^{\circ\ddagger}\mathbf{v}_+ \equiv \zeta^*$ on $\zeta = \mathbf{b}^\circ\mathbf{v}_+$, and thus it can be completed to become a Hilbert $\star$-algebra $\mathfrak{k}$.

Since $l(\mathbf{e}_i) = 0$ for all $i \in J$ corresponding to $\mathbb{E}[m_i] = 0$, the matrices $\mathbf{e}_i = \mathbf{e}_{-i}^\ddagger$ form a reflect-Hermitian basis in $\mathfrak{b}^\circ = \mathfrak{b}/\mathfrak{d}$ corresponding to the reflect-real basis $e_i^\circ = \mathbf{e}_i\mathbf{v}_+ = e_{-i}^{\circ *}$ in the Hilbert algebra $\mathfrak{k}$. This basis may be overcomplete in $\mathfrak{k}$, but it is normal if standard in the sense that the Hermitian reflect-symmetric Gram matrix $\boldsymbol{\sigma} = \left[l\left(\mathbf{e}_i^\ddagger\mathbf{e}_j\right)\right]$ is orthoprojective, $\boldsymbol{\sigma} = \boldsymbol{\pi} = \boldsymbol{\pi}^\dagger\boldsymbol{\pi}$. It is orthonormal, $\boldsymbol{\pi} = \mathbf{1}$, iff the standard basis $\{\mathbf{e}_i\}$ is linearly independent in $\mathfrak{b}^\circ$.

2.2. *Implementation of noncommutative Itô $\star$-algebras*

2.2.1. *Quantum matrix-implemented HP-Itô algebra*

The non-orthogonal triangular matrix representation $\mathbf{e}_w$, $\mathbf{e}_m$ for $\mathbf{e}_w = \mathbf{e}_w^\ddagger$ and $\mathbf{e}_m = \mathbf{e}_m^\ddagger$ in the same pseudo-Euclidean space $\mathbb{k} = \mathbb{C}\oplus\mathbb{C}\oplus\mathbb{C}$ generates a noncommutative Itô $\star$-algebra $\mathfrak{b}(\mathbb{C})$ with $\mathbf{e}_w\mathbf{e}_m \neq \mathbf{e}_m\mathbf{e}_w$. This is four-dimensional $\star$-algebra of linear spans $\mathbf{b} = \kappa\mathbf{e}_\emptyset + \kappa_\circ\mathbf{e}_- + \kappa^\circ\mathbf{e}^+ + \kappa_\circ^\circ\mathbf{e}$ in the

Hermit-symmetric basis of the canonical triangular matrix basis of $\mathbf{e}_{-}^{+} = \mathbf{e}_{\emptyset}$,

$$\mathbf{e}_{-} = \begin{bmatrix} 0\,1\,0 \\ 0\,0\,0 \\ 0\,0\,0 \end{bmatrix}, \ \mathbf{e}^{+} = \begin{bmatrix} 0\,0\,0 \\ 0\,0\,1 \\ 0\,0\,0 \end{bmatrix}, \ \mathbf{e} = \begin{bmatrix} 0\,0\,0 \\ 0\,1\,0 \\ 0\,0\,0 \end{bmatrix}.$$

Here $\mathbf{e} = \mathbf{e}^{\ddagger}$ is the selfadjoint projector and $\mathbf{e}^{+} = \mathbf{e}_{-}^{\ddagger}$ is the *pseudo-creation* matrix with pseudo-adjoint $\mathbf{e}_{-}$ having the *pseudo-annihilation* property $\mathbf{e}_{-}\mathbf{v}_{+} = 0$ on the pseudo-vacuum state vector as nonzero column $\mathbf{1}_{+} = \left[\delta_{+}^{\iota}\right]^{\iota=-,\circ,+}$ of zero pseudo-norm with respect to the pseudo-Euclidean metric $(\mathbf{k}|\mathbf{k}) = k_{-}k^{-} + k_{\circ}k^{\circ} + k_{+}k^{+}$, where $k^{\iota} = k_{-\iota}^{*}$. As it was discovered in Refs,[1,10] the quadruple $\boldsymbol{b} = \left(\mathbf{b}_{\mu}^{\nu}\right)_{\mu=-,\circ}^{\nu=+,\circ}$ of the canonical matrices $\mathbf{b}_{\circ}^{\circ} = \mathbf{e}$, $\mathbf{b}_{\circ}^{+} = \mathbf{e}^{+}$, $\mathbf{b}_{-}^{\circ} = \mathbf{e}_{-}$ and $\mathbf{b}_{-}^{+} = \mathbf{e}_{\emptyset}$, given by the algebraic combinations

$$\mathbf{e}_{-} = \mathbf{e}_{w}\mathbf{e}_{m} - \mathbf{e}_{\emptyset}, \quad \mathbf{e}^{+} = \mathbf{e}_{m}\mathbf{e}_{w} - \mathbf{e}_{\emptyset}, \quad \mathbf{e} = \mathbf{e}_{m} - \mathbf{e}_{w}$$

of the noncommuting matrices $\mathbf{e}_{w}$ and $\mathbf{e}_{m}$ with $\varepsilon = 1$, realize the quantum stochastic Itô table

$$\mathbf{e}_{-} \cdot \mathbf{e}^{+} = \mathbf{e}_{\emptyset}, \quad \mathbf{e}_{-} \cdot \mathbf{e} = \mathbf{e}_{-}, \quad \mathbf{e} \cdot \mathbf{e} = \mathbf{e}, \quad \mathbf{e} \cdot \mathbf{e}^{+} = \mathbf{e}^{+},$$

which we call vacuum multiplication table if all otherwise products are equal zero. It is known as Hudson-Parthasarathy (HP) table[19] in terms of the canonical QS differentials of the Bosonic annihilation $\mathrm{E}_{-}^{t} \equiv \mathrm{A}_{-}^{\circ}(t)$, the adjoint creation $\mathrm{E}_{t}^{+} \equiv \mathrm{A}_{\circ}^{+}(t)$ and the self-adjoint exchange $\mathrm{E}_{t} \equiv \mathrm{A}_{\circ}^{\circ}(t) = \mathrm{E}_{t}^{*}$ processes in Fock space $F = \Gamma(K)$ over $K = L^{2}(\mathbb{R}_{+})$ which has all differential products zero except $(\mathrm{dE})^{2} = \mathrm{dE}$ and

$$\mathrm{dE}_{-}\mathrm{dE} = \mathrm{dE}_{-}, \ \mathrm{dEdE}^{+} = \mathrm{dE}^{+}, \ \mathrm{dE}_{-}\mathrm{dE}^{+} = \mathrm{d}t\mathrm{I}. \tag{1}$$

2.2.2. *The general affine and Heisenberg Itô algebras*

The above HP-Itô algebra is the scalar case $\mathfrak{k} = \mathbb{C}$ of the general *vacuum Itô $\star$-algebra* over a Hilbert space $\mathfrak{k}$ of ket-vectors $k^{\circ} = k_{\circ}^{*}$ as Hermitian adjoints to the bra-vectors $k_{\circ} \in \mathfrak{k}^{*}$. Such $\star$-algebra consists of all linear combinations $\boldsymbol{K} \centerdot \boldsymbol{b} := \sum_{\mu,\nu} \kappa_{\nu}^{\mu}\mathbf{b}_{\mu}^{\nu} \equiv \kappa_{\nu}^{\mu}\mathbf{b}_{\mu}^{\nu}$ given in the canonical basis $\boldsymbol{b} = \left(\mathbf{b}_{\mu}^{\nu}\right)$ by the quadruples $\boldsymbol{K} = \left(\kappa_{\nu}^{\mu}\right)_{\nu=+,\circ}^{\mu=-,\circ}$ with $\kappa_{+}^{-} \in \mathbb{C}$, $\kappa_{+}^{\circ} \in \mathfrak{k}$, $\kappa_{\circ}^{-} \in \mathfrak{k}^{*}$ and $\kappa_{\circ}^{\circ} \in \mathbb{B}(\mathfrak{k})$. These quadruples arranged into the triangular matrices $\mathbf{K} = [\kappa_{\nu}^{\mu}]_{\nu=-,\circ,-}^{\mu=-,\circ,+} \equiv \mathrm{K}(\boldsymbol{b})$ with $\kappa_{\nu}^{+} = 0 = \kappa_{-}^{\mu}$ for all μ, ν form a nonunital $\star$-algebra $\mathfrak{b}(\mathfrak{k})$ with respect to the usual block-matrix product $\mathbf{K}^{\ddagger}\mathbf{K}$ and unusual involution $\mathbf{K}^{\ddagger} = \boldsymbol{K}^{\star} \centerdot \boldsymbol{b} \equiv \mathbf{K}^{\ddagger}$ introduced in Ref.[1] Note that for a Hilbert space $\mathfrak{k}$ the involution $\mathbf{K} \mapsto \mathbf{K}^{\ddagger}$ is described by the Hermitian conjugation $\boldsymbol{K}^{\star} = \boldsymbol{K}^{\dagger}$

of the quadratic block-matrices $\boldsymbol{K} = (\kappa_\nu^\mu)_{\nu=+,\circ}^{\mu=-,\circ}$. Each such quadruple $\boldsymbol{K}$ defines a pair of noninvertible *adjoint affine transformations*

$$\mathrm{K} : \xi \oplus \lambda \mapsto (\zeta + \kappa_\circ^\circ \xi) \oplus (\kappa + \eta^* \xi),$$
$$\mathrm{K}^\ddagger : \xi \oplus \lambda \mapsto (\eta + \kappa_\circ^{\circ *} \xi) \oplus (\overline{\kappa} + \zeta^* \xi)$$

given on the vector space $\mathfrak{v} = \mathfrak{k} \oplus \mathbb{C}$ of $\xi \in \mathfrak{k}$ and $\lambda \in \mathbb{C}$ by a scalar $\kappa_+^- = \kappa$, a column-vector $\kappa_+^\circ = \zeta \in \mathfrak{k}$, a row-vector $\kappa_\circ^- = \eta^* \in \mathfrak{k}^*$ and a matrix $\kappa_\circ^\circ \in \mathbb{B}(\mathfrak{k})$ as adjointable linear transformation of $\mathfrak{k}$. The algebra $\mathfrak{b}(\mathfrak{k})$ defines the Lie group of the invertible affine transformations $\mathrm{T} = \mathrm{I} \oplus \mathrm{K}$ implemented on the pseudo-Hilbert space $\mathbb{k} = \mathbb{C} \oplus \mathfrak{v}$ by the invertible pseudo-adjointable block-matrices

$$\mathbf{T} = \begin{bmatrix} 1 & \eta^* & \kappa \\ 0 & \mathrm{T}_\circ^\circ & \zeta \\ 0 & 0 & 1 \end{bmatrix}, \ \mathbf{T}^\ddagger = \boldsymbol{I}\mathbf{T}^\dagger \boldsymbol{I}, \ \boldsymbol{I} = \begin{bmatrix} 0 & 0 & 1 \\ 0 & \mathrm{I}_\circ^\circ & 0 \\ 1 & 0 & 0 \end{bmatrix},$$

where the "falling identity" is metric tensor which should not be mixed with the identity matrix $\mathbf{I} = \boldsymbol{I}^2$ in the decomposition $\mathbf{T} = \mathbf{I} + \mathbf{K}$ implementing to the unitalization $\mathfrak{s} = \imath + \mathfrak{b}$ of $\mathfrak{b} = \mathfrak{b}(\mathfrak{k})$. Both $\mathbf{T}$ and $\mathbf{T}^\ddagger$ leave invariant $\mathfrak{v} = \mathfrak{k} \oplus \mathbb{C}$ inserted as hyperplane $\mathfrak{v}_1 = 1 \oplus \mathfrak{v}$ into $\mathbb{k} = \mathbb{C} \oplus \mathfrak{k} \oplus \mathbb{C}$. The nonunital $\star$-algebra $\mathfrak{b} = \mathfrak{b}(\mathfrak{k})$ of all such transformations described by block-matrices $\boldsymbol{K} = \mathbf{K}\boldsymbol{I}$ with $\boldsymbol{K}^\star = \mathbf{K}^\ddagger \boldsymbol{I}$ and the associative product $\boldsymbol{K} \cdot \boldsymbol{K}^\star = \boldsymbol{K}\boldsymbol{I}\boldsymbol{K}^\star$ is called *vacuum* (respectively, *Heisenberg vacuum*) Itô algebra. Its pure state $l(\mathrm{K}) = (\mathbf{v}_+ | \mathbf{K} \mathbf{v}_+) = \kappa$ is called *pseudo-vacuum* as defined by the vector $\mathbf{v}_+ = (1, 0, 0)^\ddagger$ having the vacuum property $\mathbf{b}_\iota^\kappa \mathbf{v}_+ = 0$ for all $\kappa \neq +$ and zero pseudo-norm $(\mathbf{v}_+ | \mathbf{v}_+) = 0$ (but $\mathbf{v}_+^\dagger \mathbf{v}_+ = (\mathbf{v}_+ | \boldsymbol{I} \mathbf{v}_+) = 1$.)

The general (finite-dimensional if $\mathfrak{k} = \mathbb{C}^d$) Itô algebra described as a nonunital associative $\star$-subalgebra $\mathfrak{b} \subseteq \mathfrak{b}(\mathfrak{k})$ on the (d + 1-dimensional) vector space $\mathfrak{v} = \mathfrak{k} \oplus \mathbb{C}$ was introduced in Refs.[9,18] It replaces the role of Lie algebra for a $\star$-submonoid $\mathfrak{s} \subseteq \mathrm{I} \oplus \mathfrak{b}(\mathfrak{k})$ of not necessarily invertible pseudo-adjointable affine transformations $\mathrm{S} \in \mathfrak{s}$ of $\mathfrak{v}$. In particular, the nilpotent of second order Itô $\star$-subalgebra of the triples $(\kappa_\circ^-, \kappa_+^-, \kappa_+^\circ)$ defining the only three nonzero entries in the block-matrix representation $\mathbf{H} = [\kappa_\nu^\mu]$ corresponding to $\kappa_\circ^\circ = 0$, is the Heisenberg $\star$-algebra as a Lie algebra of the Heisenberg $\star$-group of the invertible affine transformations $\mathbf{S} = \mathbf{I} + \mathbf{H}$ with $\mathrm{S}_\circ^\circ = \mathrm{I}_\circ^\circ$. It is the vacuum Itô $\star$-algebra $\mathfrak{b}_0(\mathfrak{k})$ of quantum Brownian motion over the vector space $\mathfrak{k}$.

2.2.3. *Quantum Banach Itô algebras and B^*-algebras*

An infinite dimensional Itô $\star$-algebra $\mathfrak{a}$ is called B*-algebra if it is complete with respect to the quadruple $\|\mathbf{a}\|_{\nu=+,\circ}^{\mu=-,\circ}$ of the operator seminorms $\|\mathbf{a}\|_{\nu}^{\mu} = \|\mathbf{a}_{\nu}^{\mu}\|$ given by the triangular block-matrix representation on $\Bbbk$ with $\|\mathbf{a}_{+}^{-}\| = |\mathbf{a}_{+}^{-}|$ since $\mathbf{a}_{+}^{-} \in \mathbb{C}$. These algebras were introduced by Belavkin[17] who classified them and also studied their operator representations, and we will shall call them after his name.

In general, the quadruple of $\star$-seminorms $\|\mathbf{a}\|_{\nu}^{\mu}$ on an abstract $\star$-algebra $\mathfrak{a}$ is called B*-norm if it separates the algebra $\mathfrak{a}$ in the sense $\|\mathbf{a}\|_{\nu=+,\circ}^{\mu=-,\circ} = 0 \Rightarrow \mathbf{a} = 0$ and satisfies the four inequalities $\|\mathrm{ab}\|_{\nu}^{\mu} \leq \|\mathbf{a}\|_{\circ}^{\mu} \|\mathbf{b}\|_{\nu}^{\circ}$ for all $\mathbf{a}, \mathbf{b} \in \mathfrak{a}$ with the $\star$-property $\|\mathbf{a}^{\star}\|_{-\nu}^{\mu} = \|\mathbf{a}\|_{-\mu}^{\nu}$ and two equalities

$$\|\mathbf{a}^{\star}\mathbf{a}\|_{+}^{-} = \|\mathbf{a}^{\star}\|_{\circ}^{-} \|\mathbf{a}\|_{+}^{\circ}, \quad \|\mathbf{a}^{\star}\mathbf{a}\|_{\circ}^{\circ} = \|\mathbf{a}^{\star}\|_{\circ}^{\circ} \|\mathbf{a}\|_{\circ}^{\circ} \quad \forall \mathbf{a} \in \mathfrak{a}$$

Note that the first of these equalities defines $\|\mathbf{a}\|_{+}^{\circ}$ as a Hilbert module seminorm, and the second defines $\|\mathbf{a}\|_{\circ}^{\circ}$ as a C*-seminorm. norm. A $\star$-algebra $\mathfrak{a}$ with a B*-norm is called B*-algebra if it is complete, i.e. is Banach space with respect to any if the norms $\max_{\mu,\nu} \|\mathbf{a}\|_{\nu}^{\mu}$ or $\sum_{\mu,\nu} \|\mathbf{a}\|_{\nu}^{\mu}$. In particular, if all $\|\mathbf{a}\|_{\nu}^{\mu} = 0$ except $\|\mathbf{a}\|_{+}^{\circ} = \|\mathbf{a}^{\star}\|_{\circ}^{-}$, the B*-algebra is a Hilbert module, and if all $\|\mathbf{a}\|_{\nu}^{\mu} = 0$ except $\|\mathbf{a}\|_{\circ}^{\circ} = \|\mathbf{a}^{\star}\|_{\circ}^{\circ}$, the B*-algebra is any C*-algebra.

It can be easily shown that the operator seminorms on any Itô $\star$-algebra form a B*-norm, and that an abstract B*-algebra with $\|\mathbf{a}\|_{+}^{-} = |l(\mathbf{a})|$ for a linear $\star$-positive functional $l : \mathfrak{a} \to \mathbb{C}$ normalized as $l(\mathfrak{d}) = 1$ on a unique death element $\mathbf{e}_{\emptyset}^{\star} = \mathbf{e}_{\emptyset}$ is an Itô B*-algebra in the sense that it has the unique up to a pseudo-unitary equivalence operator $\star$-representation $\pi = \|\pi_{\nu}^{\mu}\|_{\nu=+,\circ}^{\mu=-,\circ}$ in the quadruples of $\mathrm{A}_{\nu}^{\mu} = \pi_{\nu}^{\mu}(\mathbf{a})$ on the complex Minkowski space $\Bbbk = \mathbb{C} \oplus \mathfrak{k} \oplus \mathbb{C}$ with the minimal Hilbert component $\mathfrak{k}$.[17] Moreover, every Itô B*-algebra $\mathfrak{a}$ has a unique orthogonal decomposition $\mathfrak{b}_0 + \mathfrak{b}_1 + \mathfrak{d}$, up to the $\star$-ideal $\mathfrak{d} = \mathbb{C}\mathbf{e}_{\emptyset}$, into the quantum Brownian Itô B*-algebra $\mathfrak{b}_0$ and the quantum Lévy B*-algebra $\mathfrak{b}_1$ called in the thermal, e.g. classical case respectively Wiener-Itô and Poisson-Itô B*-algebra.[18]

2.2.4. *The modular Itô algebras and dual B^*-algebras*

Let us now consider a general noncommutative Itô B*-algebra,[18] i.e. a Banach subalgebra $\mathfrak{b} \subseteq \mathfrak{b}(\mathfrak{k})$ of block-matrices $\mathbf{b} = [b_{\nu}^{\mu}]$ indexed by $\mu, \nu = -, \circ +$ with $b_{\nu}^{\mu} = 0$ if $\mu = +$ or $\nu = -$ and involution $b \mapsto b^{\star}$ represented by $\mathbf{b}^{\ddagger} = [\bar{b}_{\mu}^{\nu}] = \bar{\mathbf{b}}^{\intercal}$, where $\bar{b}_{\mu}^{-\nu} = b_{-\mu}^{\nu *}$. The linear functional $l : b \mapsto b_{+}^{-}$ describes the 'vacuum' state on the maximal Itô algebra $\mathfrak{b}(\mathfrak{k})$ of quadruples $(b_{\nu}^{\mu})_{\nu=+,\circ}^{\mu=-,\circ}$ with independent values $b_{+}^{-} \in \mathbb{C}$, $b_{+}^{\circ} \in \mathfrak{k}$, $b_{\circ}^{-} \in \mathfrak{k}^{*}$

and $b_\circ^\circ \in \mathfrak{L}(\mathfrak{k})$ for a Hilbert space $\mathfrak{k}$. The Itô algebra $\mathfrak{b}$ is called modular (non-vacuum, or thermal) if the state

$$l(b) = (\mathbf{v}_\emptyset|\mathbf{b}\mathbf{v}_\emptyset) := \mathbf{v}_\emptyset^\ddagger \mathbf{b}\mathbf{v}_\emptyset = b_+^-,$$

induced in the canonical representation by the 'vacuum' vector $\mathbf{v}_\emptyset = (\delta_+^\mu)$ with the covector $\mathbf{v}_\emptyset^\ddagger = (1,0,0) = \mathbf{v}_\emptyset^\dagger \boldsymbol{I}$, has the faithfulness property $(\mathbf{b}^\ddagger\mathbf{b})_+^- = 0 \Rightarrow b = 0$ on $\mathfrak{b}_\circ = \{b - l(b)\,d : b \in \mathfrak{b}\}$, as it is always assumed by default in the commutative case. This makes from the quotient $\mathfrak{h}/\mathfrak{d} \simeq \mathfrak{b}_\circ$ a left Hilbert $\star$-algebra dense in the Hilbert subspace $\mathfrak{k}_\circ \subseteq \mathfrak{k}$ generated by $\mathfrak{b}\mathbf{v}_\emptyset$ with respect to the scalar product

$$(b|b) := (\mathbf{b}\mathbf{v}_\emptyset|\mathbf{b}\mathbf{v}_\emptyset) = (\mathbf{b}^\ddagger\mathbf{b})_+^- \equiv b_+^{\circ *} b_+^\circ \geq 0 \quad \forall b \in \mathfrak{b}_\circ \,.$$

Let $\mathfrak{k}$ be separable and $J_\circ = J_\circ^\star$ be an index set with involution $j = i^\star \Leftrightarrow i = j^\star$ for a $\star$-symmetric basis

$$e_{J_\circ} := (e_j)_{j=J_\circ}, \quad e_i^\star = e_{i^\star} \in \mathfrak{b}_\circ \ \forall i,j \in J_\circ \,.$$

A basis $e_{J_\circ} = (e_j)_{j\neq\emptyset}$ in $\mathfrak{b}_\circ$ is called *standard* if the Hermitian semipositive Gram matrix $\boldsymbol{\kappa} = [\kappa_{i,j}]$ of

$$\kappa_{i,j} := e_i^{\circ *} e_j^\circ = l(e_i^\star \cdot e_j) \equiv \sigma_{i^\star,j} \tag{2}$$

for the basic vectors $e_i^\circ = (\mathbf{e}_i)_+^\circ \in \mathfrak{k}_\circ$ has quasiinverse $\tilde{\boldsymbol{\kappa}} = \boldsymbol{\kappa}_\star^*$ as its $\star$-conjugation $\tilde{\kappa}^{i,j} = \kappa_{i^\star,j^\star}^*$, the complex conjugation of $\boldsymbol{\kappa}_\star = [\sigma_{i,j^\star}]$ corresponding to the Hermitian adjoint covariance matrix $\boldsymbol{\sigma}_\star = [\sigma_{i^\star,j^\star}] = \boldsymbol{\sigma}^\dagger$.

In general the Gram matrix $\boldsymbol{\kappa}$ of a modular basis has quasiinverse $\tilde{\boldsymbol{\kappa}}$ with a common $\star$-real support $\tilde{\boldsymbol{\kappa}}\boldsymbol{\kappa} = \boldsymbol{\pi} = \boldsymbol{\kappa}\tilde{\boldsymbol{\kappa}}$ defined as the minimal $\star$-real orthoprojection matrix $\boldsymbol{\pi} = [\pi_j^i] = \boldsymbol{\pi}_\star^*$ such that

$$\boldsymbol{\kappa}\boldsymbol{\pi} = \boldsymbol{\kappa}, \quad \boldsymbol{\pi}\tilde{\boldsymbol{\kappa}} = \tilde{\boldsymbol{\kappa}}, \quad \boldsymbol{\pi}^2 = \boldsymbol{\pi} = \boldsymbol{\pi}^\dagger, \quad \boldsymbol{\pi}^\dagger = \boldsymbol{\pi}^{*\intercal}. \tag{3}$$

If the basis is not overcomplete, then $\boldsymbol{\pi} = \mathbf{1}$ such that for the standard basis $\boldsymbol{\kappa}^* = \boldsymbol{\kappa}_\star^{-1}$, otherwise $\boldsymbol{\kappa}_\star^*\tilde{\boldsymbol{\gamma}} = \boldsymbol{\gamma}\tilde{\boldsymbol{\kappa}}$, where $\tilde{\boldsymbol{\gamma}} = \tilde{\boldsymbol{\gamma}}_\star^*$ is quasiinverse to a $\star$-real Hermitian matrix $\boldsymbol{\gamma} = [\gamma_{i,j}] = \boldsymbol{\gamma}^\dagger$ with $\tilde{\boldsymbol{\gamma}}\boldsymbol{\gamma} = \boldsymbol{\pi} = \boldsymbol{\gamma}\tilde{\boldsymbol{\gamma}}$. One can take semipositve such $\boldsymbol{\gamma}$, uniquely defined as the geometric mean

$$\boldsymbol{\gamma} = \boldsymbol{\kappa}^{\frac{1}{2}} \left(\tilde{\boldsymbol{\kappa}}^{\frac{1}{2}} \boldsymbol{\kappa}_\star^* \tilde{\boldsymbol{\kappa}}^{\frac{1}{2}}\right)^{\frac{1}{2}} \boldsymbol{\kappa}^{\frac{1}{2}}$$

of $\boldsymbol{\kappa}_\star^*$ and $\boldsymbol{\kappa}$. The semipositive matrix $\tilde{\boldsymbol{\kappa}}$ is dual Gram matrix

$$\tilde{\kappa}^{i,j} := \bar{e}_\circ^i \bar{e}_\circ^{j*} = l(\bar{e}^i \cdot \bar{e}^{j*}) \equiv \tilde{\sigma}^{i^\star,j}$$

for adjoint covectors $\bar{e}_\circ^i = (\bar{\mathbf{e}}^i)_\circ^- \in \mathfrak{k}_\circ^*$ such that $\overline{e_i} = \gamma_{i,j}\bar{e}^j$ is the modular $\star$-conjugation $\overline{e_{i^\star}} = \overline{e_i}^\star$ represented by $\overline{\mathbf{e}_i} = \mathrm{J}^*\mathbf{e}_i\mathrm{J}$ in terms of an antilinear isometry $\mathrm{J}^* = s\mathrm{J} = \mathrm{J}^{-1}$, where $s = s^{\ddagger} = \bar{s}$ is any signature $s^2 = 1$ commuting with $\mathfrak{b}$. The dual $\star$-Hermitian matrix $\tilde{\boldsymbol{\sigma}} = \left[\tilde{\sigma}^{i,j}\right] = \tilde{\boldsymbol{\sigma}}_\star^\dagger$ is defined by the covariances $\tilde{\sigma}^{i,j} = l\left(\tilde{e}^i \cdot \tilde{e}^j\right)$ of $\tilde{e}^i = \tilde{\gamma}^{i,j}\widetilde{e_j}$ like $\boldsymbol{\sigma} = [\sigma_{i,j}] = \boldsymbol{\sigma}_\star^\dagger$ by (2), where $\widetilde{e_i} = \overline{e_i^\star}$ represented by $\widetilde{\mathbf{e}_i} = \mathrm{J}^*\mathbf{e}_i^\ddagger\mathrm{J}$.

In the classical case we have always $\boldsymbol{\sigma}_\star^* = \boldsymbol{\sigma}$ due to the usual symmetricity $\boldsymbol{\sigma} = \boldsymbol{\sigma}^\intercal$ following from the commutativity $e_i \cdot e_j = e_j \cdot e_i$. Therefore in the standard basis $\tilde{\boldsymbol{\kappa}} = \boldsymbol{\kappa}$ corresponding to orthonormality $\boldsymbol{\kappa} = \mathbf{1}$ of the basis $e_{J_\circ}^\circ$ in $\mathfrak{k}_\circ$, or, if the classical standard basis is overcomplete, $\boldsymbol{\kappa}$ satisfies the projectivity condition $\boldsymbol{\kappa} = \boldsymbol{\pi} = \overline{\boldsymbol{\kappa}}$ corresponding to the normality of overcomplete basis in $\mathfrak{k}_\circ$. Thus, the classical standard basis is always selfdual in the sense that the completeness condition

$$e_j^\circ \bar{e}_\circ^j = \mathrm{I}_\circ^\circ = \bar{e}_j^\circ e_\circ^j \tag{4}$$

in the Hilbert space $\mathfrak{k}_\circ$ is satisfied for $\bar{e}^i = e^{i^*}$ i.e. $\tilde{e}^i = e^i$ implying symmetricity $\widetilde{\mathbf{e}_i} = \mathbf{e}_i$ of the matrix representation $\mathfrak{b} \subset \mathfrak{b}\left(\mathfrak{k}_\circ\right)$ for the abelian $\mathfrak{b}$ in any $\star$-Hermitian basis.

For the general modular algebra the standard basis in $\mathfrak{k}_\circ$ cannot be selfdual, unless the Itô state l is tracial,

$$l\left(e_i^\star \cdot e_k\right) = l\left(e_k \cdot e_i^\star\right) \quad \forall i.k \in J_\circ \,,$$

which does not imply the commutativity $e_i^\star \cdot e_k = e_k \cdot e_i^\star$. However, one can prove that the *dual Itô algebra* $\bar{\mathfrak{b}} = \tilde{\mathfrak{b}}$ generated in $\mathfrak{b}\left(\mathfrak{k}_\circ\right)$ by the basis $\mathbf{e}_J$ of the transposed matrices $\widetilde{\mathbf{e}}_i = \overline{\mathbf{e}_{i\star}}$, commutes in the standard representation with the modular Itô algebra $\mathfrak{b}$ since $\bar{\mathbf{e}}^i\mathbf{e}_k = \mathbf{e}_k\bar{\mathbf{e}}^i$ due to $\pi_{k\star}^{i^*} = \pi_k^{i*}$ for $\pi_k^i = \left(\bar{\mathbf{e}}^i\mathbf{e}_k\right)_+^- = \bar{e}_\circ^i e_k^\circ$ and $\mathbf{m}_{k\star}^{i^*} = \mathbf{m}_k^{i\ddagger}$ for the 'martingale' part $\mathbf{m}_k^i = \bar{\mathbf{e}}^i\mathbf{e}_k - \pi_k^i\mathbf{d}$ of $\bar{\mathbf{e}}^i\mathbf{e}_k$ in the following multiplication table

$$\mathbf{e}_i\mathbf{e}_k = \sigma_{i,k}\mathbf{d} + \varepsilon_{i,k}^j\mathbf{e}_j,\ \bar{\mathbf{e}}^i\mathbf{e}_k = \pi_k^i\mathbf{d} + \mathbf{m}_k^i,\ \bar{\mathbf{e}}^i\bar{\mathbf{e}}^k = \bar{\sigma}^{i,k}\mathbf{d} + \bar{\varepsilon}_j^{i,k}\bar{\mathbf{e}}^j. \tag{5}$$

Here $\sigma_{i,k} = \sigma_{k^\star,i^\star}^*$ such that $\bar{\boldsymbol{\sigma}} = \bar{\boldsymbol{\sigma}}_\star^\dagger \equiv \tilde{\boldsymbol{\sigma}}_\star$ in the standard basis is $\boldsymbol{\sigma}^* = \boldsymbol{\sigma}_\star^\intercal$, $\boldsymbol{\pi} = \left[\pi_k^i\right] = \tilde{\boldsymbol{\pi}}$ is $\star$-symmetric orthoprojector, $\varepsilon_{i,k}^j = \varepsilon_{k^\star,i^\star}^{j^*{*}}$ are associative structural coefficients, $\bar{\boldsymbol{\varepsilon}} = \bar{\boldsymbol{\varepsilon}}_\star^\dagger \equiv \tilde{\boldsymbol{\varepsilon}}_\star$ coincide for the standard basis with $\boldsymbol{\varepsilon}^* = \boldsymbol{\varepsilon}_\star^\intercal$. Note that $\boldsymbol{\gamma}_\star\bar{\boldsymbol{\varepsilon}}_i\boldsymbol{\gamma} = \gamma_{i,j}\boldsymbol{\varepsilon}^{j*}$, including $\boldsymbol{\gamma}_\star\bar{\boldsymbol{\sigma}}\boldsymbol{\gamma} = \boldsymbol{\sigma}^*$ as the case $\bar{\boldsymbol{\varepsilon}}_\emptyset = \bar{\boldsymbol{\sigma}}$ for $i = \emptyset$, where $\boldsymbol{\varepsilon}^*$ coincides in the commutative case $\boldsymbol{\varepsilon} = \boldsymbol{\varepsilon}^\intercal$ with $\boldsymbol{\varepsilon}^\dagger = \left[\varepsilon_{k,i}^{j*}\right] = \boldsymbol{\varepsilon}_\star$.

The components of the quantized vector-processes $\hat{e}_j\left(t\right) = \Lambda\left(t, \mathbf{e}_j\right)$ and $\check{e}^j\left(t\right) = \Lambda\left(t, \bar{\mathbf{e}}^j\right)$, defined on the Fock-Guichardet space $F = \Gamma\left(K\right)$ over $K = \mathfrak{k} \otimes L^2\left(\mathbb{R}_+\right) \equiv L_{\mathfrak{k}}^2$ as in Refs.[8,10] in general do not commute with their adjoints but commute with their tensor-independent increments as in the classical Lévy case. However, the dual quantum standard Lévy processes $\hat{e}_{J_\circ}(t)$, $\check{e}^{J_\circ}(t)$ do not coincide as in the classical case but are *maximally entangled* having mutually commuting maximally correlated $\star$-Hermitian components $\hat{e}_{i^\star} = \hat{e}_i^*$, $\check{e}^{i^\star} = \check{e}^{i*}$. They generate the dual W*-filtration of maximal mutually commuting W*-algebras $\mathbb{B}_0^t$ and $\widetilde{\mathbb{B}}_0^t = \mathrm{J}^*\mathbb{B}_0^t\mathrm{J}$ on the Fock filtration over $K_0^t = L^2[0,t]$ which are antiisomorphic by the transposition $\widetilde{\mathrm{B}} = \overline{\mathrm{B}}^*$, and both have vacuum vector as a separating cyclic vector common in all $F_0^t = \Gamma(K_0^t)$ inducing a faithful complex-conjugated white noise thermostate consistent on all $\mathbb{B}_0^t$ and $\widetilde{\mathbb{B}}_0^t$.

3. Quantum Stochastic Calculus and Differential Equations

Here we give a brief survey of the universal stochastic calculus[8,10] based on the Belavkin's Itô B*-algebra approach to the universal quantum stochastic analysis.[9] Although this approach was developed for also nonadapted quantum stochastic processes for which the universal quantum Itô formula was established in Ref,[8] we shall restrict our consideration to only adapted processes. We shall give the functional universal Itô formula in a very compact form applicable for any B*-algebra even of infinite dimensionality, although in the noncommutative case it can be justified and computed only for the polynomials or analytic functions, or at least the Fourier transforms of such functions. It is interesting to note that this formula gives also an "explicit" answer for the classical stochastic trajectories with noncommutative matrix values which seams is unknown in classical stochastics, and its nonadapted extension still has the same form both for Wiener and Poisson case.

3.1. *Noncommutative stochastic integration*

3.1.1. *Quantum adapted processes and martingales*

Let $\mathbb{B}^r \subseteq \mathbb{B}^t$ for $r < t \in \mathbb{T}$ denote the increasing unital operator subalgebras which are assumed to be modular, or contain weakly dense modular subalgebras $\mathbb{L}_\star^r \subseteq \mathbb{L}_\star^t$ as preduals of their L^1-completions $\mathbb{L}^r \subseteq \mathbb{L}^t$ with respect to the symmetric pairings $\langle \mathrm{q}\cdot\mathrm{p}\rangle_\mu^r = \mu^r(\mathrm{q}\cdot\mathrm{p})$ of $\mathbb{B}^r =$ and $\mathbb{L}^r$ induced by the reference weights $\mu^r = \mu^t|\mathbb{L}^r$. An adapted $\mathbb{B}$-process with respect to the *filtration* $(\mathbb{B}^t)$ is described by a function $\mathrm{Q}(t) = \mathrm{q}^t \in \mathbb{B}^t$ mapping a totally ordered set $\mathbb{T}$ into the corresponding $*$-subalgebras of $\mathbb{B} := \cup\mathbb{B}^t$. The embeddings $\iota^t(r) : \mathbb{L}_\star^r \hookrightarrow \mathbb{L}_\star^t$ are assumed to be regular such that they are adjointable, $\iota^t(r) := \mu^r(t)_\star \equiv \mu_\star^t(r)$ in the sense

$$(\mathrm{q}|\mu^r(t,\mathrm{p}))_\mu^r := \langle \mathrm{q}^*\cdot\mathrm{p}\rangle_\mu^t \equiv \left(\iota^t(r,\mathrm{q})\,|\mathrm{p}\right)_\mu^t \quad \forall \mathrm{q}\in\mathbb{B}^r, \mathrm{p}\in\mathbb{L}^t$$

with respect to the symmetric pairing given by μ. Then the adjoint comorphisms $\mu^r(t) := \iota^t(r)^\star_\mu$ form a *forward hemigroup* of surjections $\mu^r(t) : \mathbb{L}^t \to \mathbb{L}^r$ in the sense

$$\mu^r(s) \circ \mu^s(t) = \mu^r(t) \ \ \forall r < s < t,$$

such that $\mu^r = \mu^r(t) \circ \mu^t$ for all *localizations* $\mu^t : \mathbb{L} \to \mathbb{L}^t$ at $t \geq r$ defined on $\mathbb{L} = \mathbb{B}^\star$ as the comorphisms onto $\mathbb{L}^t$ adjoint to the embeddings $\iota(t) : \mathbb{B}^t \hookrightarrow \mathbb{B}$. If each $\mathbb{B}^t$ is invariant with respect to the involution $\dagger = \dagger_\mu$, then, obviously $\dagger^t = \dagger|\mathbb{B}^t$, and the localizations are simply restrictions $\mu^r(t) = \mathsf{E}^r|\mathbb{L}^t$ of the conditional expectations $\mathsf{E}^r : \mathbb{L} \to \mathbb{L}^r$ as the positive projections onto $\mathbb{L}^r$ embedded into the L^1-completion $\mathbb{L}$ of $\mathbb{B} \supseteq \mathbb{B}^r$.

An $\mathbb{L}$-adapted process $\mathrm{Z}(t) = \mathrm{z}^t \in \mathbb{L}^t$ is called L^1-martingale if

$$\mu_r\left(t, \mathrm{z}^t\right) = \mathrm{z}^r \ \ \left(\text{or } \mathsf{E}^r\left[\mathrm{z}^t\right] = \mathrm{z}^r\right) \forall r < t,$$

and is called global martingale if $\mathrm{z}^t = \mu^t(\mathrm{z})$ (or $\mathrm{z}^t = \mathsf{E}^t[\mathrm{z}]$). Every positive normalized L^1-martingale $t \mapsto \mathrm{p}^t$ defines a state $\pi^t(\mathrm{q}) = \langle \mathrm{q} \cdot \mathrm{p}^t \rangle$ on $\mathbb{B}$ which is called global if p^t is the global martingale given by a positive normalized $\mathrm{p} \in \mathbb{L}$.

From now on we assume the invariance of all $\mathbb{B}^t \subseteq \mathbb{B}$ with respect to left (and right) modular involutions with respect to the symmetric pairing. Then the weight $\mu(\mathrm{p}) = \langle 1 \cdot \mathrm{p} \rangle$ on $\mathbb{L}$ admits the *compatible conditional expectations* $\mathsf{E}^t : \mathbb{L} \mapsto \mathbb{L}^t$ as positive projections such that $\mathsf{E}^s \circ \mathsf{E}^t = \mathsf{E}^s$ $\forall$ $r, t \in \mathbb{R}_+$ onto $\mathbb{L}^t = \mathbb{M}^t_\star \subseteq \mathbb{L}$:

$$\mathsf{E}^s_\mu\left(1^t\right) = 1^s, \ \mathsf{E}^t_\mu(\mathrm{a}^*\mathrm{p}\mathrm{a}) = \mathrm{a}^*\mathsf{E}^t_\mu(\mathrm{p})\,\mathrm{a} \ \ \forall \mathrm{p} \in \mathbb{L}, \mathrm{a} \in \mathbb{B}^t.$$

Let $\mathcal{B}^t$ denote the $*$-algebras of *adapted* C-processes as continuous functions $\mathrm{Q} : r \to \mathrm{Q}(r) \in \mathcal{B}^r$ mapping $r \in [0, t[$ into the modular subalgebras $\mathcal{B}^r \subseteq \mathcal{B}^t$ such that $\mathcal{B}^s \preceq \mathcal{B}^t$ if $s < t$. Then the dual space $\mathcal{L} = \mathcal{B}^\star_\mu \equiv \mathcal{L}_\mu$ to $\mathcal{B} = \curlyvee_{t>0}\mathcal{B}^t$ with respect to the integral pairing

$$\langle \mathrm{Q} \cdot \mathrm{P} \rangle_\mu := \int \langle \mathrm{Q}(t) \cdot \mathrm{P}(t) \rangle \, dt \ \ \forall \mathrm{Q} \in \mathcal{B}, \mathrm{P} \in \mathcal{L}_\mu$$

consists of all locally integrable QS adapted processes $\mathrm{P}(t) \in \mathbb{L}^t$ dominated by $\mathrm{I}(t) = 1^t$.

An adapted QS process $\mathrm{Z} : t \mapsto \mathrm{Z}(t) \in \mathbb{L}^t$ is called L^1-process if $\mathrm{Z} \in \mathcal{L}$, *locally bounded* (*contractive*) if $\mathrm{Z}^*\mathrm{Z} \leq p\mathrm{I}$ for a positive $p\mathrm{I}$ (for $p = 1$). The process Z is called *L^1-martingale (supermartingale)* if

$$\mathsf{E}^r[\mathrm{Z}(t)] = \mathrm{Z}(r) \ \ \left(\mathsf{E}^r[\mathrm{Z}(t)] \leq \mathrm{Z}(r)\right) \ \ \forall r < t.$$

Absolutely continuous QS processes are described as the indefinite integrals $\mathrm{Y}(t) = \mathrm{Y}_0 + \int_0^t \mathrm{L}(s)\,\mathrm{d}s$ of L^1-process $\mathrm{L} \in \mathcal{L}$. They have finite variation norm $\int_0^t \|\mathrm{L}(s)\|_\star \,\mathrm{d}s$ for each $t < \infty$.

3.1.2. *Quantum stochastic integrators and integrands*

We consider adapted *quantum Itô L^1-processes* Z defined as the *special semimartingales* by

$$\mathrm{Z}(t) - \mathrm{Z}(r) = \int_r^t \boldsymbol{K}(s) \centerdot \mathrm{d}\boldsymbol{B}(s) \equiv \imath_r^t(\boldsymbol{K}).$$

Here $\boldsymbol{K} = (\mathrm{K}_i)_{i\in\mathbb{J}_\circ}$ are adapted integrands indexed by a separable set $\mathbb{J}_\circ$ with an isolated point $\emptyset \in \mathbb{J}_\circ$ invariant under an involution $\star : \mathbb{J}_\circ \to \mathbb{J}_\circ$ such that $J_\circ^\star = J_\circ$ for $J_\circ = \mathbb{J}_\circ \setminus \emptyset$ and

$$\imath_r^t(\boldsymbol{K}) = \int_r^t \mathrm{K}_i(s)\,\mathrm{dB}^i(s) + \int_r^t \mathrm{K}_\emptyset(s)\,\mathrm{d}s,$$

where $\mathrm{B}^i(t) \in \mathbb{L}^t$ are not necessarily canonical QS martingales with QS differentials forming a basis for an Itô algebra for each t. The finite variation part

$$\int_r^t \mathrm{K}_\emptyset(s)\,\mathrm{d}s := \int_r^t \epsilon^s(\mathrm{dZ}(s)) \equiv \epsilon_r^t[\mathrm{Z}(t) - \mathrm{Z}(r)]$$

is given by the deterministic integrator $\mathrm{B}^\emptyset(t) = t1 = \mathrm{B}^\emptyset(t)^*$ defining it as absolutely continuous process

$$\epsilon_0^t[\mathrm{Z}](t) = \mathrm{Z}_0 + \int_0^t \mathrm{K}_\emptyset(s)\,\mathrm{d}s$$

for $\mathrm{K}_\emptyset(\cdot) :\in \mathcal{L}$. We assume that the martingale basis is $\star$-Hermitian in the sense

$$\mathrm{B}_i^\star(t) := \mathrm{B}_{i^\star}(t,)^* = \mathrm{B}_i(t) \quad \forall i \in J_\circ$$

Let us consider a quadruple $\boldsymbol{B} = \left(\mathrm{A}_\mu^\nu\right)_{\mu=-,\circ}^{\nu=+,\circ} \equiv \boldsymbol{A}$ of the *canonical quantum stochastic (QS) integrators* $\mathrm{A}_\mu^\nu(t) = \Lambda\left(t, \mathrm{e}_\mu^\nu\right) \equiv \hat{e}_\mu^\nu$ of annihilation, A_-°, creation $\mathrm{A}_\circ^+$, number exchange $\mathrm{A}_\circ^\circ = \mathrm{E}$ and conservation $\mathrm{A}_-^+(t) = t\mathrm{I}$ processes up to time t as defined by Hudson-Parthasarathy[19] in the Bosonic Fock space $F = \Gamma(K)$ over the Hilbert space $K = \mathfrak{k} \otimes L^2(\mathbb{R})$ of $\mathfrak{k}$-norm square-integrable functions $\varphi : \mathbb{R} \mapsto \mathfrak{k}$. The correspondence $\Lambda(t)$ with the canonical matrix basis $\boldsymbol{e} = \left(\mathrm{e}_\mu^\nu\right)_{\mu=-,\circ}^{\nu=+,\circ}$ for the matrix Itô $\star$-algebra $\mathfrak{b}(\mathfrak{k})$, noticed by Belavkin in Ref,[1] realizes the pseudo-Poisson matrix table $\mathrm{e}_\mu^\iota \mathrm{e}_\kappa^\nu = \delta_\kappa^\iota \mathrm{e}_\mu^\nu$ as the HP multiplication differential table 1 which can be written in the Belavkin form simply as $\mathrm{dA}^{\iota}_{\mu}\mathrm{dA}^{\nu}_{\kappa} = \delta^{\iota}_{\kappa}\mathrm{dA}^{\nu}_{\mu}$, and that the matrix pseudo-adjointness $\mathbf{e}^{\iota\ddagger}_{-\kappa} = \mathbf{e}^{\kappa}_{-\iota}$ corresponds to the Hermitian adjointness $\mathrm{A}^{\iota*}_{-\kappa} = \mathrm{A}^{\kappa}_{-\iota}$ on the Hilbert space F. Moreover, linear spans $\boldsymbol{\kappa}\,\boldsymbol{.}\,\boldsymbol{B}(t) := \kappa^{\mu}_{\nu}\mathrm{A}^{\nu}_{\mu}(t)$ of these QS-integrators determine on F a Lie $\star$-representation $\hat{\kappa} = \Lambda(t,\mathrm{K})$ of K implemented as the triangular block-matrices $\mathbf{K} = \boldsymbol{\kappa I}$ as the spans $\kappa^{\mu}_{\nu}\mathbf{e}^{\nu}_{\mu} \equiv \boldsymbol{\kappa}\,\boldsymbol{.}\,\boldsymbol{e}$ such that $\Lambda(t,\mathrm{K})^{*} = \Lambda(t,\mathrm{K}^{\ddagger})$ and

$$\left[\Lambda\left(t,\mathrm{K}^{\ddagger}\right),\Lambda\left(t,\mathrm{K}\right)\right]_{-} = \Lambda\left(t,\left[\mathrm{K}^{\ddagger},\mathrm{K}\right]_{-}\right)\ \forall \mathrm{K}\in\mathfrak{b}\left(\mathfrak{k}\right) \tag{1}$$

for the Lie product given by usual commutators $[\mathrm{X},\mathrm{Z}]_{-} = \mathrm{XZ}-\mathrm{ZX}$. The forward QS differentials $\Lambda(\mathrm{d}t,\mathrm{K}) = \mathrm{d}\Lambda(t,\mathrm{K})$ have the Pseudo-Poisson multiplication table

$$\Lambda\left(\mathrm{d}t,\mathrm{K}\right)\Lambda\left(\mathrm{d}t,\mathrm{K}^{*}\right) = \left(\boldsymbol{\kappa}\star\boldsymbol{\kappa}\right)\boldsymbol{.}\boldsymbol{A}\left(\mathrm{d}t\right) = \Lambda\left(\mathrm{d}t,\mathrm{KK}^{\ddagger}\right), \tag{2}$$

where $\boldsymbol{\kappa}\star\boldsymbol{\kappa} = \boldsymbol{\kappa}\cdot\boldsymbol{\kappa}^{\star}$ is the Itô product $\boldsymbol{\kappa}\cdot\boldsymbol{\kappa}^{\star} = \boldsymbol{\kappa I \kappa}^{\dagger}$ defied in the triangular matrix representation $\mathbf{K} = \boldsymbol{\kappa I}$ simply as $\boldsymbol{\kappa I \kappa}^{\dagger}\,\boldsymbol{I} = \mathbf{KK}^{\ddagger}$.

A quantum process $\mathrm{X}(t)$ is said to be *QS-differentiable* if its increments are the QS integrals

$$\mathrm{X}(t) - \mathrm{X}(0) = \int_{0}^{t}\Lambda\left(\mathrm{d}r,\mathrm{K}(r)\right) \equiv \Lambda\left(t,\mathrm{K}\right)$$

as defined in Refs[1,8] even for nonadapted $\mathrm{X}(t)$. Here $\mathrm{K}(t)$ is the QS integrand defined by only four nonzero, in general operator-valued entries $(\mathrm{K}^{\mu}_{\nu}(t))^{\mu=-,\circ}_{\nu=+,\circ}$ in the triangular block-matrix representation $\mathbf{K}(t) = \mathrm{K}^{\mu}_{\nu}(t)\otimes\mathbf{e}^{\mu}_{\nu}$ of $\boldsymbol{K} = (\mathrm{K}^{\mu}_{\nu})^{\mu=-,\circ}_{\nu=+,\circ}$ giving the quadruple of the QS derivatives $\mathrm{K}^{\mu}_{\nu}(t)$ of $\mathrm{X}(t)$ as the entries to $\boldsymbol{D}^{t}_{\mathrm{X}} = \mathbf{K}(t)\boldsymbol{I} \equiv \boldsymbol{K}(t)$. If the process $\mathrm{X}(t)$ is adapted in the sense that $\mathrm{X}(t)\in\mathbb{A}^{t}$, or affiliated to the operator algebra $\mathbb{B}^{t} = \mathbb{B}(F^{t}_{[0}\otimes\mathrm{I}_{t})$ generated by $\{\boldsymbol{A}(r) : r\le t\}$, then its QS derivatives are also adapted process, $\mathrm{K}^{\mu}_{\nu}(t)\vdash\mathbb{B}^{t}$, having zero entries $\mathrm{K}^{\mu}_{-} = 0 = \mathrm{K}^{+}_{\nu}$ for all μ and ν. Quite weak sufficient conditions of QS-integrability for the operator-valued valued processes $\mathrm{K}^{\mu}_{\nu}(t)$ affiliated to the W*-algebra $\mathbb{B}^{t}$ and also satisfying (1) and (2) were established in Refs.[8,10]

3.1.3. *The associative quantum stochastic covariation*

From now on we shall consider only adapted QS-differentiable processes $\mathrm{X}(t)$ defined by an initial value $\mathrm{X}(0)\in\mathbb{B}(\mathfrak{h})\otimes\mathrm{I}_{0}\equiv\mathbb{B}_{0}$ on the Hilbert space $\mathfrak{h}\equiv F^{0}_{[0}$, and the quadruple $\boldsymbol{D}_{\mathrm{X}} = (\mathrm{D}^{\mu}_{\nu})^{\mu=-,\circ}_{\nu=+,\circ}$ of its adapted QS derivatives $D^{\mu}_{\nu}(t)$. The product Itô formula

$$\mathrm{d}(XX^*) = (\mathrm{d}X^*)X + \mathrm{d}X\mathrm{d}X^* + X(\mathrm{d}X^*)$$

defines $X(t)X(t)^* - X(0)X(0)^*$ as the *adapted* QS integral

$$\Lambda(t, \boldsymbol{K}) := \int_0^t K^{\mu}_{\nu}(s)\, A^{\nu}_{\mu}(\mathrm{d}s) \equiv \boldsymbol{K}^t_0 \,.\, \boldsymbol{B} \tag{3}$$

of $\boldsymbol{K} = X\boldsymbol{D}^*_X + \boldsymbol{D}_X \star \boldsymbol{D}_X + \boldsymbol{D}_X X^*$ in terms of the QS Itô $\star$-product $\boldsymbol{D} \star \boldsymbol{D} := \boldsymbol{D} \cdot \boldsymbol{D}^\star \equiv \boldsymbol{DID}^*$ as the block-matrix product $\mathbf{D}\mathbf{D}^\ddagger \boldsymbol{I}$ of $\mathbf{D} = \boldsymbol{DI}$ and $\mathbf{D}^\ddagger = \boldsymbol{I}\mathbf{D}^\star\boldsymbol{I}$, provided that the products in $D^{*\mu}_{-\iota}X + D^{*\mu}_{\circ}D^{\circ}_{-\iota} + D^{*\mu}_{-\iota}X$ are well-defined as QS-integrable functions of t.[8,9] The positive increasing QS process

$$[X|X](t) = \int_0^t (\boldsymbol{D}_X \star \boldsymbol{D}_X)(r) \,.\, \boldsymbol{B}(\mathrm{d}r) = \Lambda\left(t, \mathbf{D}_X\mathbf{D}^\ddagger_X\right), \tag{4}$$

symbolically defined as the integral of the Itô $\star$-product $\mathrm{d}[X|X] = \mathrm{d}X\mathrm{d}X^*$, is called the Hermitian QS variation of the process $X(t)$.

Note that the bilinear bracket $[X; Z] = (\boldsymbol{D}_X \cdot \boldsymbol{D}_Z) \,.\, \boldsymbol{B}$ as the Hermitian covariation of X and $Y = Z^*$ obtained from $[Z|Z] = (\boldsymbol{D}_Z \star \boldsymbol{D}_Z) \,.\, \boldsymbol{B}$ by polarization

$$[X|Y] = \sum_{n=0}^{3} \frac{i^n}{4}[Z_n|Z_n], \quad Z_n = X + i^n Y,$$

satisfies all the properties of the classical covariation bracket except the symmetricity since in general $\boldsymbol{D}_X \cdot \boldsymbol{D}_Z \neq \boldsymbol{D}_Z \cdot \boldsymbol{D}_X$. Moreover, due to the property

$$(\boldsymbol{K} \,.\, \boldsymbol{B}) \star (\boldsymbol{L} \,.\, \boldsymbol{B}) := [\boldsymbol{K} \,.\, \boldsymbol{B} | \boldsymbol{L} \,.\, \boldsymbol{B}] = (\boldsymbol{K} \star \boldsymbol{L}) \,.\, \boldsymbol{B}, \tag{5}$$

it is simply the Itô product, $X \cdot Z = [X; Z]$, which is obviously associative, $(X \cdot Y) \cdot Z = X \cdot (Y \cdot Z)$ as

$$(\boldsymbol{K} \,.\, \boldsymbol{B}) \cdot (\boldsymbol{L} \,.\, \boldsymbol{B}) \cdot (\boldsymbol{M} \,.\, \boldsymbol{B}) = (\boldsymbol{K} \cdot \boldsymbol{L} \cdot \boldsymbol{M}) \,.\, \boldsymbol{B}, \tag{6}$$

due to the associativity of the matrix product $\mathbf{KLM}\boldsymbol{I} = \boldsymbol{K} \cdot \boldsymbol{L} \cdot \boldsymbol{M} = \boldsymbol{KILIM}$. We may also use the notation $X \star Y = [X|Y]$ instead of cdot $X \cdot Z$ with $Z = Y^*$, not to be mixed with the QS integration dot $X \,.\, Z$, and avoid use of classical square bracket notation $[X; Z]$ keeping it for another bilinear form $[X, Z] = [X|\bar{Z}]$, not to be mixed with the commutator $[X, Z]_-$ defining the antisymmetric covariation

$$[X; Z]_-(t) := (X \cdot Z - Z \cdot X)(t) = [X(t), Z(t)]_- - [X(0), Z(0)]_- .$$

3.1.4. *The universal quantum Itô formula*

The associativity of the noncommutative QS covariation give rise to the effective noncommutative calculus based on the adapted QS Itô formula

$$\mathrm{d}\,(\mathrm{XZ})\,(t) = \left(\mathrm{X}\,(t)\,\boldsymbol{D}_{\mathrm{Z}}^{t} + \boldsymbol{D}_{\mathrm{X}}^{t}\cdot\boldsymbol{D}_{\mathrm{Z}}^{t} + \boldsymbol{D}_{\mathrm{X}}^{t}\mathrm{Z}\,(t)\right)\boldsymbol{.}\,\boldsymbol{B}\,(\mathrm{d}t) \qquad (7)$$

as derived by Hudson and Parthasarathy for the product of two QS adapted integrals in Ref.[19] More efficiently it is written as

$$\mathrm{d}\,(\mathrm{XZ})\,(t) = \Lambda\left(\mathrm{d}t, \mathbf{J}_{\mathrm{X}}^{t}\mathbf{J}_{\mathrm{Z}}^{t} - \mathbf{X}\,(t)\,\mathbf{Z}\,(t)\right) \qquad (8)$$

in terms of the ampliation $\mathbf{X}\,(t) = \mathrm{X}\,(t)\otimes\mathbf{I}$ and the *QS germ* $\mathbf{J}_{\mathrm{X}}^{t} = \mathbf{X}\,(t) + \mathbf{K}\,(t) \equiv \mathbf{G}\,(t)$ for $\mathrm{X}\,(t)$ at the time t which was introduced in Refs.[10,14] Notice that this formula is universal also for the classical stochastic calculus unifying the Itô-Wiener and the Itô-Poisson calculi in one and admitting their noncommutative extensions to operator-valued classical stochastic trajectories $\mathrm{X}\,(\omega)\,, \mathrm{Z}\,(\omega)$ with noncommuting values at each t.

The germ form of Itô formula admits also a functional extension which was made by Belavkin in Refs:[8,11]

$$\mathrm{d}f\left[\mathrm{X}\,(t)\right] = \Lambda\left(\mathrm{d}t, f\left[\mathbf{G}\,(t)\right] - f\left[\mathbf{X}\,(t)\right]\right) = \left(\boldsymbol{F}\,(t) - f\left[\mathrm{X}\,(t)\otimes\boldsymbol{I}\right]\right)\boldsymbol{.}\boldsymbol{B}\,(\mathrm{d}t) \qquad (9)$$

This formula is valid for any polynomial or analytic function f for which $\mathbf{F}\,(t) = \boldsymbol{F}\,(t)\,\boldsymbol{I}$ can be evaluated at each t as a triangular block-matrix $f\,(\mathbf{G}) = \mathbf{F}$ of the germ $\mathbf{F}\,(t) = \mathbf{J}_{\mathrm{Y}}^{t}$ for $\mathrm{Y}\,(t) = f\left[\mathrm{X}\,(t)\right]$ with $\mathrm{F}_{-}^{-} = f\,[\mathrm{X}] = \mathrm{F}_{+}^{+}$. Since all classical Itô algebras are commutative subalgebras of $\mathfrak{b}\,(\mathfrak{k})$, this formula contains both classical cases of diffusive and counting Itô formula. As it was proved in Ref,[8] this universal Itô formula remains unchanged even in the general case of nonadapted QS processes, with only difference that ampliation $\mathrm{Y}\,(t)\otimes\mathbf{I}$ of $\mathrm{Y}\,(t) = f\left[\mathrm{X}\,(t)\right]$ should be replaced by the function $\mathbf{Y}\,(t) = f\left[\mathbf{X}\,(t)\right]$ of the germ $\mathbf{X}\,(t) = \mathbf{J}_{\mathrm{X}}^{t}$ for the left limit $\mathrm{X}\,(t_{-}) = \mathrm{X}\,(t)$ at the same t.

3.2. *Quantum stochastic dynamics and equations*

3.2.1. *Quantum object-output stochastic processes*

Let $\mathbb{M}_{0}^{t]} = \mathbb{M}^{o}\bar{\otimes}\mathbb{M}_{0}^{t}$ denote the W*-product algebra generated on the Hilbert product $H_{0}^{t]} = \mathfrak{h}\otimes H^{t}$ by the present (quantum object) algebra $\mathbb{M}^{o}\subseteq\mathbb{B}\,(\mathfrak{h})$, say the matrix algebra as operator algebra on $\mathfrak{h} = \mathbb{C}^{d}$, and the past (classical output) algebra $\mathbb{M}_{0}^{t}\subseteq\mathbb{B}\,(H_{0}^{t})$, say generated by multiplication operators $Y_{0}^{t} : g_{0}^{t}\mapsto y_{0}^{t}g_{0}^{t}$ for adapted $\mathbb{C}$-valued functionals $y_{0}^{t}\in M_{0}^{t}$, $g_{0}^{t}\in H_{0}^{t}$ of classical Lévy process trajectories $v_0^t = \{v(r) : r \in [0,t)\}$ such that $M_0^t \subseteq L_Q^\infty(\Upsilon, \mathfrak{B})$, $H_0^t \subseteq L_Q^2(\Upsilon, \mathfrak{B})$ on the Lévy probability space $(\Upsilon, \mathfrak{B}, Q)$. The latter example has the split property $\mathbb{M}_0^{t+s} = \mathbb{M}_0^t \bar{\otimes} \mathbb{M}_t^s$ on $H_0^{t+s} = H_0^t \otimes H_t^s$ for every t and $s > 0$, and we shall assume this infinite divisibility not only for the Abelian algebras $\mathbb{M}_0^t \simeq M_0^t$ but also for the general W*-algebras $\mathbb{M}_0^t$, say generated by the dual quantum Brownian or Lévy processes

$$\hat{e}_j(t) := \Lambda(t, \mathbf{e}_j) \equiv \mathrm{B}_j(t), \quad \check{e}^j(t) := \Lambda\left(t, \tilde{\mathbf{e}}^j\right) \equiv \bar{\mathrm{B}}^j(t) \quad \forall j \in J_\circ$$

given by the canonical bases $\mathbf{e}_{J_\circ}$, $\tilde{\mathbf{e}}^{J_\circ}$ respectively for the dual noncommutative Itô algebras $\mathfrak{b}$ and $\widetilde{\mathfrak{b}}$.

Every QS-differentiable *adapted* process $\mathrm{Z}(t) \vdash \mathbb{M}_0^{t]} \otimes I_t$ has increments $\mathrm{Z}(t) - \mathrm{Z}(r) = \boldsymbol{K}_r^t \centerdot \boldsymbol{A}$ written in terms of $Z\boldsymbol{I} := \mathbf{I} \otimes Z \equiv \boldsymbol{I}\mathrm{Z}$ as in the canonical QS integral form $\int_r^t (\boldsymbol{J}(s) - \mathrm{Z}(s)\boldsymbol{I}) \centerdot \mathrm{d}\boldsymbol{A}(s)$:

$$= \int_r^t \left(\mathrm{J}_\nu^\mu(s) - \mathrm{Z}(s)\delta_\nu^\mu\right) \mathrm{dA}_\mu^\nu(s) \equiv \Lambda_r^t(\mathbf{J}_\mathrm{Z} - \mathrm{Z}\mathbf{I})$$

of the *QS derivative* $\boldsymbol{K}(t) = \boldsymbol{D}_\mathrm{Z}^t$ defining the *QS germ-matrix* $\mathbf{J}_\mathrm{Z}^t = \mathbf{Z}(t) + \boldsymbol{D}_\mathrm{Z}^t \boldsymbol{I}$, where $\mathbf{Z} := \mathbf{I} \otimes \mathrm{Z} \equiv \mathrm{Z}\mathbf{I}$ is the ampliation $[\mathrm{Z}\delta_\nu^\mu]$ of Z. The QS dynamics is described by a hemigroup

$$\widetilde{\alpha}_{t_0}(t) = \widetilde{\alpha}_{t_0}(r) \circ \widetilde{\alpha}_r(t) \quad \forall t_0 < r < t$$

of QS monomorphic transformations $\widetilde{\alpha}_{t_0}(t) : \mathbb{M}_0^{t]} \bar{\otimes} \mathrm{N}_r \to \mathbb{M}_0^{r]} \bar{\otimes} \mathrm{N}_r$

$$\widehat{\mathrm{Z}}_r(t) = \mathrm{V}_r(t)\,\mathrm{Z}(t)\,\mathrm{V}_r(t)^* \equiv \widetilde{\alpha}_r(t, \mathrm{Z}(t)),$$

where $\mathrm{V}_r(t) : H_0^{t]} \otimes F_t \to H_0^{r]} \otimes F_r$ is a unitary or an isometric evolution $\mathrm{V}_r(t) = \mathrm{I} + \Lambda_r^t(\mathrm{VL})$ such that $\widetilde{\alpha}_r(\mathrm{Z}^*\mathrm{Z}) = \widetilde{\alpha}_r(\mathrm{Z})^* \widetilde{\alpha}_r(\mathrm{Z})$ for any $\mathrm{Z} \in \mathbb{M}_0^{t]}$. We shall call the isometric W*-monomorphisms $\widetilde{\alpha}_{t_0}(t)$ *semimorphisms* as they preserve all algebraic operations except,.maybe, the identity $\mathrm{I}_0^{t]} \in \mathbb{M}_0^{t]}$ which is mapped into the decreasing orthoprojectors $\mathrm{P}_r(t) = \mathrm{V}_r(t)\,\mathrm{V}_r(t)^* \geq \mathrm{P}_r^{t+s]}$ $\forall s > 0$ of the quantum object survival from r up to t which is said to be *stable* only if $\mathrm{P}_r(t) = \mathrm{I}_0^{r]}$ for all t and $r < t$.

3.2.2. *Quantum stochastic Schrödinger equation*

The input – output semiunitary transformations $\mathrm{V} = \{\mathrm{V}_r(t) : r \leq t\}$ form a linear *hemigroup*

$$\mathrm{V}_{t_0}(t) = \mathrm{V}_{t_0}(r) \circ \mathrm{V}_r(t) \quad \forall t_0 < r < t.$$

as the representation of a small category of the totally ordered set $\mathbb{R}_+$. Since the seminal paper[19] it has been main object of study in quantum stochastics as a resolving family for the Hudson-Parthasarathy (HP) QS differential equation

$$\mathrm{dV}(t) = \mathrm{V}(t)\,\Lambda(\mathrm{d}t, \mathbf{S} - \mathbf{I})\,\mathrm{V}(t) = \boldsymbol{L}(t) \centerdot \boldsymbol{A}(\mathrm{d}t)$$

written in the canonical Hudson-Parthasarathy basis $\boldsymbol{A} = [\mathrm{A}_\mu^\nu]_{\mu=-,\circ}^{\nu=+,\circ}$ as

$$\mathrm{dV}(t) = \mathrm{V}(t)\left(\mathrm{L}_+^-\mathrm{d}t + \mathrm{L}_\circ^-\mathrm{dA}_-^\circ + \mathrm{L}_\circ^\circ\mathrm{dA}_\circ^\circ + \mathrm{L}_+^\circ\mathrm{dA}_\circ^+\right).$$

Here $\boldsymbol{L} = [\mathrm{L}_\kappa^\iota]_{\kappa=+,\circ}^{\iota=-,\circ} \equiv \mathbf{L}\boldsymbol{I}$ is the quadruple of QS logarithmic derivatives defining the QS germ $\mathbf{G}(t) = \mathrm{V}(t)\,\mathbf{S}(t) \equiv \mathbf{J}_\mathrm{V}^t$ for the Hemigroup V by $\mathbf{S} = \mathbf{I} + \mathbf{L}$. The following theorem gives necessary HP unitarity conditions in a compact germ form. They are also sufficient for the existence and uniqueness of the semiunitary solutions under the appropriate integrability conditions[8,9] for the operator-valued adapted function $\mathbf{L}(t)$.

Theorem 3.1. *The solution of HP equation is unitary,* $\mathrm{V}^* = \mathrm{V}^{-1}$ *(isometry,* $\mathrm{V}^*\mathrm{V} = \mathrm{I}$*) iff the germ is pseudo-unitary,* $\mathbf{G}^\ddagger = \mathbf{G}^{-1}$ *(pseudo-isometry* $\mathbf{G}^\ddagger\mathbf{G} = \mathbf{I}$*), i.e. iff the logarithmic germ* $\mathbf{S}(t) = \mathrm{V}(t)^*\,\mathbf{J}_\mathrm{V}^t$ *is pseudo-unitary (pseudo-isometry). This implies the relations*

$$\mathrm{S}_\circ^{\circ *}\mathrm{S}_\circ^\circ = \mathrm{I}_\circ^\circ, \quad \mathrm{S}_\circ^{\circ *}\mathrm{S}_+^\circ + \mathrm{S}_\circ^{-*} = 0,$$
$$\mathrm{S}_+^{\circ *}\mathrm{S}_\circ^\circ + \mathrm{S}_\circ^- = 0, \quad \mathrm{S}_+^- + \mathrm{S}_+^{\circ *}\mathrm{S}_+^\circ + \mathrm{S}_+^{-*} = 0.$$

for $\mathrm{S}_\circ^\circ = \mathrm{I}_\circ^\circ + \mathrm{L}_\circ^\circ$, $\mathrm{S}_+^\circ = \mathrm{L}_+^\circ$, $\mathrm{S}_\circ^- = \mathrm{L}_\circ^-$, $\mathrm{S}_+^- = \mathrm{L}_+^-$.

3.2.3. *Quantum stochastic Langevin equation*

The QS Heisenberg equation for the maximal W*-algebra $\mathbb{N}_t^s = \mathbb{B}(F_t^s)$ is generated by the vacuum noise increments $\mathrm{A}_\mu^\nu(t') - \mathrm{A}_\mu^\nu(t)$, $t' \in [t, t^s)$ of the canonical HP basis $\mathrm{B}_\mu^\nu = \mathrm{A}_\mu^\nu$. It may actually be driven by a smaller Itô $\star$-subalgebra than $\mathfrak{b}(\mathfrak{k})$, corresponding to a product W*-algebra $\mathbb{N}_0^t \subseteq \mathbb{B}(F_t^s)$ of quantum noise on Fock space $F_t^s = \Gamma(K_t^s)$ for $K_t^s = L_\mathfrak{k}^2(t, t^s]$, say generated by the classical Langevin forces $\check{f}_k(t')$ given by an Abelian Itô $\star$-subalgebra for each $t^s = t + s > t$ on F_t^s. It is usually described by the Langevin QS equation

$$\mathrm{d}\widehat{\mathrm{Z}}_r(t) = \Lambda\left(\mathrm{d}t, \gamma_r\left(t, \mathbf{J}_\mathrm{Z}^t\right) - \widehat{\mathbf{Z}}_r(t)\right), \quad \widehat{\mathrm{Z}}_r(r) = \mathrm{Z}(r).$$

where the QS germ $\gamma(t) = \mathbf{J}_{\widetilde{\alpha}}^t = \widetilde{\alpha}(t) \circ \varsigma$ for $\widetilde{\alpha}$ with $\varsigma(\mathbf{J}) = \mathbf{SJS}^\ddagger$,

$$\gamma_r\left(t, \mathbf{J}_\mathrm{Z}^t\right) = \mathbf{G}_r(t)\,\mathbf{J}_\mathrm{Z}^t\mathbf{G}_r(t)^\ddagger = \widehat{\varsigma}_r\left(t, \widehat{\mathbf{J}}_\mathrm{Z}^t\right),$$

is obtained by applying quantum Itô product formula

$$\mathrm{d}\left(\mathrm{VZV}^*\right)=\Lambda\left(\mathrm{d}t,\mathbf{GJ}_{\mathrm{Z}}^{t}\mathbf{G}^{\ddagger}-\mathbf{VZV}^{\ddagger}\right).$$

Theorem 3.2.[10] *The solution* $\widehat{\mathrm{Z}}(t)=\widetilde{\alpha}(t,\mathrm{Z}(t))$ *of the QS Heisenberg equation is homomorphic (unital,* $\widetilde{\alpha}_r(t,\mathrm{I})=\mathrm{I}$*) iff all germs* $\gamma_r(t)$ *are pseudo-homomorphic (unital):*

$$\gamma_r\left(t,\mathbf{J}^{\ddagger}\mathbf{J}\right)=\gamma_r(t,\mathbf{J})^{\ddagger}\gamma_r(t,\mathbf{J}),\quad\left(\gamma_r(t,\mathbf{I})=\mathbf{I}\right).$$

In particular, $\mathbf{J}_{\mathrm{X}}^{t}=\mathbf{I}^{t}\mathrm{x}\equiv\mathbf{X}(t)$ *gives Langevin equation, and the case* $\mathbf{J}_{\mathrm{Y}}^{t]}=\mathbf{J}_{\mathrm{Y}}^{t}1$ *with* $\gamma_r\left(t,\mathbf{J}_{\mathrm{Y}_t}^{t}\right)=\widehat{\mathrm{Y}}_r(t)\,\mathbf{I}+\gamma_r\left(t,\mathbf{D}_{\mathrm{Y}}^{t}\right)$ *gives the output equation:*

$$\mathrm{d}\widehat{\mathrm{X}}=\Lambda\left(\mathrm{d}t,\widehat{\varsigma}\left(t,\widehat{\mathbf{X}}\right)-\widehat{\mathbf{X}}(t)\right),\ \mathrm{d}\widehat{\mathrm{Y}}=\Lambda\left(\mathrm{d}t,\widehat{\varsigma}\left(t,\mathbf{D}_{\widehat{\mathrm{Y}}}^{t}\right)\right).$$

4. Quantum Hidden Markov Dynamics and Belavkin QS Flows

In this Section we consider nonstationary QS hidden Markov dynamics described by Heisenberg singular interaction transformations induced by semiunitary QS hemigroups of propagators satisfying the generalized HP-Schrödinger equation. Each such dynamics gives a micro-model of a reduced quantum macro-dynamics as a dilation to nonstochastic quantum dynamics. In Markov case the nonstochastic is described by semigroups, or in nonstationary case by hemigroups of completely positive (CP) maps on the quantum object algebra $\mathtt{A}$ preserving the identity $\mathtt{I}\in\mathtt{A}$, or in Schrödinger picture preserving the trace trace on $\mathbb{A}=\mathbb{B}(\mathfrak{h})$ but mixing the pure quantum states. More generally, the hidden Markov QS mixing dynamics corresponds to a hemigroup of completely positive (CP) contractions $\mathbb{A}_0^{t]}\to\mathbb{A}_0^{r]}$, $\forall r\le t$ on a filtration of increasing object-input W*-algebras $\mathbb{A}_0^{t]}$ as the quantum expected dynamics conditioned by $\mathbb{A}_0^{t]}$. Given a sufficiently large filtration $\{\mathbb{B}^t\}$ of the output W*-algebras $\mathbb{B}^t$ commuting with $\mathbb{A}_0^{t]}$ such deterministic hemigroups can be completely unravelled to a nonmixing QS hemigroup describing the Belavkin QS master dynamics as a QS contractive CP flow on the object-output state space $\mathbb{B}_\star^{t]}$. The corresponding unraveling QS master equation,[12] known as the Belavkin filtering equation, was in general derived in Refs[13,15] following the scheme suggested for the commutative output algebras $\mathbb{B}^t$ in Refs.[2,6]

First we obtain the closed deterministic characteristic equation for QS dynamics assuming its semi-Markovianity with respect to the increasing family of *input* W*-algebras $\mathbb{A}^{t]}\subseteq\mathbb{B}\left(H^{t]}\right)$ on a consistent family of cyclic product state-vectors $\psi_0^{t]} \in H^{t]}$. It is described by backward in time completely positive mixing mappings from the larger $\mathbb{M}^{t+s]}$, $s > 0$ into the smaller W*-algebras $\mathbb{M}^{t]} = \mathbb{A}^{t]} \vee \mathbb{B}^{t]}$ generated by $\mathbb{A}^{t]}$ and increasing *output* W*-subalgebras $\mathbb{B}^{t]} \subseteq \mathbb{B}\left(H^{t]}\right)$ commuting with $\mathbb{A}^{t]}$ on the Hilbert spaces $H^{t]}$ generated by $\mathbb{A}^{t]}\psi_0^{t]}$. Then we derive the reduced QS master equation on the increasing preduals $\mathbb{B}_\star^{t]} \subseteq \mathbb{A}_\intercal^{t]}$ of modular conjugated input subalgebras $\overline{\mathbb{B}}^{t]} \subseteq \mathbb{A}^{t]}$. The solution of this equation defines the evolution of the output state densities $\hat{\rho}_t \in \mathbb{B}_\star^{t]}$ which results from a noncommutative analog of Girsanov transformation for an input quantum white noise with respect to the initial vector state. It has the generalized object-output Lindblad form coinciding in the Markovian case with the standard Lindblad equation on a fixed object algebra. We present the Belavkin QS dynamics for the general case of quantum white noise given by a thermal Lévy process described by a modular B*-Itô algebra $\mathfrak{b} \subseteq \mathfrak{b}(\mathfrak{k})$. To be concrete, we make it in a reflection-symmetric bases and introduce also the dual algebra which is obtained in the dual bases simply by a complex reflect-conjugation on the underlying Hilbert space $\mathfrak{k}$.

4.1. *Quantum hidden Markov dynamical hemigroups*

4.1.1. *The generalized Lindblad equation*

Let $\mathbb{M}^{t]} = \mathbb{A}^{t]} \vee \mathbb{B}^{t]}$ be generated by the input W*-algebras $\mathbb{A}^{t]} = \mathbb{A}^{\circ}\bar{\otimes}\mathbb{A}^{t}$ on the Hilbert spaces $H^{t]} = \mathfrak{h} \otimes H^{t}$ with an initial cyclic state-vector $\psi_0^{t]} = \eta \otimes \psi_0^{t}$ and an output modular algebras $\mathbb{B}^{t]} = \mathbb{B}^{\circ}\bar{\otimes}\mathbb{B}^{t}$ from the commutants $\widetilde{\mathbb{A}^{t]}} = \mathrm{J}^{*}\mathbb{A}^{t]}\mathrm{J}$ of $\mathbb{A}^{t]}$ on $H^{t]}$. We assume that $(\mathbb{A}^{t}, \mathbb{B}^{t})$ is increasing W*-product system on the components $H^{t} = F_0^{t}$ of Fock space $F_0 = F_0^{t} \otimes F_t$ and take the vacuum $\delta_\emptyset^{t} \in F_0^{t}$ as an initial product state vector ψ_0^{t} separating $\mathbb{B}^{t}$. A QS adapted dynamics $\widetilde{\alpha}_r(t) : \mathbb{M}^{t]} \to \mathbb{M}^{r]}\bar{\otimes}\mathbb{N}_r$ is called $\mathbb{A}^{t]}$-*Markov* on vacuum state vectors $\delta_\emptyset \in F_t$ for the decreasing future quantum noise W*-algebras $\mathbb{N}_t \subseteq \mathcal{B}(F_t)$ if

$$\phi_t(t^s, \mathrm{Z}) = \mathrm{E}_t(t^s)\,\widetilde{\alpha}_t(t^s, \mathrm{Z})\,\mathrm{E}_t(t^s) \in \mathbb{A}^{t]}\ \forall t > r,\ \mathrm{Z} \in \mathbb{A}^{t}\bar{\otimes}\mathbb{M}_t^{s]}, \tag{1}$$

where $\mathrm{E}_t(t^s) = \delta_\emptyset^{s*}$ is the vacuum projection of $H^{t]} \otimes F_t^s$ onto $H^{t]}$ as Hermitian adjoint to the injection $\delta_\emptyset^s : \psi^{t]} \mapsto \psi^{t]} \otimes \delta_\emptyset^s$ of $\psi_t \in H^{t]}$ into $H^{t]} \otimes F_t^s$. Then the family of CP maps $\phi_t(t^s)$ is output compatible, $\left[\phi_t(t^s, \mathrm{Z}), \mathbb{B}^{t]}\right] = 0$, and can be lifted to a hemigroup of CP maps $\mathbb{M}^{t+s]} \to \mathbb{M}^{t]}$ satisfying the $\mathbb{B}^{t}$-modularity condition

$$\phi_t(t^s, \mathrm{Y}_t\mathrm{Z}\mathrm{Y}_t^{*}) = \mathrm{Y}\phi_t(t^s, \mathrm{Z})\,\mathrm{Y}^{*}\quad \forall \mathrm{Y} \in \mathbb{B}^{t}, \mathrm{Z} \in \mathbb{M}^{t+s]}, \mathrm{Y}_t = \mathrm{Y} \otimes \mathrm{I}_t^{s]}.$$

It is resolving family for the generalized Lindblad equation

$$\frac{\mathrm{d}}{\mathrm{d}t}\phi_r\left(t,\mathrm{Z}\left(t\right)\right)=\phi_r\left(t,\lambda\left(t,\mathbf{J}_{\mathrm{Z}}^t\right)\right),\ \phi_r\left(r,\mathrm{Z}\right)=\mathrm{Z}\in\mathbb{M}^{r]}. \tag{2}$$

defined by the form-generator $\lambda(t,\mathbf{J}):=\varsigma_+^-(t,\mathbf{J})$ commuting with $\mathbb{B}^{t]}$ on the germs $\mathbf{J}\in\mathfrak{G}^{\circ}\bar{\otimes}\mathbb{A}^t$ of the process $\mathrm{Z}(t^s)\in\mathbb{A}^t\bar{\otimes}\mathbb{M}_t^{s]}$ at each t. Here $\mathfrak{G}^{\circ}=\mathbb{M}^{\circ}\bar{\otimes}\mathfrak{g}$ is the initial germ algebra defined as the $\mathbb{M}^{\circ}$-envelope of the $\star$-semigroup $\mathfrak{g}=\mathbf{1}\oplus\mathfrak{m}$ unitizing the Itô $\star$-algebra $\mathfrak{M}(t)=\mathfrak{a}\vee\mathfrak{b}\equiv\mathfrak{m}$ of the mutually commutative Itô subalgebras $\mathfrak{a},\mathfrak{b}\subset\mathfrak{b}(\mathfrak{k})$ generating respectively $\mathbb{A}^t=\imath_0^t(\mathfrak{A}^{\otimes})$ and $\mathbb{B}^t=\imath_0^t(\mathfrak{B}^{\otimes})$ by multiple QS integrals $\imath_0^t(\mathfrak{M}^{\otimes})$ of $\mathfrak{M}^{\otimes}=\oplus_{n=0}^{\infty}\mathfrak{M}^{\otimes n}$ as defined for $\mathfrak{M}=\int_{t>0}^{\oplus}\mathfrak{M}(t)\,\mathrm{d}t$ in Refs.[8,9] Every such completely bounded generator $\lambda(t)$ can be uniquely lifted to the $\mathbb{B}^{t]}$-modular map $\mathfrak{G}^{\circ}\bar{\otimes}\mathbb{M}^{t]}\to\mathbb{M}^{t]}$.

Proposition 4.1. *Let $\mathbf{J}(t)=\mathbf{J}_{\mathrm{Z}}^t$ be the germ of a QS process $\mathrm{Z}:t\mapsto\mathbb{M}_t$ with $\mathrm{J}_+^-(t)=0$ for all t. Then the $\mathbb{A}^{t]}$-Markov generator $\lambda(t)$ for a QS hemigroup of $\widetilde{\alpha}_r(t,\mathrm{Z}(t))=\mathrm{V}_r(t)\,\mathrm{Z}(t)\,\mathrm{V}_r(t)^*$ has the generalized Lindblad structure $\lambda(t,\mathbf{J})=\left(\mathbf{SJS}^{\ddagger}\right)_+^-$ on $\mathfrak{G}^{\circ}\bar{\otimes}\mathbb{M}^{t]}$ written in terms of the logarithmic QS derivatives $\mathrm{L}_\nu=\mathrm{S}_\nu^-$ of the QS evolution $\mathrm{V}_r(t)$ as*

$$\lambda\left(t,\mathbf{J}_{\mathrm{Z}}^t\right)=\left(\mathrm{L}_+\mathrm{Z}+\mathrm{Z}\mathrm{L}_+^*+\mathrm{L}_\circ\mathrm{J}_+^\circ+\mathrm{L}_\circ\mathrm{J}_\circ^\circ\mathrm{L}_\circ^*+\mathrm{J}_\circ^-\mathrm{L}_\circ^*\right)(t)\,. \tag{3}$$

Note that the condition $\mathrm{J}_+^-=0$ corresponds to the property of $\mathrm{Z}(t)$ to be a local vacuum martingale $\mathrm{M}(t)$: $\mathrm{E}_r(t)\,\mathrm{M}(t)\,\mathrm{E}_r(t)=\mathrm{M}(r)$, and $\lambda(t,\mathbf{J}_{\mathrm{Z}}^t)=\lambda(t,\mathbf{J}_{\mathrm{M}}^t)+\mathrm{J}_+^-(t)$ for any vacuum semimartingale $\mathrm{Z}(t)=\mathrm{M}(t)+\int_0^t\mathrm{J}_+^-\mathrm{d}r$. In particular, if $\mathbf{D}=\mathbf{J}-\mathbf{X}=0$, λ reads simply as the Lindbladian on $\mathrm{X}=\mathrm{Ix}$,

$$\lambda(\mathbf{X})=\mathrm{L}_+\mathrm{X}+\mathrm{X}\mathrm{L}_+^*+\mathrm{L}_\circ\mathrm{X}\mathrm{L}_\circ^*\equiv\lambda(\mathrm{x})\,. \tag{4}$$

However, it defines the $\mathbb{A}^{\circ}$-Markov dynamics only if $\lambda(t,\mathbf{J})\in\mathbb{A}^{\circ}$ for $\mathbf{J}\in\mathfrak{G}^{\circ}$, e.g. if $\mathrm{L}_\nu(t)=\mathrm{I}\mathrm{L}_\nu(t)$ with $\mathrm{L}_\nu(t)\in\mathbb{A}^{\circ}$.

For the output processes $\mathrm{Z}(t)=\mathrm{Y}_t1$ having the QS derivative $\mathbf{D}=\mathbf{1b}(t)\,\mathrm{Y}_t$ with $\mathbf{b}(t)\in\mathbb{B}\ \forall\, t\geq 0$,

$$\lambda\left(\mathbf{J}_{\mathrm{Y}_t}^t\right)=\left(\lambda(1)+\mathrm{L}_\circ b_+^\circ+\mathrm{L}_\circ b_\circ^\circ\mathrm{L}_\circ^*+b_\circ^-\mathrm{L}_\circ^*+b_+^-\right)\mathrm{Y}_t.$$

Here $\mathrm{Y}_t\in\mathbb{B}^{t]}$ is a local QS semimartingale, e.g. a local QS martingale characterized by $b_+^-=0$, say $\mathrm{Y}_t=\mathrm{W}_t(\mathbf{b})\,\mathrm{y}$ with any $\mathrm{y}\in\mathbb{B}$ of zero expectation and a QS exponential martingale $\mathrm{W}_t(\mathbf{b})\in\mathbb{B}^t$ as the normal-ordered Weyl exponents $\mathrm{W}_t(\mathbf{b})=:\exp\left[\Lambda_0^t(\mathbf{b})\right]:$. The compatibility of $\lambda(\mathbf{J})$ with $\mathbb{B}^{t]}$ on the QS logarithmic germs $\mathbf{J}=\mathbf{J}_{\mathrm{Y}}^t\mathrm{Y}_t^{-1}=\mathbf{1}+\mathbf{b}(t)$ is ensured by $\mathrm{L}_\nu\in\mathbb{A}^{t]}$.

In the following we assume that $\mathfrak{b}=\widetilde{\mathfrak{a}}$ is dual to a modular Itô B*-algebra in the standard representation $\mathfrak{a}\subseteq\mathfrak{b}(\mathfrak{k})$.

4.1.2. *Unraveling the hidden Markov dynamics*

Let us consider a QS hemigroup $\{\check{\mathrm{F}}_r(t)\}$ of propagators $H^{r]}\otimes F_r \to H^{t]}\otimes F_t$ satisfying the QS evolution equation

$$d\check{\mathrm{F}}_r(t) = \check{\mathrm{F}}_r(t)\left(\mathrm{L}_+(t)\,dt + \mathrm{L}_k(t)\,d\check{e}^k_t\right) = \Lambda\left(dt, \check{\mathrm{F}}_r(t)\,\mathbf{L}(t)\right)\quad \check{\mathrm{F}}_r(r) = \mathrm{I}^{r]}. \tag{5}$$

with $\mathrm{L}_k = \mathrm{L}_\circ e^\circ_k$ and $\check{e}^k(t) = \Lambda\left(t, \bar{\mathbf{e}}^k\right)$. Applying the QS Itô formula to $\check{\phi}_r(t, \mathrm{Z}(t)) = \check{\mathrm{F}}_r(t)\,\mathrm{Z}(t)\,\check{\mathrm{F}}_r(t)^*$, we obtain a QS evolution equation

$$d\check{\phi}(t, \mathrm{Z}(t)) = \Lambda\left(dt, \check{\phi}\left(t, \boldsymbol{\zeta}\left(t, \mathbf{J}^t_{\mathrm{Z}}\right) - \mathbf{X}(t)\right)\right)\quad \mathrm{Z}(t)\in \mathbb{M}^{t]} \tag{6}$$

describing in terms of $\boldsymbol{\zeta}(t, \mathbf{J}) = \mathbf{C}(t)\,\mathbf{J}\mathbf{C}(t)^{\ddagger}$ a QS *entangling process* $\check{\phi}(t) : \mathbb{M}^{t]} \to \mathbb{M}^{\circ}\bar{\otimes}\mathbb{A}^t_\star$ by the hemigroup $\{\check{\phi}_r(t)\}$ of *simple* CP transformations $\check{\phi}_r(t) : \mathbb{M}^{t]} \to \mathbb{M}^{r]}\bar{\otimes}\mathbb{A}^{t-r}_{r\star}$. They are defined by the logarithmic QS germ $\mathbf{C}(t) = \mathbf{I} + \mathbf{L}(t)$ of $\check{\mathrm{F}}_r(t)$ given by the logarithmic QS derivative

$$\mathbf{L}(t) = \mathrm{L}_+(t)\,\mathbf{d} + \sum_{k\in J_\circ} \mathrm{L}_k(t)\,\bar{\mathbf{e}}^k. \tag{7}$$

In general the QS transformations $\check{\phi}_r(t)$ map $\mathbb{M}^{t]}$ not into a W*-algebra but into the kernels[15] which are affiliated to $\mathbb{M}^{r]}\bar{\otimes}\mathbb{A}^{t-r}_r$ since, as it follows from the next Lemma, they satisfy the *quasi-Markovianity* condition

$$\phi_t(t^s, \mathrm{Z}) = \mathrm{E}_t(t^s)\,\check{\phi}_t(t^s, \mathrm{Z})\,\mathrm{E}_t(t^s) \in \mathbb{A}^{t]}\ \forall t > r,\ \mathrm{Z}\in \mathbb{A}^t\bar{\otimes}\mathbb{M}^{s]}_t$$

following from the condition (1) for $\widetilde{\alpha}_r(t, \mathrm{Z})$. This quasi-Markovianity, defined as the Markovianity with respect to an increasing algebras $\mathbb{A}^{t]}$ in place of a single object algebra $\mathbb{A}^{\circ}$, and also the uniqueness of the solutions to (5), it is sufficient to assume that only the operator-functions $\mathrm{L}_+(t)$ and $\mathrm{L}_\circ(t)$ are respectively locally L^1 and L^2-integrable[8,10] with values in $\mathbb{A}^{t]}$, or $\mathbb{A}^{t]}$-adapted $\mathrm{L}_\nu(t) \vdash \mathbb{A}^{t]}$ as affiliated to $\mathbb{A}^{t]}$.

Lemma 4.1. *Let the $\mathbb{A}^{t]}$-Markov hemigroup $\{\phi_r(t)\}$ be defined as the unique solution of the generalized Lindblad equation (2) on the increasing $\mathbb{M}^{t]}$ with the generator (3) $\lambda(\mathbf{J}) = \varsigma^-_+(\mathbf{J})$. Then it is unraveled by the $\mathbb{A}^{t]}$-Markov QS hemigroup of simple CP transformation $\check{\phi}_r(t)$ in the sense that it is the conditional expectation*

$$\phi_r(t, \mathrm{Z}(t)) = \epsilon_r\left[\check{\phi}_r(t, \mathrm{Z}(t))\right] := \mathrm{E}_r\check{\phi}_r(t, \mathrm{Z}(t))\,\mathrm{E}_r$$

corresponding to the vacuum-induced white noise thermostates $\epsilon_r[\mathrm{X}] = \delta^{r*}_\emptyset \mathrm{X}\delta^r_\emptyset$ on the algebras $\mathbb{A}_r$ generated by the independent increments of the input quantum Lévy noise $\check{\mathbf{e}}(t) - \check{\mathbf{e}}(r)$.

Proof. Since $E_r(t)E_t = E_r = E_tE_r(t)$, the vacuum expectation $\epsilon_r = \epsilon_r(t)\circ\epsilon_t$ of $d\breve{\phi}_r(Z)(t)$ is defined by the expectation $\breve{\phi}_r\left(t,\zeta_+^-(\mathbf{J}_Z^t)\right)dt$ of $d\breve{\phi}_r(Z)(t)$ written by modularity property as $\epsilon_t\left[d\breve{\phi}_r(Z)\right](t) =$

$$\breve{F}_r(t)\,\epsilon_t\left[\Lambda\left(dt,\boldsymbol{\zeta}\left(t,\mathbf{J}_Z^t\right)-\mathbf{Z}(t)\right)\right]\breve{F}_r(t)^* = \breve{\phi}_r\left(t,\zeta_+^-(t,\mathbf{J})\right)dt,$$

where $\zeta_+^-(\mathbf{J})\,dt = \epsilon_t\left[\Lambda\left(dt,\mathbf{C}(t)\,\mathbf{J}\mathbf{C}(t)^\ddagger\right)\right]$ with $J_+^- = 0$ is a generalized Lindblad generator

$$(\boldsymbol{\zeta}(\mathbf{J})-\mathbf{Z})_+^- = C_+^-Z + ZC_+^{-*} + C_\circ^- J_+^\circ + C_\circ^- J_\circ^\circ C_\circ^{-*} + J_\circ^- C_\circ^{-*} \equiv \zeta_+^-(\mathbf{J})\,. \quad (8)$$

In fact, since $C_+^- = L_+$ and $C_\circ^- = \sum_k L_k(t)\,\bar{e}_\circ^k = L_\circ$ due to the completeness relation $e_k^\circ\bar{e}_\circ^k \equiv I_\circ^\circ$ with $\bar{e}_\circ^k$, this $\zeta_+^-(t)$ coincides with the generator $\lambda(t) = \varsigma_+^-(t)$ of the $\mathbb{A}^{t]}$-Markov backward master equation vacuum induced by the isometric or unitary evolution $V_r(t)$ on $\mathbb{M}^{t]}$. Thus, expectation $\epsilon_r\circ\breve{\phi}_r(t)$ of the QS bracket evolution $\breve{\phi}_r(t)$ driven by the white thermonoise $\breve{\mathbf{e}}(t)$ must coincide by the uniqueness argument with the $\mathbb{A}^{t]}$-Markov evolution $\epsilon_r\circ\widetilde{\alpha}_r(t)$ vacuum induced by the Heisenberg evolution $\widetilde{\alpha}_r(t)$. □

4.1.3. *The Belavkin unraveling flow and equation*

The following theorem defines the structure of the generators for QS master equations induced by the general QS dynamics with respect to any output quantum Lévy process given by a modular Itô B*-algebra. This result extends the semi-quantum filtering theory[10] determined by any classical output Lévy process to fully quantum case. The noncommutative filtering in the $\mathbb{A}^\circ$-Markovian case was outlined for quantum finite-dimensional Wiener temperature noise corresponding to a Wiener-Itô modular algebra in Ref.[7]

Theorem 4.1. *Let $\mathfrak{b}$ be a modular output Lévy-Itô B*-algebra with the symmetric basis $\mathbf{e}_{J_\bullet}$ not necessarily complete in $\widetilde{\mathfrak{a}}$. Then the $\mathbb{A}^{t]}$-Markov QS evolution*

$$\breve{\phi}(t,\mathrm{x}) = \breve{F}_0(t)\,\mathrm{x}\breve{F}_0(t)^*\,, \quad t\geq 0$$

restricted to the object algebra $\mathbb{A}^\circ$ and filtered with respect to $\mathbb{B}^t$ satisfies the Heisenberg QS equation

$$d\breve{\phi}_r(t,\mathrm{x}) + \breve{\phi}(t,\kappa(t,\mathrm{x}))\,dt = \sum_{i,k\in J_\bullet}\breve{\phi}\left(t,\kappa_j^\varepsilon(t,\mathrm{x})\right)d\breve{e}^j(t)\;, \quad (9)$$

where $\kappa(\mathrm{x}) = \mathrm{Kx}+\mathrm{xK}^* - \sum_{i,k\in J_\circ} L_i\mathrm{x}\bar{\sigma}^{i,k}\hat{\rho}L_k^\star = -\lambda(\mathrm{x})$ defines the Lindbladian (4) in the Hermit-symmetric basis with $K = -L_+$, $L_k^\star = L_{k\star}^*$ and

$$\kappa_j^\varepsilon(t,\mathrm{x}) = L_j(t)\,\mathrm{x} + \mathrm{x}L_j^\star(t) + \sum_{i,k\in J_\circ} L_i(t)\,\mathrm{x}\bar{\varepsilon}_j^{i,k}L_k^\star(t)\,,$$

are the fluctuating coefficients $\boldsymbol{\kappa} = \left(\kappa_j^{\varepsilon}\right)_{j\in J_\circ}$ *with* $\bar{\varepsilon}_j^{i,k} = \varepsilon_{k,i}^j$.

Proof. By QS product Itô formula applied to $\check{\mathrm{F}}_0(t)\,\mathrm{x}\check{\mathrm{F}}_r(t)^*$ we obtain the equation (6) with $\mathrm{Z}(t) = \mathrm{I}^t\mathrm{x}$ and $\mathbf{J}_{\mathrm{Z}}^t = \mathrm{Z}(t)\,\mathbf{I}$. Taking for these the element (8) as vacuum conditional expectation with $\mathbf{C} = \mathbf{I} + \mathbf{L}$ given by (7) we obtain

$$\check{\kappa}_+^-(\mathrm{x}) = \mathrm{L}_+\mathrm{x} + \mathrm{x}\mathrm{L}_+^* + \mathrm{L}_i\mathrm{x}\left(\bar{\mathbf{e}}^i\bar{\mathbf{e}}^k\right)_+^-\hat{\rho}\mathrm{L}_k^\star = -\kappa(\mathrm{x}),$$

since $\left(\bar{\mathbf{e}}^i\bar{\mathbf{e}}^k\right)_+^- = \bar{\sigma}^{i,k}$, and by $\bar{\mathbf{e}}^i\bar{\mathbf{e}}^k = \bar{\varepsilon}_j^{i,k}\bar{\mathbf{e}}^j$ we obtain

$$\check{\kappa}(\mathrm{x})_j\,\bar{\mathbf{e}}^j = \mathrm{L}_i\bar{\mathbf{e}}^i\mathrm{x} + \mathrm{x}\bar{\mathbf{e}}^k\mathrm{L}_k^\star + \mathrm{L}_i\mathrm{x}\bar{\mathbf{e}}^i\bar{\mathbf{e}}^k\mathrm{L}_k^\star = \kappa_j^{\varepsilon}(\mathrm{x})\,\bar{\mathbf{e}}^j \qquad \square$$

4.2. *Reduced state dynamics and QS master equations*

4.2.1. *The standard pairing and marginalization*

Let $\mathbb{A}$ be a C*-algebra with a vector state $\omega(\mathrm{A}) = (\upsilon|\mathrm{A}\upsilon) \equiv \upsilon^*\mathrm{A}\upsilon$ on the Hilbert space H generated by $\mathbb{A}\upsilon$ and $\mathbb{B} \subseteq \mathbb{A}'$ be a modular subalgebra of $\mathcal{B}(H)$ commuting with $\mathbb{A}$, with the dual subalgebra $\widetilde{\mathbb{B}} \subseteq \mathbb{A}$ represented on the Hilbert subspace $H_\circ \subseteq H$ generated by $\mathbb{A}'\upsilon$. Let $(\mathfrak{a}, l)$ be a B*-Itô $\star$-algebra with

$$l(a) = (\mathbf{v}_\emptyset|\mathbf{a}\mathbf{v}_\emptyset) \equiv \mathbf{v}_\emptyset^\ddagger\mathbf{a}\mathbf{v}_\emptyset = a_+^-$$

in the canonical matrix representation $\mathfrak{a} \subseteq \mathfrak{b}(\mathfrak{k})$ of the Hilbert space $\mathfrak{k}$ generated by the action of $\mathfrak{a}$ on the vector $\mathbf{v}_\emptyset = \left(\delta_+^\mu\right)$, and $\mathfrak{b} \subseteq \mathfrak{a}'$ be a modular Itô subalgebra of the commutant Itô algebra $\mathfrak{a}' = \left\{c_+^-\mathbf{d} + \mathbf{c}_\circ\right\}$ with the dual subalgebra $\widetilde{\mathfrak{b}} \subseteq \mathfrak{b}(\mathfrak{k}_\circ)$ of $\mathrm{J}^*\mathfrak{a}'\mathrm{J} = \mathfrak{a}_+^-\mathbf{d} + \mathfrak{a}_\circ$ represented in $\mathfrak{b}(\mathfrak{k}_\circ)$ of the Hilbert subspace $\mathfrak{k}_\circ \subseteq \mathfrak{k}$ generated by $\mathfrak{a}'$ on $\mathbf{v}_+$. We denote by $\mathfrak{R} = \mathfrak{r}\bar{\otimes}\mathbb{A}$ and $\mathfrak{S} = \mathfrak{s}\bar{\otimes}\mathbb{B}$ the germ algebras given respectively by $\mathfrak{r} = 1 \oplus \mathfrak{a}$ and $\mathfrak{s} = 1 \oplus \mathfrak{b}$. represented in $\mathfrak{b}(\mathfrak{k})$ by matrices $\mathbf{I} + \mathbf{a}$ and $\mathbf{I} + \mathbf{b}$ with $\mathbf{a} \in \mathfrak{a}$, $\mathbf{b} \in \mathfrak{b}$ such that $\mathfrak{S} \subseteq \mathfrak{R}'$. Let us define the standard $(\mathfrak{R}, \mathfrak{R}_\mathsf{T})$-*pairing*

$$\langle \mathbf{R}, \mathbf{S}\rangle := (\mathbf{v}_+|\mathbf{R}\mathbf{S}\mathbf{v}_+) = \left(\upsilon|\mathrm{R}_\circ^-\mathrm{S}_+^\circ\upsilon\right) \equiv \left\langle \mathrm{R}_\circ^-, \mathrm{S}_+^\circ\right\rangle \quad \forall \mathbf{R} \in \mathfrak{R}, \mathbf{S} \in \mathfrak{R}_\mathsf{T}, \tag{10}$$

induced by the state $l \otimes \omega$, where $\mathbf{v}_+^\ddagger = (\upsilon^*, 0, 0)$, by extending this bilinear from $\mathbf{S} \in \mathfrak{R}'$ on the completion $\mathfrak{R}_\mathsf{T} = \mathfrak{R}'_\star$ of the commutant $\mathfrak{R}'$ represented on $\mathfrak{k}_\circ \otimes H_\circ$ as the dual space to the normed B*-algebra $\mathfrak{R}$. Note the decomposition $\mathfrak{R}_\mathsf{T} = \mathbb{A}_\mathsf{T} \oplus \dot{\mathfrak{R}}_\mathsf{T}$ corresponding to $\mathfrak{R} = \mathfrak{R}_+^- \oplus \mathfrak{R}_\circ$ such that $\langle \mathbf{R}, \mathbf{S}\rangle$ for $\mathbf{R} = \mathrm{R}_+^-\mathbf{d} + \mathbf{R}_\circ$ and $\mathbf{S} = \mathrm{Y}\mathbf{I} + \dot{\mathbf{S}}$ can be written as

$$\left\langle \mathbf{R}_\circ, \dot{\mathbf{S}}\right\rangle + \langle \mathrm{R}_+^-, \mathrm{Y}\rangle = \langle \mathrm{X}, \mathrm{P}_+^-\rangle + \left\langle \dot{\mathbf{R}}_\circ, \dot{\mathbf{S}}^\circ\right\rangle + \langle \mathrm{R}_+^-, \mathrm{Y}\rangle = \langle \mathrm{X}, \mathrm{P}_+^-\rangle + \left\langle \dot{\mathbf{R}}, \mathbf{S}^\circ\right\rangle.$$

Here $\mathbf{R}_\circ = \mathrm{X} \oplus \dot{\mathbf{R}}_\circ \in \mathfrak{R}_\circ$ is the germ-martingale, $(\mathbf{R}_\circ)_+^- = 0$, of X, uniquely defined by the decomposition $\mathbf{S} = \mathrm{S}_+^- \mathbf{d} + \mathbf{S}^\circ$.

It can be easily seen that the standard pairing has the modularity property

$$\left\langle \mathbf{CRC}^\ddagger, \mathbf{S} \right\rangle = \left(\mathbf{v}_+ | \mathbf{CRSC}^\ddagger \mathbf{v}_+ \right) = \langle \mathbf{R}, \mathbf{C}'\mathbf{SC}'^* \rangle \quad \forall \mathbf{S} \in \mathfrak{R}_\mathsf{T}, \mathbf{R} \in \mathfrak{R}_\star^\star$$

for any $\mathbf{C} \in \mathfrak{R}$ having a *transposed* $\mathbf{C}' \in \mathfrak{R}'$ uniquely defined by

$$\mathbf{v}_+^\ddagger \mathbf{RC}' = \mathbf{v}_+^\downarrow \mathbf{CR} = \mathbf{v}_+^\downarrow \mathbf{C}'\mathbf{R} \quad \forall \mathbf{R} \in \mathfrak{R}$$

as $\mathbf{C}' := \mathrm{J}^* \mathbf{C}^\dagger \mathrm{J} \equiv \mathbf{C}^{\flat\ddagger}$ by the unitary conjugation $\mathrm{J}^* = \mathrm{J}^{-1}$. Here $\dagger$ denotes the left involution $(\mathbf{X} \cdot \mathbf{C}^\dagger \mathbf{Z}) = \left(\mathbf{XL}^\ddagger \cdot \mathbf{Z} \right)$ with respect to the *modular pairing* $(\mathbf{X} \cdot \widetilde{\mathbf{Y}}) = \langle \mathbf{X}, \mathbf{Y} \rangle$ for $\widetilde{\mathbf{X}} = \overline{\mathbf{X}}^\ddagger, \mathbf{Y} \in \mathfrak{R}'$, which is symmetric due to $\mathbf{JXJ}^* = \overline{\mathbf{X}} = \mathrm{J}^* \mathbf{XJ}$:

$$(\mathbf{X} \cdot \mathbf{Z}) := (\overline{\mathbf{X}} \mathbf{v}_+ | \mathbf{Z} \mathbf{v}_+) = (\overline{\mathbf{Z}} \mathbf{v}_+ | \mathbf{X} \mathbf{v}_+) \equiv (\mathbf{Z} \cdot \mathbf{X}) \quad \forall \mathbf{X}, \mathbf{Z} \in \mathfrak{R}.$$

A positive normalized element $\hat{\rho} \in \mathbb{A}_\mathsf{T} := \mathbb{A}'_\star$ is called the *covariant density* of the *regular state* $\rho(\mathrm{X}) = \langle \mathrm{X}, \hat{\rho} \rangle$ on $\mathbb{A}$, which is *absolutely continuous* with respect to ω in the sense $\omega(\mathrm{A}^*\mathrm{A}) = 0 \Rightarrow \rho(\mathrm{A}^*\mathrm{A}) = 0$. A *simple conditionally CP transformation* $\lambda(\mathbf{R}) = (\mathbf{CRC})_+^-$ from $\mathfrak{R}$ into $\mathbb{A}$ such that $\lambda(\widetilde{\mathfrak{S}}) \subseteq \mathbb{B} := \mathbb{B}_\mathsf{T}^\star$ transforms $\mathbb{B}_\mathsf{T}$ into $\mathfrak{S}_\star$ by

$$\langle \mathbf{R}, \lambda_\mathsf{T}(\hat{\rho}) \rangle = \rho \circ \lambda(\mathbf{R}) = \langle \lambda(\mathbf{R}), \hat{\rho} \rangle \quad \forall \mathbf{R} \in \mathfrak{R}, \hat{\rho} \in \mathbb{B}_\mathsf{T}$$

if it is given by the transposable $\mathbf{C} \in \mathfrak{R}$. One can show that λ_T is also conditionally CP such that each $\lambda_\mathsf{T}(\hat{\rho})$ is also conditionally absolutely continuous with respect to $l \otimes \omega$. It is defined on $\mathfrak{R}_\circ$ by the decomposition $\lambda_\mathsf{T}(\hat{\rho}) = \hat{\rho}\mathbf{I} + \varkappa(\hat{\rho}) \in \dot{\mathfrak{S}}_\star$, where

$$\varkappa(\hat{\rho}) = \mathbf{L}'\hat{\rho} + \hat{\rho}\mathbf{L}'^\ddagger + \mathbf{L}'\hat{\rho}\mathbf{L}'^\ddagger \tag{11}$$

in terms of $\mathbf{L} = \mathbf{C} - \mathbf{I}$ is decomposed as $\varkappa(\hat{\rho}) = \varkappa^\emptyset(\hat{\rho})\,\mathbf{d} + \varkappa^\circ(\hat{\rho})$.

4.2.2. *The dual and reduced QS dynamics*

Let $\mathbb{A}^t$ be the increasing W*-algebras generated by quantum Itô-Lévy noises $\breve{e}^j(r)$ for $r \in [0, t)$ and $j \in J$ on Fock space $F = \Gamma\left(L^2_\mathfrak{k}\right)$. The modular subalgebras $\widetilde{\mathbb{B}}^t = \mathrm{J}^* \mathbb{B}^t \mathrm{J}$ are defied on $F_\circ = \Gamma\left(L^2_{\mathfrak{k}_\circ}\right)$ as the dual to $\mathbb{B}^t$ generated by $\hat{e}_i = \Lambda(\mathbf{e}_i)$ for $i \in J_\bullet \subseteq J_\circ$ with the Weyl operator basis $\{\mathrm{W}_t(\mathbf{b}) : \mathbf{b} \in \mathfrak{b}\}$, where $\mathfrak{b}$ is the modular Itô algebra spanned by $\mathbf{e}_\emptyset$ and $\mathbf{e}_{J_\bullet}$. It is given by the solutions to

$$d\mathrm{W}_t(\mathbf{b}) = \mathrm{W}_t(\mathbf{b})\,\Lambda(dt, \mathbf{b}), \quad \mathrm{W}_0(\mathbf{b}) = \mathrm{I}.$$

This solution can be written in the normal form in terms of the logarithmic matrix $l^\circ_\circ = \ln(\mathrm{I}^\circ_\circ + b^\circ_\circ)$ as.

$$\mathrm{W}_t(\mathbf{b}) = \mathrm{e}^{\int_0^t b^i_+(r)\mathrm{dA}^+_i}\mathrm{e}^{\int_0^t l^i_k(r)\mathrm{dA}^k_i}\mathrm{e}^{\int_0^t b^-_k(r)\mathrm{dA}^k_-}\mathrm{e}^{\int_0^t b^-_+(r)\mathrm{d}r}.$$

If the QS dynamics is a $\mathbb{A}^{t]}$-Markov process on the state vector $\upsilon = \eta\otimes\delta_\emptyset \in \mathfrak{h}\otimes F$ with the vacuum $\delta_\emptyset \in F_t$, the dual state dynamics $\varphi_t(r)$ form a forward hemigroup defied on $\mathbb{A}^{r]}_\intercal$ by the CP maps $\varphi_t(r)$ predual to $\phi_r(t)$:

$$\langle \mathrm{X}_t, \varphi_t(r,\hat{\rho})\rangle_t = \langle \phi_r(t,\mathrm{X}), \hat{\rho}\rangle_r \quad \forall \mathrm{X}\in\mathbb{A}^{t]},\ \hat{\rho}\in\mathbb{A}^{r]}_\intercal.$$

Here and below we shall take the vacuum $(\mathbb{A}^{t]}, \mathbb{A}^{t]}_\intercal)$-pairings

$$\langle \mathrm{X}, \hat{\rho}\rangle := \left\langle \eta\otimes\delta^t_\emptyset | \mathrm{X}\hat{\rho}\left(\eta\otimes\delta^t_\emptyset\right)\right\rangle \equiv \langle \mathrm{X},\hat{\rho}\rangle_t,\ \forall \mathrm{X}\in\mathbb{A}^{t]}, \hat{\rho}\in\mathbb{A}^{t]}_\intercal.$$

Let $\mathbb{B}^{t]}_\star$ denote the preduals to the increasing W*-subalgebras $\overline{\mathbb{B}}^{t]} = \overline{\mathbb{B}}^\circ\bar{\otimes}\overline{\mathbb{B}}^t$ of $\mathbb{A}^{t]}$ generated an initial modular subalgebra $\widetilde{\mathbb{B}}^\circ \subseteq \mathbb{A}^\circ$ and the Weyl basis $\mathrm{W}_t(\mathbf{a})$, $\mathbf{a}\in\widetilde{\mathfrak{b}}\subseteq\mathfrak{a}$. If the QS dynamics is also Markov with respect to $\overline{\mathbb{B}}^{t]} \subseteq \mathbb{A}^{t]}$ for all t, the dual dynamics can be reduced to a hemigroup of $\vartheta_t(r) : \mathbb{B}^{r]}_\star \to \mathbb{B}^{t]}_\star$. It is given by the marginalization $\iota_\intercal$ of the reference $\lambda = \omega|\mathbb{A}$ to $\overline{\mu} = \omega|\overline{\mathbb{B}}$ defining the CP maps $\vartheta_t(r)\circ\iota_\intercal = \iota_\intercal\circ\varphi_t(r)$ as the preduals to the restrictions

$$\vartheta^\intercal_r(t) := \phi_r(t)\,|\overline{\mathbb{B}}^{t]} \equiv \vartheta_t(r)^\intercal.$$

In particular, the reduced evolution $\hat{\rho}_t = \vartheta_t(\hat{\varrho})$ from the initial state ϱ on $\overline{\mathbb{B}}^\circ \subseteq \mathbb{A}^\circ$ is defined by

$$\langle \mathrm{X}_t, \vartheta_t(\hat{\varrho})\rangle_\mu = \langle \phi_0(t,\mathrm{X}_t), \hat{\varrho}\rangle_\lambda, \quad \forall \mathrm{X}_t\in\overline{\mathbb{B}}^{t]},\ \hat{\varrho}\in\overline{\mathbb{B}}^\circ_\star.$$

`The reduced states` $\hat{\rho}_t$ can be described on $\mathrm{X}_t = \mathrm{W}_t(\mathbf{a})\,\mathrm{x}$ with $\mathrm{x}\in\overline{\mathbb{B}}$ by the characteristic function

$$C_t(\mathbf{a},\mathrm{x}) = \langle \mathrm{W}_t(\mathbf{a})\,\mathrm{x}, \hat{\rho}_t\rangle = \langle \phi_0(t,\mathrm{X}_t), \hat{\varrho}\rangle$$

satisfying the equation

$$\mathrm{d}C_t(\mathbf{a},\mathrm{x}) = \mathrm{d}\langle \mathrm{X}_t(\mathbf{a}), \hat{\rho}_t\rangle \ \text{ for } \ C_t(\mathbf{a},\mathrm{x}) = \langle \mathrm{X}_t(\mathbf{a}), \hat{\rho}_t\rangle.$$

Note that since $\vartheta_t(r)$ mapping $\mathbb{B}^{r]}_\star$ into the increasing state spaces $\mathbb{B}^{t]}_\star$, they can not satisfy a deterministic evolution equation which should involve the increments $\mathrm{d}\hat{e}_{J_\bullet}$ generating $\mathrm{d}\mathbb{B}^t_\star$. Thus they should satisfy a QS evolution equation $\mathrm{d}\hat{\rho}_t = \varkappa(\hat{\rho}) \cdot \mathrm{d}\hat{\mathrm{e}}$ with $\mathrm{d}\hat{\mathrm{e}}_\emptyset = \Lambda(\mathrm{d}t, \mathbf{d}) = \mathrm{d}t\mathrm{I}$ and the innovation $\{\mathrm{d}\hat{\mathrm{e}}_j : j \in J_\bullet\}$ such that by Itô formula

$$\langle \mathrm{dX}_t, \hat{\rho}_t\rangle + \langle \mathrm{X}_t, \mathrm{d}\hat{\rho}_t\rangle + \langle \mathrm{dX}_t, \mathrm{d}\hat{\rho}_t\rangle = \mathrm{d}C_t(\mathbf{a}, \mathbf{x}).$$

It is characterized on $\mathrm{X}_t = \mathrm{W}_t(\mathbf{a})\mathbf{x}$ with $\mathbf{a} \in \widetilde{\mathfrak{b}}$, $\mathbf{x} \in \widetilde{\mathbb{B}}^\circ$ by the equation

$$\mathrm{d}C_t(\mathbf{a}, \mathbf{x}) = \mathrm{d}\langle \mathrm{X}_t(\mathbf{a}), \hat{\rho}_t\rangle \text{ for } C_t(\mathbf{a}, \mathbf{x}) = \langle \mathrm{X}_t(\mathbf{a}), \hat{\rho}_t\rangle,$$

where $\mathrm{d}\langle \mathrm{X}_t(\mathbf{a}), \hat{\rho}_t\rangle = 0$ in the martingale case $a_+^- = 0$, $\langle \mathrm{X}_t, \mathrm{d}\hat{\rho}_t\rangle = \langle \mathrm{X}_t, \varkappa^\emptyset(\hat{\rho}_t)\rangle\, dt$ and

$$\langle \mathrm{dX}_t, \mathrm{d}\hat{\rho}_t\rangle = \sum_{i\in J_\circ} \langle \mathrm{X}_t\mathbf{a}, \varkappa^i(\hat{\varrho})\mathbf{e}_i\rangle\, \mathrm{d}t \equiv \langle \mathrm{X}_t\mathbf{a}, \varkappa^\circ(\hat{\varrho})\rangle\, \mathrm{d}t \;\; \forall \mathbf{a} \in \widetilde{\mathfrak{b}}. \qquad (12)$$

4.2.3. *The Belavkin modular master equation*

The following theorem defines the structure of the generators for QS master equations induced by the general QS dynamics with respect to any output quantum Lévy process given by a modular Itô B*-algebra. This result extends the semi-quantum unraveling dynamics[3] determined by any classical output Lévy process to the quantum Levy-Itô temperature process. The proof of the theorem in the general case is similar to the proof of the semi-quantum filtering theorem[5] for the commutative processes $\hat{e}_i = \check{e}_{i^\star}$ but noncommutative initial algebra $\mathbb{B}$. It was outlined in Ref[7] for fully quantum diffusive case of noncommuting self-adjoint Wiener processes $\hat{e}_i = \hat{e}_i^*$ corresponding to quantum finite-dimensional Wiener-Itô nilpotent in second order modular algebra.

Theorem 4.2. *(Main) Let $\{\vartheta_r^\intercal(t)\}$ be the quantum hemigroup dynamics on the increasing W*-subalgebras $\overline{\mathbb{B}}^{t]} \subseteq \mathbb{A}^{t]}$ generated by $\widetilde{\mathbb{B}}^\circ \subseteq \mathbb{A}^\circ$ and the modular Lévy-Itô B*-algebra $\widetilde{\mathfrak{b}} \subseteq \mathfrak{a}$ with a symmetric basis $\{\mathbf{d}, \overline{\mathbf{e}}_i : i \in J_\bullet\}$ indexed by $J_\bullet = J_\bullet^\star \subseteq J_\circ$. Assume that it is induced by the hemigroup $\{\phi_r(t)\}$ having the generator (3) with $\mathbb{A}^{t]}$-adapted transposable coefficients $\mathrm{L}_+^\iota(t)$. Then the reduced states $\hat{\rho}(t) = \vartheta_t(\hat{\rho}) \in \mathbb{B}_\star^{t]}$ satisfy the QS stochastic equation*

$$\mathrm{d}\hat{\rho}(t) + \varkappa(t, \hat{\rho}(t))\, \mathrm{d}t = \sum_{j\in J_\bullet} \varkappa^j(t, \hat{\rho}(t))\, \mathrm{d}\hat{e}_j(t) \;\; \forall \hat{\rho}(0) \in \mathbb{B}_\star^\circ \qquad (13)$$

with QS modular noises $\hat{e}^j(t) = \Lambda(t, \mathbf{e}^j)$ and fluctuating coefficients given by

$$\varkappa^j(t, \hat{\rho}) = L^{j^\star}(t)\,\hat{\rho} + \hat{\rho}L^{j*}(t) + \sum_{i,k\in J_\circ} L^{k^\star}(t)\,\varepsilon_{i,k}^j\hat{\rho}L^{k*}(t)$$

including $\varkappa^{\emptyset} = -\varkappa$ *with* $L^{\emptyset} = -K = \mathrm{L}_{+}^{-\prime}$ *and* $L^{i} = (\mathbf{L}')_{\circ}^{-}\bar{e}_{\circ}^{i*} = \mathrm{L}_{+}^{\circ\prime}\bar{e}_{\circ}^{i*}$.

Proof. The predual generator $\varkappa : \mathbb{B}_{\star}^{t]} \to \dot{\mathfrak{S}}^{t]}$ for the generalized Lindbladian written for an $\mathbb{A}^{t]}$-martingales $\mathrm{Z} = \mathrm{X}$ in the form (3) is found by modularity in (11) for the $\mathbb{A}^{t]}$-adapted transposable $\mathbf{C} = \mathbf{I} + \mathbf{L}$ such that $\mathbf{C}'(t)$ commutes with $\mathfrak{R}^{t]}$. This allows the expansions

$$\mathbf{L}' = \sum_{i \in J_{\circ}} \mathbf{e}_{i\star} L^{i}, \quad \mathbf{L}'^{\ddagger} = \sum_{i \in J_{\circ}} \mathbf{e}_{i} L^{i*},$$

where $L^{i} = \mathrm{L}_{+}^{\circ\prime}\bar{e}_{\circ}^{i*}$, $L^{i*} = \bar{e}_{\circ}^{i}\mathrm{L}_{+}^{\circ\flat}$ are determined by the completeness as

$$\begin{aligned}\left\langle \mathrm{X}^{*}\bar{\mathbf{e}}^{j}, \mathbf{L}'^{\ddagger}\right\rangle &= \left(\mathrm{X}v|\left(\bar{\mathbf{e}}^{j}\mathbf{L}'^{\ddagger}\right)_{+}^{-}v\right) = \left(\bar{\mathbf{e}}^{j}\mathbf{e}_{i}\right)_{+}^{-}\left(\mathrm{X}v|L^{i*}v\right) \\ &= \pi_{i}^{j}\bar{e}_{\circ}^{i}\left(\mathrm{X}v|\mathrm{L}_{+}^{\circ\flat}v\right) = \left(\mathrm{X}v|L^{j*}v\right)\end{aligned}$$

since $\pi_{j}^{i} = \left(\bar{\mathbf{e}}^{i}\mathbf{e}_{j}\right)_{+}^{-} = \bar{e}_{\circ}^{i}e_{j}^{\circ}$ is either δ_{j}^{i} or a an idempotent kernel such that $\pi_{j}^{i}\bar{e}_{\circ}^{j} = \bar{e}_{\circ}^{i}$ in the case of overcompleteness of the basis $\left\{\bar{e}_{\circ}^{i}\right\}$ in $\mathfrak{k}_{\circ}^{*}$. In the similar way we obtain $L^{j^{*}} = \left(\mathbf{L}'\bar{\mathbf{e}}^{j}\right)_{+}^{-}$ and

$$\begin{aligned}\left\langle \mathrm{X}^{*}\bar{\mathbf{e}}^{j}, \mathbf{L}'\hat{\rho}\mathbf{L}'^{\ddagger}\right\rangle &= \left(\mathrm{X}v|\left(\mathbf{L}'\hat{\rho}\bar{\mathbf{e}}^{j}\mathbf{L}'^{\ddagger}\right)_{+}^{-}v\right) \\ &= \left(\mathbf{e}_{i}\bar{\mathbf{e}}^{j}\mathbf{e}_{k}\right)_{+}^{-}\left(\mathrm{X}v|L^{i^{*}}\hat{\rho}L^{k*}v\right) = \left(\mathrm{X}v|L^{i^{*}}\varepsilon_{i,k}^{j}\hat{\rho}L^{k*}v\right).\end{aligned}$$

since $\left(\mathbf{e}_{i}\bar{\mathbf{e}}^{j}\mathbf{e}_{k}\right)_{+}^{-} = \left(\bar{\mathbf{e}}^{j}\mathbf{e}_{i}\mathbf{e}_{k}\right)_{+}^{-} = \varepsilon_{i,k}^{m}\left(\bar{\mathbf{e}}^{j}\mathbf{e}_{m}\right)_{+}^{-} = \pi_{m}^{j}\varepsilon_{i,k}^{m} = \varepsilon_{i,k}^{m}$. Therefore,

$$\varkappa^{j}(\hat{\rho}) = L^{j^{*}}\hat{\rho} + \hat{\rho}L^{j*} + L^{i^{*}}\varepsilon_{i,k}^{j}L^{k*}$$

in (11) determined for $j \in J_{\bullet}$ by (12) for any $\mathrm{X} \in \mathbb{B}^{t]}$ and $\mathbf{a} \in \widetilde{\mathfrak{b}}$ spanned by $\bar{\mathbf{e}}^{J_{\bullet}}$. This formula also determines $\varkappa^{\emptyset}(\hat{\rho})$ with $\varepsilon_{i,k}^{\emptyset} = \sigma_{i.k}$ by

$$\left(\mathrm{X}_{t}v|\varkappa^{\emptyset}(\hat{\rho}_{t})v\right) = \left(\mathrm{X}_{t}v|\left(\mathbf{L}'\hat{\rho} + \hat{\rho}\mathbf{L}'^{\ddagger} + \mathbf{L}'\hat{\rho}\mathbf{L}'^{\ddagger}\right)_{+}^{-}v\right)$$

with $(\mathbf{L}'\hat{\rho})_{+}^{-} = L^{\emptyset}\hat{\rho}$, $\left(\hat{\rho}\mathbf{L}'^{\ddagger}\right)_{+}^{-} = \hat{\rho}L^{\emptyset*}$ given by $L^{\emptyset} = (\mathbf{L}')_{+}^{-} = \mathrm{L}_{+}^{-\prime}$. This completes the proof. □

Note that if $\mathbf{L}(t) \in \widetilde{\mathfrak{S}}^{t]}$ corresponding to the span $L^{\circ} = e_{i}^{\circ}L^{i}$ of $(\mathbf{L}')_{+}^{\circ}$ by the family $e_{J_{\bullet}}^{\circ}$ of Itô basis $\mathbf{e}_{J_{\bullet}}$ with $\mathbb{B}^{t]}$-adapted $L^{i}(t)$, the $\mathbb{B}^{t]}$-Markov QS evolution (13) is completely unraveled by the hemigroup of QS propagators $\hat{V}_{r}(t)$ satisfying the *QS unraveling equation*

$$\mathrm{d}\hat{V}_{r}(t) + \hat{V}_{r}(t)K(t)\,\mathrm{d}t = \hat{V}_{r}(t)L^{i}(t)\,\mathrm{d}\hat{e}_{i}, \quad \hat{V}_{r}(t) = \mathrm{I}.$$

This can be easily seen by applying the Itô formula to the solution of (13) for $t > r$ with $\hat{\rho}(r) = \hat{\rho}$ in the form $\vartheta_t(t, \hat{\rho}) = \hat{V}_r(t)\, \hat{\rho} \hat{V}_r(t)^*$. In particular, this is the case if $\mathbf{e}_{J_\bullet} = \mathbf{e}_{J_\circ}$ is complete basis in the maximal modular Itô algebra $\mathfrak{a}$ generating $\mathfrak{k}_\circ$ corresponding to $\mathbb{A}^{t]}$-Markovianity of the dual dynamics $\vartheta^\intercal = \phi$. Such unraveling is predetermined by hemigroup Markovianity on the W*-algebras $\overline{\mathbb{B}}^{t]}$. If the basis $\mathbf{e}_{J_\bullet}$ of $\mathfrak{b}$ is not complete in $\mathfrak{k}_\circ$, the QS dynamics ϑ may not completely unravel the quantum Markov dynamics $\vartheta^\intercal$. However it can be extended to an $\mathbb{A}^{t]}$-Markov unravelled dynamics ϕ. by a choice of the complementary basis $\mathbf{e}_{J_\circ \setminus J_\bullet}$. The extended unraveling evolution satisfies the QS equation of the same form as (9) with fluctuating coefficients $\varkappa^j$ indexed by the complete set $J_\circ = J$. Note that such raveling is not unique and can be chosen classical by taking a commutative Itô algebra $\mathfrak{c} \subset \mathfrak{b}\left(\mathfrak{k}_\bullet^\perp\right)$ which is always modular with the basis $\overline{\mathbf{e}}_j = \mathbf{e}_{j^\star}$ in the orthogonal complement $\mathfrak{k}_\bullet^\perp = \mathfrak{k} \ominus \mathfrak{k}_\bullet$ corresponding to the classical Lévy-Itô noise $\breve{e}_j = \hat{e}_{j^\star}$ for $j \in J \setminus J_\bullet$ independent of $\breve{e}_{J_\bullet} = \overline{\hat{e}_{J_\bullet}}$. The incompletely unraveled $\overline{\mathbb{B}}^{t]}$-Markov dynamics $\vartheta^\intercal$ is then represented as a conditional expectation of $\mathbb{A}^{t]}$-Markov dynamics ϕ over the noise indexed by $J_\circ \setminus J_\bullet$. which has obviously the same form as (9) $\phi_r(t, \mathrm{Z}) = \epsilon_r\left[\vartheta_r^\intercal(t, \mathrm{Z})\right]$.

If ϕ is unital, the density operators $\hat{\rho}(t)$ are normalized on $\overline{\mathbb{B}}^{t]}$ to the positive martingale $\hat{\pi}(t) = \langle \mathrm{I}, \hat{\rho}(t) \rangle$ describing the statistics of the complementary processes $\mathbf{y}_j(t) = \mathrm{V}_0(t)\, \hat{e}_\circ(t)\, \mathrm{V}_0(t)^*$, otherwise $\hat{\pi}(t)$ is supermartingale normalized to the probability $\pi(t) = \langle \hat{\pi}(t) \rangle_\emptyset$ of the object survival up to t.

4.2.4. *The general QS completely positive dynamics*

Belavkin has also generalized the notion of conditional complete positivity (CCP) for QS flows on an operator W*-algebra $\mathbb{B}$[12] in terms of the QS germ introduced in Ref.[14] He found the general form for conditionally bounded generators of such flows extending the Lindblad form to the general quantum stochastic master equations. In particular, if the QS generator is completely dissipative, it corresponds to the contractive CP dynamics on the operator algebra $\mathbb{B}$, the case which was also studied by Lindsay and Parthasarathy in Ref.[21] Belavkin also characterized[13,15] the nondissipative CCP QS generators on the generalized operator spaces given by densely defined sesquilinear forms commuting with an operator algebra $\mathbb{A}$. In particular, the preadjoint dynamics to normal operator contractive completely positive QS dynamics as a rule is not operator-contractive, but well defied as contractive in the dual to the operator norm. For example, the predual

QS dynamics described by the general Belavkin quantum stochastic equation is obviously CP and predually contractive on the state densities $\varrho \in \mathbb{B}_\star$ as the generalized operators commuting with $\overline{\mathbb{B}}$.

The general structure of completely bounded CCP sesquilinear germ-form for any QS contractive CP dynamics on $\mathbb{B}_\star$ can be obtained by duality from the structure of the CCP germ-form on $\mathbb{B}$ derived by Belavkin in Refs:[13,15]

$$\mathrm{d}\hat{\rho}_t = \left(\mathrm{C}^{\mu}_{\kappa}\hat{\rho}_t\mathrm{C}^{\ddagger\iota}_{\mu} - \hat{\rho}_t\delta^{\iota}_{\kappa}\right)\mathrm{dA}^{\kappa}_{\iota}, \quad \hat{\rho}_t = \hat{\rho}\otimes I_0$$

Here $\mathrm{C}^{\ddagger\iota}{}_{-\mu} = \mathrm{C}^{\mu *}_{-\iota}$ is an adapted, given by the QS germ $\mathbf{C}(t) = \mathbf{J}^t_{\mathrm{V}}$ as a triangular matrix $\mathbf{C} = [\mathrm{C}^{\mu}_{\kappa}]^{\mu=-,\circ,+}_{\kappa=-,\bullet,+}$ indexed by $\kappa = k \in J_\bullet, +$, $\mu = -, j \in J_\circ$ with $J_\circ \supseteq J_\bullet$ for also for $\nu = \circ, +$. The martingale *normalization condition* is $\mathrm{S}^{-}_{\mu}\mathrm{S}^{\ddagger\mu}_{+} = \mathrm{O}$ in terms of left transposed $\mathrm{S}^{\iota}{}_{-\mu} = \widetilde{\mathrm{C}}^{\mu}{}_{-\iota}$ with respect to standard pairing. It describes a QS object-output *entangling* process continuously in time. This is the most general QS equation preserving complete positivity and normalization in the martingale sense. Denoting $\mathrm{K}^{-}_{\iota} = -\mathrm{C}^{-}_{\iota} \equiv \mathrm{K}_{\iota}$ such that $\mathrm{C}^{\ddagger\iota}_{+} = -\mathrm{K}^{*}_{\iota}$, it can be written as

$$\mathrm{d}\hat{\rho}_t + 2\Re\left[\mathrm{K}_{\iota}\hat{\rho}_t\mathrm{dA}^{\iota}_{-}\right]\mathrm{d}t = \left(\sum_{j\in J_\circ}\mathrm{C}^{j}_{\kappa}\hat{\rho}_t\mathrm{C}^{j*}_{-\iota} - \hat{\rho}_t\delta^{\iota}_{\kappa}\right)\mathrm{dA}^{\kappa}_{\iota}.$$

The generator of the Belavkin master equation for the modular B*-algebra of the quantum Lévy-Itô noise $\hat{v}^t_\bullet$ can obviously be written in this form by representing the temperature quantum noise in the right hand side of the Belavkin modular master equation as $\varkappa^\bullet \mathrm{d}\hat{v}^t_\bullet = \varkappa^\bullet \mathbf{e}_\bullet \cdot \mathrm{d}\boldsymbol{B}$. In general the full quantum vacuum noise as the quadruple $\boldsymbol{B} = (\mathrm{A}^{\kappa}_{\iota})$ can induce non-temperature noise if the derived above by duality QS form-generator for the most general master equation has nonmodular form.

Acknowledgements This work was supported by British Council PMI2-RC156 grant for UK-Korea Research Co-operation and completed while the first author visited Korean Institute of Advanced Studies (KIAS) in April, 2010. The first author would like to thank to all members of the KIAS, specially to Professors Jaeseong Heo, Un Cig Ji and Jaewan Kim for their warm hospitality. J.H. was partially supported by a grant No. R01-2007-000-20064-0 from the Korea Science and Engineering Foundation (KOSEF). U.C.J. was partially supported by Mid-career Researcher Program through NRF grant funded by the MEST (No. R01-2008-000-10843-0).

References

1. V. P. Belavkin. A new form and $*$-algebraic structure of quantum stochastic integrals in fock space. In *Rendiconti del Seminario Matematico e Fisico di Milano LVIII*, pages 177–193, 1988.
2. V. P. Belavkin. Non-demolition measurements, nonlinear filtering and dynamic programming of quantum stochastic processes. In A.Blaquiere, editor, *Proc of Bellmann Continuum Workshop 'Modelling and Control of Systems', Sophia–Antipolis 1988*, volume 121 of *Lecture notes in Control and Inform Sciences*, pages 245 265, Berlin–Heidelberg–New York–London–Paris–Tokyo, 1988. Springer–Verlag.
3. V. P. Belavkin. Non-demolition stochastic calculus in fock space and nonlinear filtering and control in quantum systems. In *Proc of Fourteenth Winter School in Theor Phys, Karpacz 1988*, Stochastic Methods in Mathematics and Physics, pages 310–324, Singapore, 1989. World Scientific.
4. V. P. Belavkin. Stochastic calculus of quantum input-output processes and non-demolition filtering. In *Reviews on Newest Achievements in Science and Technology*, volume 36 of *Current Problems of Mathematics*, pages 29–67. VINITI, Moscow, 1989. Translation in: J. Soviet Math. 56 (1991) No 5, 2525-2647.
5. V. P. Belavkin. A posterior Schrödinger equation for continuous non-demolition measurement. *J of Math Phys*, 31(12):2930–2934, 1990.
6. V. P. Belavkin. Stochastic equations of quantum filtering. In B.Grigelionis et al., editor, *Prob. Theory and Math. Statistics*, volume 1, pages 91–109. "Mokslas", Vilnius, 1990.
7. V. P. Belavkin. Continuous non-demolition observation, quantum filtering and optimal estimation. In *Quantum Aspects of Optical Communication, Proceedings, Paris 1990*, volume 379 of *Lecture notes in Physics*, pages 151–163, Berlin, 1991. Springer.
8. V. P. Belavkin. A quantum nonadapted Ito formula and stochastic analysis in Fock scale. *J of Funct Analysis*, 102(2):414–447, 1991.
9. V. P. Belavkin. Chaotic states and stochastic integrations in quantum systems. *Usp. Mat. Nauk*, 47:47–106, 1992. Translation in: Russian Math. Surveys, No 1 pp. 53-116 (1992).
10. V. P. Belavkin. Quantum stochastic calculus and quantum nonlinear filtering. *Journal of Multivariate Analysis*, 42(2):171–201, 1992.
11. V. P. Belavkin. Quantum functional ito formula. In *Quantum Probability and Related Topics*, volume 8, pages 81–85. World Scientific, 1993.
12. V. P. Belavkin. On stochastic generators of completely positive cocycles. *Russ Journ of Math Phys*, 3(4):523–528, 1995.
13. V. P. Belavkin. On the general form of quantum stochastic evolution equation. In I. M. Davies at al, editor, *Stochastic Analysis and Applications*, pages 91–106. World Scientific, 1996.
14. V. P. Belavkin. Positive definite germs of quantum stochastic processes. *C. R. Acad. Sci. Paris*, 322(1):385–390, 1996.
15. V. P. Belavkin. Quantum stochastic positive evolutions: Characterization, construction, dilation. *Commun. Math. Phys.*, 184:533–566, 1997.

16. V. P. Belavkin. On quantum ito algebras and their decompositions. *Letters in Mathematical Physics*, 45:131–145, 1998.
17. V. P. Belavkin. Quantum ito b*-algebras, their classification and decomposition. In *Quantum Probability*, volume 43, pages 63–70. Banach Center Publications, 1998.
18. V. P. Belavkin. Infinite dimensional ito algebras of quantum white noise. In *Trends in Contemporary Infinite Dimensional Analysis and Quantum Probability*, pages 57–80. Instituto Italiano di Cultura, Kyoto, 2000.
19. R. L. Hudson and K. R. Parthasarathy. Quantum Itôs formula and stochastic evolutions. *Communications in Mathematical Physics*, 93(3):301–323, 1984.
20. K. Itô. On stochastic differential equations. *Mem. Amer. Math. Soc.*, 4:1–51, 1951.
21. J. M. Lindsay and K. R. Parthasarathy. On the generators of quantum stochastic flows. *Journal of Functional Analysis*, 158(2):521–549, 1998.

SPECTRAL ANALYSIS FOR TWISTED WAVEGUIDES

P. BRIET*

Centre de Physique Théorique
CNRS-Luminy, Case 907
13288 Marseille, France
E-mail: briet@cpt.univ-mrs.fr

In this note we consider a twisted quantum wave guide i.e. a domain of the form $\Omega_\theta := r_\theta\omega \times \mathbb{R}$ where ω is a connected open and bounded subset of $\mathbb{R}^2$ and $r_\theta = r_\theta(x_3)$ is a rotation by an angle $\theta(x_3)$ depending on the longitudinal variable x_3. We are interested in the spectral analysis of the Dirichlet Laplacian acting in Ω_θ. The results discussed here come from a joint paper between : *P.Briet, H. Kovarik, G. Raikov, E.Soccorsi, Comm. P.D.E., (2009).*

Keywords: Schrödinger operators, waveguide, spectral analysis.

1. Introduction.

The mathematical analysis of the guided propagation of electromagnetic waves have received considerably attention in the last decades. It is essentially motivated by important physical applications as e.g. in nano-sciences. One important question is the influence of the defects of the guide on the propagation, see e.g. the monograph.[1]

In this work we study this problem from a spectral point of view i.e. the study of bound states of the corresponding Derichlet Laplace operator inducing by the deformation of the surface of the guide.[2–4,8–10]

The case of locally bent guides was intensively studied see e.g.[2,3,5–7] Here we consider the pure twisting effect.[8,9] Let us precise the mathematical formulation of this problem.

Denote $\Omega = \omega \times \mathbb{R}$ where ω be an open bounded and connected domain in $\mathbb{R}^2$. Furthermore we suppose that ω contains the origin of $\mathbb{R}^2$. For $\mathbf{x} = (x_1, x_2, x_3) \in \Omega$ we write $\mathbf{x} = (x_t, x_3)$ where $x_t = (x_1, x_2)$. Let $\theta \in C^2(\mathbb{R})$ with bounded first and second derivatives and define the twisted domain

$$\Omega_\theta = \{r_\theta(x_3)(\mathbf{x}),\ \mathbf{x} \in \Omega\}$$

where r_θ is the following matrix-valued function

$$r_\theta(x_3) = \begin{pmatrix} \cos\theta(x_3) & \sin\theta(x_3) & 0 \\ -\sin\theta(x_3) & \cos\theta(x_3) & 0 \\ 0 & 0 & 1 \end{pmatrix}.$$

We consider the Dirichlet Laplace operator $-\Delta^D$ in $\mathrm{L}^2(\Omega_\theta)$ i.e. the selfadjoint operator in $\mathrm{L}^2(\Omega_\theta)$ defined in the quadratic form sense by

$$Q_\theta[\psi] = \int_{\Omega_\theta} |\nabla\psi(\mathbf{x})|^2 d\mathbf{x}, \quad \psi \in \mathrm{D}(Q_\theta) = \mathrm{H}_0^1(\Omega_\theta).$$

Here H_0^1 stands for the usual sobolev space. Of course if ω is an open disk in $\mathbb{R}^2$ then Ω is twisting invariant, in this last case we can see easily from arguments developed in the section below that the spectrum of $-\Delta^D$ satisfy

$$\sigma(-\Delta^D) = \sigma_{ac}(-\Delta^D) = [\mu_0, +\infty)$$

where μ_0 is the first eigenvalue of the operator $-\Delta_t$ in $\mathrm{L}^2(\omega)$.

In a more general situation it is convenient to use the transform

$$(\mathcal{U}\psi)(\mathbf{x}) = \psi\left(r_\theta^{-1}(x_3)(\mathbf{x})\right), \quad \psi \in \mathrm{L}^2(\Omega_\theta).$$

It is easy to see that $\mathcal{U}$ is a unitary operator from $\mathrm{L}^2(\Omega_\theta)$ onto $\mathrm{L}^2(\Omega)$. Note also that $\mathcal{U}(\mathrm{H}_0^1(\Omega_\theta)) = \mathrm{H}_0^1(\Omega)$.

In $\mathrm{L}^2(\Omega)$ introduce the self-adjoint operator

$$H_\theta := \mathcal{U}(-\Delta^D)\mathcal{U}^{-1} = -\Delta_t - (\dot\theta(x_3)\partial_\tau + \partial_3)^2, \tag{1}$$

where $\Delta_t = \partial_1^2 + \partial_2^2$ is the transverse Laplacian, $\partial_\tau := x_1\partial_2 - x_2\partial_1$ and $\dot\theta$ denotes the derivative of θ. The operator H_θ is associated with the closed quadratic form on $\mathrm{H}_0^1(\Omega)$

$$q_\theta[\psi] = \int_\Omega \left(|\nabla_t\psi|^2 + |(\dot\theta(x_3)\partial_\tau + \partial_3)\psi|^2\right) d\mathbf{x}. \tag{2}$$

Here we use the shorthand $\nabla_t = (\partial_1, \partial_2)$.

In this note we consider the following model already introduced in.[8] Let $\dot\theta = \beta - \varepsilon(x_3)$ where $\beta \in \mathbb{R}$, the function $\varepsilon \in C^1(\mathbb{R})$ satisfies:

$$\lim_{|x_3|\to\infty} \varepsilon(x_3) = 0. \tag{3}$$

We will see below that if $\varepsilon = 0$ then the spectrum of the corresponding operator is purely a.c.. Our goal is to study the spectral effects as e.g. occurrence of discret eigenvalues after the perturbation ε is turned on.

2. The constant twisting.

In this section we consider the simple case corresponding to $\varepsilon = 0$ and $\beta \in \mathbb{R}$. So that from (1)

$$H_\beta := H_{\dot\theta} = -\Delta_t - (\beta\partial_\tau + \partial_3)^2, \tag{4}$$

Due to the translational invariance in the x_3-direction, the operator H_β is unitarily equivalent to

$$\hat{H}_\beta := \int_{\mathbb{R}}^{\oplus} h_\beta(p)dp, \quad h_\beta(p) := -\Delta_t - (\beta\partial_\tau + ip)^2, \quad p \in \mathbb{R}, \tag{5}$$

Let $\beta \in \mathbb{R}$. $h_\beta(p), p \in \mathbb{R}$ is a family of selfadjoint operators associated to the quadratic form

$$q_p[\varphi] := \int_\omega (|\nabla_t \varphi(x_t)|^2 + |(i\beta\partial_\tau - p)\,\varphi(x_t)|^2)dx_t\,, \quad \varphi \in \mathrm{H}_0^1(\omega)$$

and extends to a type B entire family of operators.[11,12]

Further, For all $p \in \mathbb{C}$ the operator $h_\beta(p)$ has a compact resolvent and then only discrete spectrum. For $p \in \mathbb{R}$ let $\{E_j(p)\}_{j=1}^{\infty}$ be the non-decreasing sequence of the eigenvalues of $h_\beta(p)$ and $\{\psi_j(x_t;p)\}_{j=1}^{\infty}$ the associated eigenfunctions.

We summarize below the main properties of *the band functions* $\{E_j(p), p \in \mathbb{R}\}_{j=1}^{\infty}$.[9,11,12]

Proposition 2.1. *The band function are continuous piece-wise analytic satisfying*

$$E_j(p) = p^2(1 + o(1)), \quad p \to \pm\infty. \tag{6}$$

Therefore, the general theory of analytically fibred operators (see e.g.[12]) implies that the spectrum of the operator H_β is purely a.c. :

$$\sigma(H_\beta) = \sigma_{\mathrm{ac}}(H_\beta) = \cup_{j\in\mathbb{N}} E_j(\mathbb{R}) = [\mathcal{E}, \infty), \tag{7}$$

with $\mathcal{E} = \mathcal{E}(\beta) := \min_{p\in\mathbb{R}} E_1(p,\beta)$.

Notice that if $\beta = 0$ the band functions are explicitly given as

$$E_j(p) = \mu_j + p^2, \quad j \in \mathbb{N},$$

where μ_j are the eigenvalues of the operator $-\Delta_t$ in $L^2(\omega)$. Evidently this implies that $\mathcal{E} = \mu_0$. For $\beta \in \mathbb{R}$ we have[9]

Proposition 2.2. *Let $\beta \in \mathbb{R}$. Then*

$$E_1(0,\beta) + (1 - \epsilon_\omega(\beta))\,p^2 \le E_1(p,\beta) \le E_1(0,\beta) + p^2, \quad p \in \mathbb{R} \tag{8}$$

where $\epsilon_\omega(\beta) := \frac{\beta^2 C_\omega}{1+\beta^2 C_\omega}$ *and* $C_\omega := \sup_{x_t\in\omega}(x_1^2 + x_2^2)$.

This gives $\mathcal{E} = E_1(0,\beta)$. In the other hand the proposition implies the existence of an effective mass, $m_e := (2E_1''(0,\beta))^{-1}$. It satisfies

$$0 < (1 - \epsilon_\omega(\beta)) \leq m_e \leq 1.$$

3. The essential spectrum.

We suppose now that $\dot{\theta} = \beta - \varepsilon(x_3)$, $\beta \in \mathbb{R}$, $\varepsilon \in C^1(\mathbb{R})$ and $\lim_{|x_3|\to\infty} \varepsilon = 0$.

In this case the operator $H_{\dot{\theta}}$ can be formally written as a perturbation of the operator H_β

$$H_{\dot{\theta}} = H_\beta + W_{\varepsilon,\beta}. \tag{9}$$

with

$$W_{\varepsilon,\beta} = 2\beta\varepsilon\partial_\tau^2 + \partial_\tau\,\varepsilon\partial_3 + \partial_3\,\varepsilon\partial_\tau - \varepsilon^2\partial_\tau^2 \tag{10}$$

We have[9]

Theorem 3.1. *Under the conditions stated above then*

$$\sigma_{ess}(H_{\dot{\theta}}) = \sigma_{ess}(H_\beta) = [\mathcal{E}, \infty), \quad \mathcal{E} = E_1(0,\beta). \tag{11}$$

Choose $\beta = 0$ then this proposition implies $\sigma_{ess}(H_{\dot{\theta}}) = [\mu_1, \infty)$. But since since in the form sense we have $H_{\dot{\theta}} \geq \Delta_t \otimes I_3 \geq \mu_1$ where I_3 is the identity operator in $L^2(\mathbb{R})$ then [a]

$$\sigma(H_{\dot{\theta}}) = \sigma_{ess}(H_{\dot{\theta}}) = [\mu_1, \infty). \tag{12}$$

In[14] it is also shown that the twisting effect prevent against eventual eigenvalues generated by small local bending of the tubes. In this sense these results are closely related to the one about the influence of a magnetic field on the spectrum in certain situation of waveguide.[15,16]

In the next section we give results about discret spectrum when $\beta \neq 0$ i.e. the twisting does not decay at the infinity. We wll show that this set is not empty even infinite under some conditions extending results of.[8]

Remark 3.1. In general the nature of the essential spectrum of the operator $H_{\dot{\theta}}$ is actually an open question, see e.g.[13] for a discussion in the case of bent tubes.

---

[a] The author thanks G.D. Raikov who pointed out this simple argument proving the absence of discrete eigenvalues in this case.

4. The discrete spectrum.

We now suppose that $\dot{\theta} = \beta - \varepsilon(x_3)$ with $\beta > 0$, $\varepsilon \in C^1$ is a positive function satisfying there exist $\alpha > 0$, $C > 0$ and $L > 0$ s.t.

$$\lim_{|x|\to\infty} |x|^\alpha \varepsilon(x) = L, \quad \text{and} \quad |\dot{\varepsilon}(x)| \leq C(1+|x|)^{-\alpha-1}, \quad x \in \mathbb{R}.$$

Let $N(H_{\dot{\theta}}; E)$, $E \in (-\infty, \mathcal{E})$, be the number of eigenvalues smaller than E of the selfadjoint operator $H_{\dot{\theta}}$ counting with their multiplicity.

We would like to study the asymptotic of $N(H_{\dot{\theta}}; \mathcal{E} - \lambda), \lambda > 0$ when $\lambda \downarrow 0$. We have

Theorem 4.1. *Under the conditions stated above. Suppose $\alpha \in (0, 2)$ then*

$$\lim_{\lambda\downarrow 0} \lambda^{\frac{1}{\alpha}-\frac{1}{2}} N(H_{\dot{\theta}}; \mathcal{E}-\lambda) = \frac{2\sqrt{m_e}}{\pi\alpha} \left(2\beta\, L\, \|\partial_\varphi \psi_1(\cdot, 0)\|^2_{\mathrm{L}^2(\omega)}\right)^{\frac{1}{\alpha}} B\left(\frac{3}{2}, \frac{1}{\alpha} - \frac{1}{2}\right)$$

where B is the usual Euler beta function.

Of course if $\|\partial_\varphi \psi_1(\cdot, 0)\|^2_{\mathrm{L}^2(\omega)} \neq 0$ this means that there exists an infinite number of eigenvalues below the bottom of essential spectrum and the Theorem above give the precise asymptotic of their distribution. Hence this result is related with the following[9]

Proposition 4.1. *Suppose that $\omega \subset \mathbb{R}^2$ is an open bounded subset of $\mathbb{R}^2$ with C^2-boundary containing the origin. Then ω is a disk centred at the origin iff $\partial_\varphi \psi_1(\cdot; 0) = 0$.*

Indeed recall that if ω is an open disk in $\mathbb{R}^2$ then $H_{\dot{\theta}}$ has only purely absolutely continuous spectrum.

Theorem 4.2. *In the same conditions as in the last theorem but with $\alpha = 2$. Then if $2\beta\, L\, \|\partial_\varphi \psi_1(\cdot, 0)\|^2_{\mathrm{L}^2(\omega)} > \frac{1}{4m_e}$,*

$$\lim_{\lambda\downarrow 0} |\ln \lambda|^{-1} N(H_{\dot{\theta}}; \mathcal{E} - \lambda) = \frac{1}{\pi} \left(2\beta\, L m_e\, \|\partial_\varphi \psi_1(\cdot, 0)\|^2_{\mathrm{L}^2(\omega)} - \frac{1}{4}\right)_+^{1/2}.$$

If $2\beta\, L\, \|\partial_\varphi \psi_1(\cdot, 0)\|^2_{\mathrm{L}^2(\omega)} \leq \frac{1}{4m_e}$, then $N(H_{\dot{\theta}}; \mathcal{E} - \lambda) = O(1), \quad \lambda \downarrow 0$.

If ω is a disk in $\mathbb{R}^2$, then $m_e = 1/2$ and $\partial_\varphi \psi_1 = 0$ corresponding to the second case.

Theorem 4.3. *Suppose that $\dot{\theta} = \beta - \varepsilon(x)$ with $\beta > 0$, $\varepsilon \in C^1$ is a positive function satisfying there exist $\alpha > 2$ and $C > 0$ s.t.*

$$0 \leq \varepsilon(x) \leq C(1+|x|)^\alpha \quad \text{and} \quad |\dot{\varepsilon}(x)| \leq C(1+|x|)^{-1-\alpha}, \quad x \in \mathbb{R},$$

then $N(H_{\dot{\theta}}; \mathcal{E}_{ess} - \lambda) = O(1), \quad \lambda \downarrow 0$.

This last theorem is linked to the results of[8] where it is proving that if ω is not the disk in $\mathbb{R}^2$ and ε is a smooth function with compact support $-\Delta_D$ has at least one discret eigenvalue.

References

1. N.E. Hurt, *Mathematical Physics of Quantum Wires and Devices* (Mathematics and its Application **506** Kluer Academic, Dordrecht, (2000).
2. P.Exner, P.Seba, *Jour. Math.Phys.* **30**, 2574–2580 (1989).
3. P.Duclos, P.Exner, *Rev. Math.Phys.* **7**, 73–102 (1995).
4. P.Exner, *J. Math. Phys.* **34** , no. 1, 23–28 (1993).
5. P.Exner, S. Vugalter *Ann. Inst. Henri Poincaré,* **65** , no. 1, 109–123(1996).
6. J.Goldstone, R.L Jaffe, *Phys.Rev. B* **45**, 14100–14107 (1992).
7. P.Duclos, P.Exner, D. Krejcirik, *Comm. Math. Phys*, **223**, 1,13–28 (2001).
8. P.Exner, H. Kovarik, *Letter Math. Phys.*, **73**, 183–192 (2005).
9. P.Briet, H. Kovarik, G. Raikov, E.Soccorsi, *Comm. P.D.E.*, **34**, 7-9,818–836 (2009).
10. G. Bouchitté, M.L. Mascarenhas, L.Trabucho, *Control, Optim. Calc. Var.*, **13** 4, 793–808 (2007),
11. T.Kato, *Perturbation Theory for Linear Operators* (Springer-Verlag New York, Inc., New York 1966).
12. M. Reed and B. Simon, *Methods of Modern Mathematical Physics IV: Analysis of operators* (Academic Press, 1978).
13. D. Krejcirik,, R.Tiedra de Aldecoa, *The nature of the essential spectrum in curved quantum waveguides.* J. Phys. A 37 **20**, 5449–5466 (2004),
14. T.Elkholm, H. Kovarik, D. Krejcirik, *Arch. Ration. Mech. Anal.* **188**, no. 2,245–264. (2008).
15. T.Elkholm, H. Kovarik, *Comm P.D.E.* **30**, no. 4-6, 539–565. (2005).
16. P.Briet, G. D.Raikov, E. Soccorsi, *Asymptot. Anal.* **58** no. 3, 127–155 (2008),.

THE DECOHERENCE-FREE SUBALGEBRA OF A QUANTUM MARKOV SEMIGROUP ON $\mathcal{B}(\mathsf{h})$

A. DHAHRI

Università di Roma II, Centro Vito Volterra,
Via Columbia 2, 00133 Roma
E-mail: adhahri@uc.cl

F. FAGNOLA

Politecnico di Milano, Dipartimento di Matematica
Piazza Leonardo da Vinci, 20133 Milano, Italy
E-mail: franco.fagnola@polimi.it
www.mate.polimi.it/qp

R. REBOLLEDO*

Laboratorio de Análisis Estocástico,
Facultad de Matemáticas,
Pontificia Universidad Católica de Chile,
Casilla 306, Santiago 22, Chile
** E-mail: rrebolle@uc.cl*
www.anestoc.cl

Let $\mathcal{T}$ be a quantum Markov semigroup on $\mathcal{B}(\mathsf{h})$ with a faithful normal invariant state ρ whose generator is represented in a generalised GKSL form $\mathcal{L}(x) = -\frac{1}{2}\sum_\ell (L_\ell^* L_\ell x - 2L_\ell^* x L_\ell + x L_\ell^* L_\ell) + i[H, x]$, with possibly unbounded H, L_ℓ. We show that the biggest von Neumann-subalgebra $\mathcal{N}(\mathcal{T})$ of $\mathcal{B}(\mathsf{h})$ where $\mathcal{T}$ acts as a semigroup of automorphisms coincides with the generalised commutator of $\{\mathrm{e}^{-itH} L_\ell \mathrm{e}^{itH}, \mathrm{e}^{-itH} L_\ell^* \mathrm{e}^{itH} \mid \ell \geq 1, t \geq 0\}$ under some natural regularity conditions.

The proof we present here does not involve dilations $\mathcal{T}$.

Keywords: Quantum Markov semigroups, convergence to a steady state, generalised Lindblad form, multiple commutators.

1. Introduction

The evolution of an Open Quantum System is described by a Quantum Markov Semigroups (QMS), namely a weakly* continuous semigroup $\mathcal{T} = (\mathcal{T}_t)_{t\geq 0}$ of normal, completely positive, identity preserving, maps $\mathcal{T}_t$ on the algebra $\mathcal{B}(\mathsf{h})$ of all bounded operators on a complex separable Hilbert space h. On the other hand, the evolution of a closed quantum system, not interacting with the environment, can be described by a one-parameter group of automorphisms $(\alpha_t)_{t\geq 0}$, with $\alpha_t(x) = \mathrm{e}^{itH}x\mathrm{e}^{-itH}$ and H self-adjoint.

Inside an Open Quantum System, sometimes, one can find a subsystem evolving like a closed quantum system where the typical effects of the interaction with the environment do not appear and the typical quantum features of the system, like quantum coherence and entanglement of quantum states are preserved.

The *decoherence-free* subalgebra $\mathcal{N}(\mathcal{T})$ (see Prop. 2.1 (3)) of $\mathcal{T}$ is then defined as the (biggest) sub*-algebra of $x \in \mathcal{B}(\mathsf{h})$ such that

$$\mathcal{T}_t(x^*x) = \mathcal{T}_t(x^*)\mathcal{T}_t(x), \quad \mathcal{T}_t(xx^*) = \mathcal{T}_t(x)\mathcal{T}_t(x^*), \tag{1}$$

for all $t \geq 0$.

Decoherence-free subalgebras play a fundamental role in the approach of Blanchard and Olkiewicz[1] (and the references therein) to the study of decoherence (see also e.g. Lidar, Chuang and Whaley[2]). Another approach, proposed by Rebolledo,[3,4] is based on reductions of QMS to Abelian subalgebras, i.e. classical Markov sub-semigroups obtained by restrictions of a QMS to an invariant Abelian subalgebra.[5]

The decoherence-free subalgebra $\mathcal{N}(\mathcal{T})$ of a norm-continuous QMS can be characterised in a simple and natural way in terms of the operators H, L_ℓ in a Gorini-Kossakowski-Sudarshan-Lindblad (GKSL) representation of the generator $\mathcal{L}$ (see Fagnola and Rebolledo[6]). Indeed, writing the generator in the GKSL form

$$\mathcal{L}(x) = i[H,x] - \frac{1}{2}\sum_{\ell\geq 1}(L_\ell^* L_\ell x - 2L_\ell^* x L_\ell + x L_\ell^* L_\ell) \tag{2}$$

with $H, L_\ell \in \mathcal{B}(\mathsf{h})$, $H = H^*$ and the series $\sum_{\ell\geq 1} L_\ell^* L_\ell$ strongly convergent, $\mathcal{N}(\mathcal{T})$ turns out to be the commutator of the set of operators

$$\{\, L_\ell,\, L_\ell^*,\, [H,L_\ell],\, [H,L_\ell^*],\, [H,[H,L_\ell]],\, [H,[H,L_\ell]],\cdots \mid \ell\geq 1\}. \tag{3}$$

The idea of the proof is sketched in Section 2.

This result was extended by the authors in Ref.[7] to QMS on $\mathcal{B}(\mathsf{h})$ with unbounded generator represented in a generalised GKSL form. In this case $\mathcal{N}(\mathcal{T})$ was characterised as the generalised commutator of the set

$$\mathcal{D}(\mathcal{T}) := \left\{\mathrm{e}^{-itH} L_\ell\, \mathrm{e}^{itH}\,,\, \mathrm{e}^{-itH} L_\ell^*\, \mathrm{e}^{itH} \;\mid\, \ell\geq 1, t\geq 0\right\} \tag{4}$$

under some technical conditions. More precisely, it was shown that $\mathcal{N}(\mathcal{T})$ consists of bounded operators $x \in \mathcal{B}(\mathsf{h})$ such that, for all $Y \in \mathcal{D}(\mathcal{T})$, possibly unbounded, the operator Yx is an extension of the operator xY defined on the domain of Y, i.e. $xY \subseteq Yx$.

Considering operators like $\mathrm{e}^{-itH} L_\ell \mathrm{e}^{itH}$ allows us to minimize the strong domain conditions that are necessary in order to make sense of multiple commutators.

The characterisation of $\mathcal{N}(\mathcal{T})$ is useful also in the study of convergence of $\mathcal{T}$ to a normal invariant state ρ. In fact, by the Frigerio and Verri[8] Th. 3.3 p. 281, if $\mathcal{T}$ has also a normal invariant state ρ which is faithful (if not the QMS can be reduced by the support projection of ρ, see e.g. Refs.[9,10]), then for any initial state ω, $\mathrm{tr}(\omega\mathcal{T}_t(x))$ converges to $\mathrm{tr}(\omega_\infty x)$ as t goes to infinity, i.e. the initial state ω converges to ω_∞, whenever $\mathcal{N}(\mathcal{T})$ coincides with the fixed point algebra $\mathcal{F}(\mathcal{T}) = \{x \in \mathcal{B}(\mathsf{h}) \mid \mathcal{T}_t(x), \forall t \geq 0\}$. Since $\mathcal{F}(\mathcal{T})$ coincides with the generalised commutator of H, L_ℓ, L_ℓ^* $(\ell \geq 1)$ we obtain as a corollary simple algebraic conditions for convergence towards an invariant state (see also Ref.[11]).

The paper is organised as follows. In Section 2 we introduce the decoherence-free subalgebra $\mathcal{N}(\mathcal{T})$, discuss its main properties and sketch its characterisation for norm-continuous QMS. Moreover, we describe weakly*-continuous QMS with generalised GKSL generator that we shall consider.

In Section 3 prove by a new method, avoiding dilations of $\mathcal{T}$ as in Ref.[7], that $\mathcal{N}(\mathcal{T})$ coincides with the generalised commutator of $\mathcal{D}(\mathcal{T})$.

Finally, in Section 4, we give an example of a QMS with on $\mathcal{B}(\mathsf{h})$ with h infinite dimensional, having a non-trivial decoherence-free subalgebra. The operators H, L_ℓ in the generalised GKSL representation of its generator are essentially components of the momentum operator; as a result $\mathcal{N}(\mathcal{T})$ consists of operators invariant under the action of the rotation group.

2. The decoherence-free subalgebra

Let h be a complex separable Hilbert space and let $(\mathcal{T}_t)_{t\geq 0}$ be a weak*-continuous semigroup of completely positive maps on $\mathcal{B}(\mathsf{h})$, the von Neumann algebra of all bounded operators on h. When the maps $\mathcal{T}_t$ are unital, i.e. $\mathcal{T}_t(\mathbb{1}) = \mathbb{1}$ we call $(\mathcal{T}_t)_{t\geq 0}$ a quantum Markov semigroup (QMS).

We recall the following well-known properties of (see e.g. Evans[12] Th. 3.1 or Dhahri, Rebolledo and Fagnola[7] Prop. 2.1).

Proposition 2.1. *Let $\mathcal{T}$ be a quantum Markov semigroup on $\mathcal{B}(\mathsf{h})$ and let*

$\mathcal{N}(\mathcal{T})$ be the set defined by (1). Then

(1) $\mathcal{N}(\mathcal{T})$ is $\mathcal{T}_t$-invariant for all $t \geq 0$,
(2) for all $x \in \mathcal{N}(\mathcal{T})$ and all $y \in \mathcal{B}(\mathsf{h})$ we have $\mathcal{T}_t(x^*y) = \mathcal{T}_t(x^*)\mathcal{T}_t(y)$ and $\mathcal{T}_t(y^*x) = \mathcal{T}_t(y^*)\mathcal{T}_t(x)$,
(3) $\mathcal{N}(\mathcal{T})$ is a von Neumann subalgebra of $\mathcal{B}(\mathsf{h})$.

In Ref.[6] we showed that, when the QMS is norm-continuous, $\mathcal{N}(\mathcal{T})$ is the multiple commutator of the set (3). We sketch the proof here

If x belongs to $\mathcal{N}(\mathcal{T})$, then, differentiating $\mathcal{T}_t(x^*x) = \mathcal{T}_t(x^*)\mathcal{T}_t(x)$ at $t = 0$, we find $\mathcal{L}(x^*x) = x^*\mathcal{L}(x) + \mathcal{L}(x^*)x$. Therefore, since $\mathcal{L}(x^*x) = x^*\mathcal{L}(x) + \mathcal{L}(x^*)x + \sum_\ell [L_\ell, x]^*[L_\ell, x]$ it follows that $[L_\ell, x] = 0$ for all ℓ and, taking the adjoint $[L_\ell^*, x] = 0$ because also x^* belongs to $\mathcal{N}(\mathcal{T})$.

Thus $\mathcal{L}(x) = i[H, x]$ and $\mathcal{L}(x) \in \mathcal{N}(\mathcal{T})$ by Proposition 2.1 (1), therefore, by the Jacobi identity,

$$0 = [L_\ell, [H, x]] = -[H, [x, L_\ell]] - [x, [H, L_\ell]] = [[H, L_\ell], x]$$

and x commutes also with $[H, L_\ell]$. The same argument for x^* shows that x commutes with $[H, L_\ell^*]$. Iterating we show that x commutes with multiple commutators in (3).

Conversely, if x belongs to the commutator of (3), $\mathcal{L}(x) = i[H, x]$ and

$$\begin{aligned}
\mathcal{L}^2(x) &= i\mathcal{L}([H, x]) \\
&= i[H, [H, x]] - \frac{1}{2}\sum_{\ell \geq 1} \left(L_\ell^*[[H, x], L_\ell] + [L_\ell^*, [H, x]]L_\ell\right) \\
&= i[H, [H, x]] - \frac{1}{2}\sum_{\ell \geq 1} \left(L_\ell^*[[H, x], L_\ell] + [L_\ell^*, [H, x]]L_\ell\right) \\
&= i[H, [H, x]]
\end{aligned}$$

because $[[H, x], L_\ell] = -[[x, L_\ell], H] - [[L_\ell, H], x] = 0$ and $[L_\ell^*, [H, x]] = -[H, [x, L_\ell^*]] - [x, [L_\ell^*, H]] = 0$.

Defining $\delta_H(x) = [H, x]$ and iterating we find $\mathcal{L}^n(x) = i^n \delta_H^n(x)$ and then $\mathcal{T}_t(x) = \mathrm{e}^{itH} x \mathrm{e}^{-itH}$ and x belongs to $\mathcal{N}(\mathcal{T})$. We find then the following

Theorem 2.1. *The decoherence-free subalgebra $\mathcal{N}(\mathcal{T})$ of a norm-continuous QMS $\mathcal{T}$ whose generator $\mathcal{L}$ is represented in the GKSL form (2) coincides with the commutator of the set of operators (3). Moreover $\mathcal{T}_t(x) = \mathrm{e}^{itH} x \mathrm{e}^{-itH}$ for all $x \in \mathcal{N}(\mathcal{T})$.*

Clearly, when the operators H, L_ℓ are bounded, the commutator of the sets (3) and (4) coincide because, on one hand,

$$(-i)^n \delta_H^n(L_\ell) = \frac{d^n}{dt^n} \mathrm{e}^{-itH} L_\ell \, \mathrm{e}^{itH} \Big|_{t=0}$$

and, on the other hand

$$\mathrm{e}^{-itH} L_\ell \, \mathrm{e}^{itH} = \sum_{n \geq 0} \frac{(-t)^n}{n!} \, \delta_H^n(L_\ell).$$

In this paper we are concerned with quantum Markov semigroups on $\mathcal{B}(\mathsf{h})$ with a formal generator representable in a generalised GKSL form by means of operators G, L_ℓ $(\ell \geq 1)$ on h with the following property:

(H-1) the operator G is the generator of a strongly continuous semigroup on h, the domain $\mathrm{Dom}(L_\ell)$ of each operator L_ℓ is contained in $\mathrm{Dom}(G)$ and

$$\langle Gv, u \rangle + \sum_{\ell \geq 1} \langle L_\ell v, L_\ell u \rangle + \langle v, Gu \rangle = 0 \tag{5}$$

for all $u, v \in \mathrm{Dom}(G)$. The operators $\mathbb{1}, L_1, L_2, \ldots$ are linearly independent.

Linear independence of the set of operators $\mathbb{1}, L_1, L_2, \ldots$ is a natural condition ensuring that the completely positive part of the generator is represented with its minimal multiplicity (see Parthasarathy[13] Sect. 30).

For each $x \in \mathcal{B}(\mathsf{h})$ we can consider the quadratic form $\pounds(x)$ with domain $\mathrm{Dom}(G) \times \mathrm{Dom}(G)$ defined by

$$\pounds(x)[v, u] = \langle Gv, xu \rangle + \sum_{\ell \geq 1} \langle L_\ell v, x L_\ell u \rangle + \langle v, xGu \rangle.$$

The hypothesis **(H-1)** allows us to construct the minimal semigroup on $\mathcal{B}(\mathsf{h})$ associated with the operators G, L_ℓ (see e.g. Refs.[14,15]). This is the weak*-continuous semigroup $(\mathcal{T}_t)_{t \geq 0}$ of completely positive maps on $\mathcal{B}(\mathsf{h})$ satisfying

$$\langle v, \mathcal{T}_t(x) u \rangle = \langle v, xu \rangle + \int_0^t \pounds \left(\mathcal{T}_s(x) \right) [v, u] ds. \tag{6}$$

It is well-known that, in spite of (5), i.e. $\pounds(\mathbb{1}) = 0$, the minimal semigroup may not be unital i.e. $\mathcal{T}_t(\mathbb{1}) < \mathbb{1}$. In this case, it is not the unique weak*-continuous semigroup of completely positive maps on $\mathcal{B}(\mathsf{h})$ satisfying (6).

The following assumptions implying that $\mathcal{T}_t(\mathbb{1}) = \mathbb{1}$ by Th. 4.4 p. 394 of Chebotarev and Fagnola[14] turn out to be useful.

(H-Markov) There exists positive self-adjoint operators $C \geq \mathbb{1}$ and Φ satisfying

(1) $\mathrm{Dom}(G) \subseteq \mathrm{Dom}(\Phi^{1/2})$, $\mathrm{Dom}(C) \subseteq \mathrm{Dom}(\Phi)$ and

$$-2\Re e \langle u, Gu \rangle = \sum_{\ell \geq 1} \langle L_\ell u, L_\ell u \rangle = \left\langle \Phi^{1/2} u, \Phi^{1/2} u \right\rangle$$

for all $u \in \mathrm{Dom}(G)$, moreover, $\|\Phi^{1/2} v\| \leq \|C^{1/2} v\|$ for all $v \in \mathrm{Dom}(C)$,

(2) $\mathrm{Dom}(G) \subseteq \mathrm{Dom}(C^{1/2})$ and $\mathrm{Dom}(G)$ is a core for $C^{1/2}$,

(3) the linear manifolds $L_\ell(\mathrm{Dom}(G^2))$ ($\ell \geq 1$) are contained in the domain of $C^{1/2}$,

(4) there exists a positive constant b such that

$$2\Re e \langle Cu, Gu \rangle + \sum_{\ell \geq 1} \left\langle C^{1/2} L_\ell u, C^{1/2} L_\ell u \right\rangle \leq b \langle u, Cu \rangle$$

for all $u \in \mathrm{Dom}(G^2)$.

In order to make sense of the operators $\mathrm{e}^{-itH} L_\ell \, \mathrm{e}^{itH}$ appearing in the definition of $\mathcal{D}(\mathcal{T})$, throughout the paper we shall also assume that

(**H-2**) there exists a linear manifold D dense in h, contained in the domains of the operators G, G^* and L_ℓ, L_ℓ^* for all $\ell \geq 1$, which is a core for G and

(a) L_ℓ is closed and D is a core for L_ℓ for all $\ell \geq 1$,

(b) the operator H defined on $u \in D$ by $Hu = i(Gu - G^*u)/2$ is essentially self-adjoint and the unitary group $(\mathrm{e}^{itH})_{t \in \mathbb{R}}$ generated by iH satisfies $\mathrm{e}^{itH}(D) \subseteq \mathrm{Dom}(G)$ for all $t \in \mathbb{R}$,

(c) the operator G_0 defined on $u \in D$ by $G_0 u = (Gu + G^*u)/2$ is essentially self-adjoint and $\mathrm{Dom}(G) \subseteq \mathrm{Dom}(G_0) \subseteq \mathrm{Dom}(L_\ell)$ for all $\ell \geq 1$.

As a consequence operators $\mathrm{e}^{-itH} L_\ell \, \mathrm{e}^{itH}$ are defined on the domain D.

3. Characterisation of $\mathcal{N}(\mathcal{T})$

We shall give here a new proof with an argument avoiding dilations.

The algebraic arguments leading to the characterisation of $\mathcal{N}(\mathcal{T})$ for norm-continuous QMS can not be extended in straightforward way because the domain of $\mathcal{L}$ may not be an algebra (see Fagnola[16]), thus $\mathcal{L}(x^*y)$ may not be defined even if both x and y belong to the domain of $\mathcal{L}$. Moreover, if the minimal QMS is Markov, we know that the domain of $\mathcal{L}$ consists of those $y \in \mathcal{B}(\mathsf{h})$ such that the quadratic form $\pounds(y)$ defines a bounded operator by Lemma 1.1 p.563 Ref.[19] but we do not know whether $\mathcal{L}(y)$ can be written as the sum $G^*y + \sum_\ell L_\ell^* y L_\ell + yG$ because the three terms of the sum may not make sense separately. The hypothesis (**H-2**) ensures that, for $y = |v\rangle\langle u|$, G^*y, yG and each $L_\ell^* y L_\ell$ are well-defined but and series converges to a bounded operator if and only if $\sum_{\ell\geq 1}\|L_\ell^* u\|^2$ converges for $u\in D$.

It is worth noticing here that convergence of the above series, together with (**H-min**) and (**H-2**) can be interpreted as sort of weak Feller property. Indeed, it was shown in Ref.[17] Th. 4.2 that the C^*-algebra of compact operators on h is $\mathcal{T}_t$-invariant when, moreover, $2\Re e\langle u, Gu\rangle+\sum_{\ell\geq 1}\|L_\ell^* u\|^2\leq b\|u\|^2$ for all $u\in D$ for some constant b independent of u.

We start then proving a weak form of the identity $\sum_{\ell\geq 1}[L_\ell, x]^*[L_\ell, y] = 0$ under this additional assumption.

Proposition 3.1. *Suppose that* ***H-1, H-Markov, H-2*** *and*

$$\sum_{\ell\geq 1}\|L_\ell^* u\|^2<\infty$$

for all $u\in D$. For all $x\in\mathcal{N}(\mathcal{T})$, $u,v\in D$ and all $y=|u'\rangle\langle v'|\in Dom(\mathcal{L})$ with $u',v'\in D$ we have

$$\begin{aligned}\pounds(x^*y)[v,u] &= \langle v, x^*\mathcal{L}(y)u\rangle\\ &\quad+\langle Gv, x^*yu\rangle+\sum_{\ell\geq 1}\langle L_\ell v, x^*L_\ell yu\rangle+\langle v, x^*Gyu\rangle. \qquad (7)\end{aligned}$$

In particular

$$\sum_{\ell\geq 1}\langle xL_\ell v, [y, L_\ell]u\rangle=\sum_{\ell\geq 1}\langle xv, L_\ell^*[y, L_\ell]u\rangle. \qquad (8)$$

Proof. By Prop. 2.1 (2), for all $t>0$ we can write

$$\begin{aligned}\langle v, (\mathcal{T}_t(x^*y)-x^*y)\,u\rangle &= \langle v, (\mathcal{T}_t(x^*)\mathcal{T}_t(y)-x^*y)\,u\rangle\\ &=\langle v, x^*(\mathcal{T}_t(y)-y)u\rangle+\langle(\mathcal{T}_t(x)-x)\,v, yu\rangle\\ &\quad+\langle(\mathcal{T}_t(x)-x)\,v, (\mathcal{T}_t(y)-y)\,u\rangle\,.\end{aligned}$$

Dividing by t and letting t tend to 0 we find that the first term in the right-hand side tends to $\langle xv, \mathcal{L}(y)u\rangle$ because y belongs to the domain of $\mathcal{L}$, the second term converges to

$$\langle xGv, yu\rangle+\sum_{\ell\geq 1}\langle L_\ell v, x^*L_\ell yu\rangle+\langle xv, Gyu\rangle$$

because both v and yu belong to Dom(G). Finally, writing $\mathcal{T}_t(y)-y$ as $\int_0^t(\mathcal{T}_s(y)-y)ds$, the third term turns out to be dominated by

$$\begin{aligned}\frac{\|(\mathcal{T}_t(x)-x)\,v\|}{t}\cdot\left\|\int_0^t\mathcal{T}_s(\mathcal{L}(y))ds\right\| &\leq\frac{\|(\mathcal{T}_t(x)-x)\,v\|}{t}\cdot\int_0^t\|\mathcal{L}(y)\|\,ds\\ &=\|(\mathcal{T}_t(x)-x)\,v\|\cdot\|\mathcal{L}(y)\|\,.\end{aligned}$$

Moreover, since $x \in \mathcal{N}(\mathcal{T})$, we have

$$\begin{aligned}\|(\mathcal{T}_t(x) - x)\, v\|^2 &= \langle v, (\mathcal{T}_t(x^*)\mathcal{T}_t(x) - \mathcal{T}_t(x^*)x - x^*\mathcal{T}_t(x) + x^*x)\, v\rangle \\ &= \langle v, (\mathcal{T}_t(x^*x) - \mathcal{T}_t(x^*)x - x^*\mathcal{T}_t(x) + x^*x)\, v\rangle\end{aligned}$$

and $\|(\mathcal{T}_t(x) - x)\, v\|$ vanishes as t tends to 0 by the weak* continuity of $\mathcal{T}$. This proves (7).

By the definition of the quadratic form $\pounds(x^*y)$, this identity reads as

$$\begin{aligned}\sum_{\ell\geq 1}\langle xL_\ell v, yL_\ell u\rangle &= \langle xv, G^*yu\rangle + \langle xv, Gyu\rangle \\ &\quad + \sum_{\ell\geq 1}\langle xv, L_\ell^* y L_\ell u\rangle + \sum_{\ell\geq 1}\langle xL_\ell v, yL_\ell u\rangle,\end{aligned}$$

and $G^*yu + Gyu = -\sum_{\ell\geq 1} L_\ell^* L_\ell u$, thus

$$\sum_{\ell\geq 1}\left(\langle xL_\ell v, yL_\ell u\rangle - \langle xL_\ell v, yL_\ell u\rangle\right) = \sum_{\ell\geq 1}\left(\langle xv, L_\ell^* y L_\ell u\rangle - \langle xv, L_\ell^* L_\ell yu\rangle\right)$$

that can be written in the form (8). □

Let k be a Hilbert space with Hilbertian dimension equal to the multiplicity of the completely positive part of $\mathcal{L}$, let $(f_\ell)_{\ell\geq 1}$ be an orthonormal basis of k and let $L : \mathsf{h} \to \mathsf{h}\otimes\mathsf{k}$ be the operator defined on the domain $\mathrm{Dom}(G_0)$ by

$$Lu = \sum_{\ell\geq 1}(L_\ell u)\otimes f_\ell.$$

Note that Lu is well defined because $\mathrm{Dom}(G_0)$ is contained in $\mathrm{Dom}(L_\ell)$ by (**H-2**) and the series $\sum_{\ell\geq 1}\|L_\ell u\|^2 = -\langle u, G_0 u\rangle$ is convergent for $u \in \mathrm{Dom}(G_0)$.

Lemma 3.1. *The operator L admits a closed extension to $(-G_0)^{1/2}$ and $\mathrm{Dom}(G_0)$ is a core for this closed extension.*

Proof. We first show that the operator L is closable. In fact, if let $(u_n)_{n\geq 0}$ is a sequence in $\mathrm{Dom}(G_0)$ such that

$$\lim_{n\to\infty} u_n = 0, \qquad \lim_{n\to\infty} Lu_n = \xi \in \mathsf{h}\otimes\mathsf{k}$$

then, for all $\eta \in \mathsf{h}\otimes\mathsf{k}$ of the form $\eta = \sum_\ell v_\ell \otimes f_\ell$ with all but a finite number of $v_\ell \in D$ equal to 0 we have

$$\langle \eta, \xi\rangle = \lim_{n\to\infty}\sum_\ell \langle v_\ell, L_\ell u_n\rangle = \sum_\ell \lim_{n\to\infty}\langle L_\ell^* v_\ell, u_n\rangle = 0.$$

It follows then by the density of those η in $\mathsf{h}\otimes\mathsf{h}\otimes\mathsf{k}$ that $\xi = 0$.

Since $\mathrm{Dom}(G_0)$ is an essential domain for $(-G_0)^{1/2}$ (see Kato[18] Th. 3.35 p. 281), for all $u \in \mathrm{Dom}(G_0)$ there exists a sequence $(u_n)_{n\geq 1}$ in $\mathrm{Dom}(G_0)$ such that

$$\lim_{n\to\infty} u_n = u, \qquad \lim_{n\to\infty} (-G_0)^{1/2} u_n = (-G_0)^{1/2} u. \tag{9}$$

We have then

$$\begin{aligned} \|Lu_n - Lu_m\|^2 &= -2\Re e \left\langle (u_n - u_m), G(u_n - u_m)\right\rangle \\ &= 2\left\|(-G_0)^{1/2}(u_n - u_m)\right\|^2. \end{aligned}$$

It follows that the sequence $(Lu_n)_{n\geq 1}$ converges. Moreover, the limit is independent of the chosen sequence because, if $(u_n)_{n\geq 1}$ and $(u'_n)_{n\geq 1}$ are two sequences satisfying (9), then

$$\lim_{n\to\infty} \|Lu_n - Lu'_n\|^2 = 2 \lim_{n\to\infty} \left\|(-G_0)^{1/2}(u_n - u'_n)\right\|^2 = 0.$$

Therefore we define $Lu = \lim_{n\to\infty} Lu_n$.

Finally we show that L is closed. Indeed, if $(u_n)_{n\geq 1}$ is a sequence in $\mathrm{Dom}((-G_0)^{1/2})$ such that $\lim_{n\to\infty} u_n = u \in \mathsf{h}$ and $\lim_{n\to\infty} Lu_n = \xi \in \mathsf{h}\otimes\mathsf{k}$, then

$$2\left\|(-G_0)^{1/2}(u_n - u_m)\right\|^2 = \|Lu_n - Lu_m\|^2.$$

Since $(-G_0)^{1/2}$ is self-adjoint, it follows that u belongs to the domain of $(-G_0)^{1/2} = \mathrm{Dom}(L)$ and

$$\|Lu - \xi\|^2 = \lim_{n\to\infty} \|Lu - Lu_n\|^2 = 2 \lim_{n\to\infty} \left\|(-G_0)^{1/2}(u - u_n)\right\|^2 = 0.$$

Finally $\mathrm{Dom}(G_0)$ is a core for the closed extension of L to $\mathrm{Dom}((-G_0)^{1/2})$ because it is a core for $(-G_0)^{1/2}$ and $2\|(-G_0)^{1/2}u\|^2 = \|Lu\|^2$ for all $u \in \mathrm{Dom}((-G_0)^{1/2})$. □

If we suppose that the series $\sum_{\ell\geq 1} \|L_\ell^* u\|$ converges for all $u \in D$, the identity (8) reads as

$$\langle (x\otimes \mathbb{1}_{\mathsf{k}})\, Lv, ((y\otimes \mathbb{1}_{\mathsf{k}})L - Ly)\, u\rangle = \langle xv, L^* \left((y\otimes \mathbb{1}_{\mathsf{k}})L - Ly\right) u\rangle \tag{10}$$

Indeed, all vectors of the form $((y\otimes \mathbb{1}_{\mathsf{k}})L - Ly)\,u$ with $y = |u'\rangle\langle v'|$ ($u', v' \in \mathrm{Dom}(G_0)$) belong to the domain of L^* because, by the hypotheses **(H-2)** for all $u, v, u', v', \in D(G_0)$ we have

$$\begin{aligned}
&|\langle Lv, ((|u'\rangle\langle v'| \otimes \mathbb{1}_{\mathsf{k}})L - L|u'\rangle\langle v'|)\, u\rangle| \\
&= \left|\sum_{\ell\geq 1} \left(\langle L_\ell v, u'\rangle\langle v', L_\ell u\rangle - \langle L_\ell v, L_\ell u'\rangle\langle v', u\rangle\right)\right| \\
&= \left|\sum_{\ell\geq 1} \left(\langle v, L_\ell^* u'\rangle\langle v', L_\ell u\rangle - \langle v, L_\ell^* L_\ell u'\rangle\langle v', u\rangle\right)\right| \\
&\leq \|v'\| \cdot \|v\| \left(\left(\sum_{\ell\geq 1} \|L_\ell^* u'\| \cdot \|L_\ell u\|\right) + \|u\| \cdot \|G_0 u'\|\right) \\
&\leq \|v'\| \cdot \|v\| \left(\left(2^{-1} \sum_{\ell\geq 1} \left(\|L_\ell^* u'\|^2 + \|L_\ell u\|^2\right)\right) + \|u\| \cdot \|G_0 u'\|\right).
\end{aligned}$$

Moreover, we would like to write the right-hand side of (10) as $\langle Lxv, ((y \otimes \mathbb{1}_{\mathsf{k}})L - Ly)\, u\rangle$. In order to this we must know that the set of $(y \otimes \mathbb{1}_{\mathsf{k}})L - Ly)\, u$ on which L^* acts is an essential domain for L^*, therefore we suppose that

(H-3) The operators L_ℓ are non-zero only for $\ell = 1, \ldots, d$ $(d \in \mathbb{N})$ and there exists a non-zero vector $u \in \mathrm{Dom}(G_0)$ which is not an eigenvector for all the L_ℓ.

This condition is obviously true when the space h is finite dimensional. Indeed, by the minimality condition in **(H-1)**, each L_ℓ is not a multiple of the identity operator, therefore the set of eigenvectors of each L_ℓ with the same eigenvalue, together with the zero vector, is a proper subspace of h and the set of $u \in \mathrm{Dom}(G_0) = \mathsf{h}$ that are eigenvectors for some L_ℓ forms then a set of Lebesgue measure 0 in h.

It is true also in several infinite dimensional cases.

Remark 3.1. (i) Suppose that the point spectrum of each L_ℓ is at most countable. Then the set of $u \in \mathsf{h}$ that are eigenvectors of some L_ℓ has empty interior.

Indeed, let $\mathcal{V}_{\ell,\lambda}$ be the subspace of h spanned by eigenvectors of L_ℓ with eigenvalue λ. Since L_ℓ is not a multiple of the identity operator, $\mathcal{V}_{\ell,\lambda}$ has empty interior. It follows then from the Baire's category theorem that the set of eigenvectors of some L_ℓ, has empty interior because it is a countable union of sets with empty interior. One can find then several vectors in h that are not eigenvectors of any L_ℓ.

(ii) Among operators arising in the applications to open quantum systems with uncountable point spectrum annihilations and creation operators on some Fock space are perhaps the most frequent. It is well-known that their spectrum of annihilation operators coincides with the whole complex plane, exponential vectors are eigenvectors but, number vectors, for instance, are not.

Under the hypothesis (**H-3**) we can prove the following

Proposition 3.2. *Suppose that (**H-1**), (**H-2**), (**H-Markov**) and (**H-3**) hold. Then the linear manifold generated by the set*

$$\{ (y \otimes \mathbb{1}_{\mathsf{k}})\, Lu - Lyu \mid y = |u'\rangle\langle v'|,\ u, u', v' \in D \} \tag{11}$$

is an essential domain for L^*.

Proof. We already showed that it is contained in the domain of L^*. We now check that it is dense in $\mathsf{h} \otimes \mathsf{k}$ by the linear independence of the operators $\mathbf{1}, L_1, L_2, \dots$. In fact, given a non-zero vector $\xi = \sum_\ell v_\ell \otimes f_\ell$ in $\mathsf{h} \otimes \mathsf{k}$ orthogonal to the set (11), then

$$\begin{aligned} 0 &= \langle \xi, (y \otimes \mathbf{1}_{\mathsf{k}})\, Lu - Lyu\rangle \\ &= \sum_{\ell \geq 1} \left(\langle v_\ell, u'\rangle \langle v', L_\ell u\rangle - \langle v_\ell, L_\ell u'\rangle \langle v', u\rangle \right) \\ &= \left\langle v', \left(\sum_{\ell \geq 1} \langle v_\ell, u'\rangle L_\ell - \sum_{\ell \geq 1} \langle v_\ell, L_\ell u'\rangle \mathbf{1} \right) u \right\rangle. \end{aligned}$$

It follows then, by the arbitrarity of v' and u in $\mathrm{Dom}(G_0)$

$$\sum_{\ell \geq 1} \langle v_\ell, u'\rangle L_\ell - \left(\sum_{\ell \geq 1} \langle v_\ell, L_\ell u'\rangle \right) \mathbf{1}$$

contradicting the linear independence of $\mathbf{1}, L_1, L_2, \dots$ for some $u' \in \mathrm{Dom}(G_0)$.

We now prove that the linear manifold generated by (11) is a core for L^* showing that vectors in $(\mathsf{h} \otimes \mathsf{k}) \times \mathsf{h}$ of the form $(\xi, L^*\xi)$ with ξ in the (algebraic) linear span of (11) are dense in the graph of L^*. Namely, if $\eta \in \mathrm{Dom}(L^*)$ is a vector such that

$$0 = \langle \eta, ((y \otimes \mathbf{1}_{\mathsf{k}})\, L - Ly)\, u\rangle + \langle L^*\eta, L^* \left((y \otimes \mathbf{1}_{\mathsf{k}})\, L - Ly \right) u\rangle \tag{12}$$

for all $y = |u'\rangle\langle v'|$ and $u', v', u \in \mathrm{Dom}(G_0)$ then $\eta = 0$.

Let $u \in \mathrm{Dom}(G_0)$ be a unit vector which is not an eigenvector for all the L_ℓ. For all $v' \in \mathrm{Dom}(G_0)$ which is orthogonal to $L_1 u, \ldots, L_d u$ but *not* orthogonal to v' we have

$$((y \otimes \mathbf{1}_\mathsf{k})\, L - Ly)\, u = -\langle v', u\rangle Lu'$$

and (12) reads as

$$0 = \langle \eta, Lu'\rangle + \langle L^*\eta, L^*Lu'\rangle = \langle L^*\eta, (\mathbf{1} + L^*L)\, u'\rangle$$

Now, since $(\mathbf{1} + L^*L)(\mathrm{Dom}(G_0)) = \mathsf{h}$ because L^*L is positive self-adjoint (Kato[18] Th. 3.24 p. 275) and -1 belongs then to the resolvent, it follows that $L^*\eta = 0$. The identity (12) implies then $\eta = 0$ by the density of the linear manifold generated by the set (11).

This completes the proof. □

It would be interesting to prove that the linear manifold generated by (11) is a core for L^* under some weaker conditions.

Proposition 3.3. *Suppose that* **(H-1)**, **(H-2)**, **(H-Markov)** *and* **(H-3)** *hold. Then all* $x \in \mathcal{N}(\mathcal{T})$ *belong to the generalised commutator of* $\{L_1, L_1^*, \ldots, L_d, L_d^*\}$.

Proof. Since the linear manifold described in Prop. 3.2 is an essential domain for L^*, identity (10) implies that xv belongs to the domain of L and we can write

$$\langle (x \otimes \mathbf{1}_\mathsf{k})\, Lv, ((y \otimes \mathbf{1}_\mathsf{k})L - Ly)\, u\rangle = \langle Lxv, ((y \otimes \mathbf{1}_\mathsf{k})L - Ly)\, u\rangle\,.$$

Therefore, by the density of vectors $((y \otimes \mathbf{1}_\mathsf{k})L - Ly)\, u$ in $\mathsf{h} \otimes \mathsf{k}$, we have $(x \otimes \mathbf{1}_\mathsf{k})\, Lv = Lxv$, i.e.

$$xL_\ell v = L_\ell xv$$

for all $v \in \mathrm{Dom}(G_0)$ and all $\ell = 1, \ldots, d$. Now, since L_ℓ is closed and $\mathrm{Dom}(G_0)$ is a core for L_ℓ, the above identity holds for all v in the domain of L_ℓ. This proves that each $L_\ell x$ is an ampliation of xL_ℓ i.e. x belongs to generalised commutator of each L_ℓ.

Finally, since also x^* belongs to $\mathcal{N}(\mathcal{T})$, by standard results on the adjoint of products of a closed and a bounded operator (see e.g. Kato[18] III.5 p.168), we have the inclusions

$$xL_\ell^* \subseteq (L_\ell x^*)^* \subseteq (x^* L_\ell)^* = L_\ell^* x$$

showing as in Refs.[7,19] that x also belongs also to the generalised commutator of each L_ℓ^*. □

We can now prove the main result of this section

Theorem 3.1. *Suppose that* **(H-1)**, **(H-2)**, **(H-Markov)** *and* **(H-3)** *hold and let* $x \in \mathcal{N}(\mathcal{T})$. *Then*

(1) $\mathcal{T}_t(x) = e^{itH} x e^{-itH}$ *for all* $t \geq 0$,
(2) x *belongs to the generalised commutator of the set of operators* $\{ e^{-itH} L_\ell e^{itH}, e^{-itH} L_\ell^* e^{itH} \mid \ell \geq 1, t \geq 0 \}$.

Proof. A $x \in \mathcal{N}(\mathcal{T})$ belongs to generalised commutator of the operators $L_1, L_1^*, \ldots, L_d, L_d^*$ by Prop. 3.3. Moreover, since $\mathcal{N}(\mathcal{T})$ is $\mathcal{T}_t$-invariant for all $t \geq 0$ by Prop. 2.1 (1), also $\mathcal{T}_t(x)$ belongs to generalised commutator of the operators $L_1, L_1^*, \ldots, L_d, L_d^*$, thus

$$\langle L_\ell^* L_\ell v, \mathcal{T}_t(x) u \rangle = \langle \mathcal{T}_t(x^*) L_\ell^* L_\ell v, u \rangle = \langle L_\ell^* \mathcal{T}_t(x^*) L_\ell v, u \rangle = \langle L_\ell v, \mathcal{T}_t(x) L_\ell u \rangle$$

for all $u, v \in D$ and also $\langle v, \mathcal{T}_t(x) L_\ell^* L_\ell u \rangle = \langle L_\ell v, \mathcal{T}_t(x) L_\ell u \rangle$. Therefore, for $x \in \mathcal{N}(\mathcal{T})$ and all $u, v \in D$, differentiating with respect to $s \in [0, t]$

$$\frac{d}{ds} \left\langle e^{i(s-t)H} v, \mathcal{T}_s(x) e^{i(s-t)H} \right\rangle = 0$$

so that $\langle v, \mathcal{T}_t(x) u \rangle = \langle e^{-itH} v, x e^{-itH} u \rangle$ and (1) is proved.

The claim (2) follows immediately from (1) since $e^{itH} x\, e^{-itH} L_\ell \subseteq L_\ell\, e^{itH} x\, e^{-itH}$ and, left multiplying by e^{-itH} and right multiplying by e^{itH}, we find $x\, e^{-itH} L_\ell\, e^{itH} \subseteq e^{-itH} L_\ell\, e^{itH} x$. □

In Ref.[7] we proved that $\mathcal{N}(\mathcal{T})$ indeed coincides with the generalised commutator of the set of operators $\{ e^{-itH} L_\ell e^{itH}, e^{-itH} L_\ell^* e^{itH} \mid \ell \geq 1, t \geq 0 \}$, also in the case of infinitely many non zero L_ℓ, under some more technical assumptions.

(H-4) There exists a linear manifold D dense in h satisfying **(H-2)** and self-adjoint operators $C \geq \mathbb{1}$, Φ satisfying **(H-Markov)** such that, moreover,

(1) $e^{itH}(D) \subseteq \mathrm{Dom}(C^{1/2})$ for all $t \in \mathbb{R}$,
(2) the domain inclusions: $\mathrm{Dom}(C) \subseteq \mathrm{Dom}(G)$, $\mathrm{Dom}(C) \subseteq \mathrm{Dom}(G^*)$, $\mathrm{Dom}(C) \subseteq \mathrm{Dom}(H)$, $\mathrm{Dom}(C) \subseteq \mathrm{Dom}(L_\ell)$, $\mathrm{Dom}(C^{1/2}) \subseteq \mathrm{Dom}(L_\ell)$, $L_\ell(\mathrm{Dom}(C^{3/2})) \subseteq \mathrm{Dom}(C)$, for all $\ell \geq 1$, $H(\mathrm{Dom}(C^k)) \subseteq \mathrm{Dom}(C)$ for some $k \geq 2$ hold,
(3) there exists positive constants b_ℓ such that $\sum_{\ell \geq 1} b_\ell^2 < \infty$ and $\|[L_\ell, C]u\|^2 \leq b_\ell^2 \|Cu\|^2$ for all $u \in \mathrm{Dom}(C^{3/2})$,
(4) there exists a positive constant b_G such that $\|[G, C]v\| \leq b_G \|C^{3/2} v\|$, $\|[G^*, C]v\| \leq b_G \|C^{3/2} v\|$, for some $k \geq 1$ and every $v \in \mathrm{Dom}(C^{3/2})$.

We refer to Ref.[7] Th. 4.1 for the proof. Note that this corrects our mistake in Ref.[19] were we claimed that $(\mathcal{N})(\mathcal{T})$ coincides with the generalised commutator of the operators L_ℓ, L_ℓ^* $(\ell \geq 1)$.

4. A rotation invariant quantum Markov semigroup

Quantum Markov semigroups with a non-trivial $\mathcal{N}(\mathcal{T})$ arise when, due to some symmetry of the system, dissipation does not affect some observables. In this section we show such an example, a QMS describing the open-system dynamics of a spin, immersed into an isotropic dissipating environment (see e.g. Khasin and Kosloff[21]) as an application of our characterisation.

Let $\mathsf{h} = L^2(\mathbb{R}^3;\mathbb{C})$ and let L_1, L_2, L_3 be the three components of the angular momentum operator

$$L_1 = -i\,(y\partial_z - z\partial_y)\,, \quad L_2 = -i\,(z\partial_x - x\partial_z)\,, \quad L_3 = -i\,(x\partial_y - y\partial_x)$$

(see e.g. Parthasarathy[13] Ch.I Sect.13, Teschl[20] Ch.8 Sect.2). These operators, defined on the the Schwarz space D of rapidly decreasing functions on $\mathbb{R}^3$ where they satisfy the commutation relations

$$[L_1, L_2] = iL_3, \quad [L_2, L_3] = iL_1, \quad [L_3, L_1] = iL_2,$$

have unique self-adjoint extensions, that we still denote by the same symbol, whose spectra are pure point coincide with $\mathbb{Z}$. Moreover, the eigenspaces of each eigenvalue are one-dimensional.

The operator $L^2 = L_1^2 + L_2^2 + L_3^2$ can also be defined as the unique self-adjoint extension of its restriction to D and its spectrum is the set $\{\, n(n+1) \,|\, n \in \mathbb{N}\,\}$ and commutes with each L_ℓ.

Let H be the self-adjoint extension of ωL_3 $(\omega \in \mathbb{R} - \{0\})$ and let G be the operator defined on the Schwarz space D by

$$G = -\frac{1}{2}\sum_{\ell=1}^{3} L_\ell^* L_\ell - iH = -\frac{1}{2}L^2 - iH.$$

It is not hard to see that G extends to the generator of a strongly continuous contraction semigroup on h with D as a core. Notice first that $G_0 := -(1/2)L^2$ is negative self-adjoint and generates an analytic semigroup. Then the operator $iH = i\omega L_3$ generates a unitary group and is relatively bounded with respect to L^2 with relative bound 0. Indeed, we have the inequalities

$$\begin{aligned}\|Hu\|^2 &= |\omega|\langle u, L_3^* L_3 u\rangle \\ &\le |\omega|\langle u, L^2 u\rangle \\ &\le 4^{-1}\varepsilon^2\|L^2u\|^2 + \omega^2\varepsilon^{-2}\|u\|^2 \\ &= \varepsilon^2\| - 2^{-1}L^2u\|^2 + \omega^2\varepsilon^{-2}\|u\|^2\end{aligned}$$

for all $u \in D$ and $\varepsilon > 0$ by the Young inequality. The elementary inequality $(a^2+b^2)^{1/2} \le a+b$ for all $a, b \ge 0$ yields then

$$\|Hu\| \le \varepsilon\| - 2^{-1}L^2u\| + |\omega|\varepsilon^{-1}\|u\|.$$

It follows that (see e.g. Kato[18] Ch. IX 3, Th. 2.7 p. 501) that G generates a strongly continuous contraction semigroup on h and the hypothesis **(H-1)** holds. The same arguments apply to G^*. Moreover, the domains of G, G^* and L^2 coincide.

The hypothesis **(H-2)** is immediately checked because D is an invariant domain for all the operators involved and the action of the operator e^{itH} of the unitary group generated by iH (rotation by an angle ωt around the z axis is explicitly given by

$$\left(\mathrm{e}^{itH}u\right)(x,y,z) = u\left(x\cos(\omega t) - y\sin(\omega t), x\sin(\omega t) + y\cos(\omega t), z\right).$$

The hypothesis **(H-Markov)** also follows easily taking $\Phi = C = L^2$ once we recall that L^2 commutes with all the L_ℓ so that the fundamental inequality of **(H-Markov)** (4) holds. Indeed we have

$$2\Re e\,\langle Cu, Gu\rangle + \sum_{\ell\ge 1}\left\langle C^{1/2}L_\ell u, C^{1/2}L_\ell u\right\rangle = 0$$

for all $u \in \mathrm{Dom}(G^2) = \mathrm{Dom}(C^2)$.

Since the spectrum of L_1, L_2, L_3 is pure point and countable, also the hypothesis **(H-3)** follows immediately from Remark 3.1.

We finally check the hypothesis **(H-4)**. Properties (3) and (4) are easily established recalling that L^2 commutes with each L_ℓ. Property (1) is trivial because we already noticed that D is an invariant domain for all the operators e^{itH}.

A straightforward computation shows that

$$\begin{aligned}\mathrm{e}^{-itH}L_1\mathrm{e}^{itH} &= \cos(\omega t)L_1 - \sin(\omega t)L_2, \\ \mathrm{e}^{-itH}L_2\mathrm{e}^{itH} &= \cos(\omega t)L_2 + \sin(\omega t)L_1\end{aligned}$$

for all $t \in \mathbb{R}$. It follows that the decoherence -free von Neumann subalgebra $\mathcal{N}(\mathcal{T})$ of the QMS is given by the generalized commutator of L_1, L_2, L_3

and clearly consists of operators on $L^2(\mathbb{R}^3;\mathbb{C})$ invariant under the action of the rotation group. This is obviously non-trivial because any multiplication operator by a rotation invariant function, for instance, belongs to $\mathcal{N}(\mathcal{T})$.

The fixed point algebra $\mathcal{F}(\mathcal{T})$, as the generalized commutator of $H = \omega L_3, L_1, L_2, L_3$, coincides with $\mathcal{N}(\mathcal{T})$.

Acknowledgments

This paper was written while FF was visiting the "Centro de Análisis Estocástico y Aplicaciones". Financial supports from "Programa Bicentenario de Ciencia y Tecnología, grant PBCT-ADI13" and MIUR PRIN 2007 "Quantum Probability and Applications to Information Theory" are gratefully aknowledged.

References

1. Ph. Blanchard and R.Olkiewicz, Decoherence as Irreversible Dynamical Process in Open Quantum Systems. In: S. Attal, A. Joye, C.-A. Pillet (eds.) *Open Quantum Systems III - Recent Developments.* Lecture Notes in Mathematics 1882 p. 117–160. (Springer, Berlin, 2006).
2. D.A. Lidar, I.L. Chuang and K.B. Whaley, Decoherence-free Subspaces for Quantum Computation, *Phys. Rev. Lett* **81**, 2594–2597 (1998).
3. R. Rebolledo, Decoherence of quantum Markov semigroups, *Ann. Inst. H. Poincaré Probab. Statist.* **41**, 349–373 (2005).
4. R. Rebolledo, A View on Decoherence via Master Equations, *Open Sys. & Information Din.* **12**, 37–54 (2005).
5. B.V. Rajarama Bhat, F. Fagnola and M. Skeide, Maximal commutative subalgebras invariant for CP-maps: (counter-)examples. *Infin. Dimens. Anal. Quantum Probab. Relat. Top.* **11**, 523 - 539 (2008).
6. F. Fagnola and R. Rebolledo, Algebraic Conditions for Convergence of a Quantum Markov Semigroup to a Steady State. *Infinite Dim. Anal. Quantum Probab. Related Topics*, **11**, 467–474 (2008).
7. A. Dhahri, F. Fagnola and R. Rebolledo, The decoherence-free subalgebra of a quantum Markov semigroup with unbounded generator. *Infinite Dim. Anal. Quantum Probab. Related Topics*, **12** (2010). To appear.
8. A. Frigerio and M. Verri, Long-time asymptotic properties of dynamical semigroups on W^*-algebras, *Math. Z.* **180**, 275–286 (1982).
9. F. Fagnola and R. Rebolledo, Subharmonic projections for a quantum Markov semigroup, *J. Math. Phys.* **43**, 1074–1082 (2002).
10. V. Umanità, Classification and decomposition of quantum Markov semigroups, *Probab. Theory Rel. Fields* **134**, 603–623 (2006).
11. L. Pantaleón Martínez and R. Quezada, The Asymmetric Exclusion Quantum Markov Semigroup. *Infinite Dim. Anal. Quantum Probab. Related Topics* **12**, 367–385(2009).
12. D.E. Evans, Irreducible quantum dynamical semigroups, *Commun. Math. Phys.* **54**, 293–297 (1977).

13. K. R. Parthasarathy, *An introduction to quantum stochastic calculus*, Monographs in Mathematics **85**, (Birkhäuser-Verlag, Basel, 1992).
14. A.M. Chebotarev and F. Fagnola, Sufficient conditions for conservativity of quantum dynamical semigroups, *J. Funct. Anal.* **153**, 382–404 (1998).
15. F. Fagnola, Quantum Markov Semigroups and Quantum Markov Flows, *Proyecciones* **18**, 1–144 (1999).
16. F. Fagnola, A simple singular quantum Markov semigroup, in *Stochastic Analysis and Mathematical Physics ANESTOC '98*, Proceedings of the Third International Workshop, (Birkäuser 2000) p. 73–88.
17. R. Carbone and F. Fagnola, The Feller property of a class of Quantum Markov Semigroups II, in *Quantum probability and infinite dimensional analysis* (Burg, 2001), 57–76, QP–PQ: Quantum Probab. White Noise Anal., 15, World Sci. Publishing, River Edge, NJ, 2003.
18. T. Kato, *Perturbation theory for linear operators.* Corrected printing of the second edition. (Springer-Verlag, Berlin, 1980).
19. F. Fagnola and R. Rebolledo, The Approach to Equilibrium of a Class of Quantum Dynamical Semigroups, *Infinite Dim. Anal. Quantum Probab. Related Topics* **1**, 561–572 (1998).
20. G. Teschl, *Mathematical Methods in Quantum Mechanics.* Graduate Studies in Mathematics **99**. (American Mathematical Society, Providence, 2009).
21. M Khasin and R Kosloff, The globally stable solution of a stochastic nonlinear Schroedinger equation, *J. Phys. A: Math. Theor.* **41**, 365203 (10pp)(2008).

A SUFFICIENT CONDITION FOR ALL INVARIANT STATES OF A QMS TO BE DIAGONAL

J. C. GARCÍA * and R. QUEZADA**

Universidad Autónoma Metropolitana, Iztapalapa Campus
Av. San Rafael Atlixco 186, Col. Vicentina
09340 Iztapalapa D.F., Mexico.
**E-mail: jcgc@xanum.uam.mx*
***E-mail: roqb@xanum.uam.mx*

L. PANTALEÓN-MARTÍNEZ

Escuela Nacional Preparatoria, Plantel 7, UNAM, Calz. de la Viga 54,
Col. Merced Balbuena. Del. Venustiano Carranza, C.P. 15810, México, D.F.
E-mail: polo3951@unam.mx

We give a sufficient condition for all elements in the set of invariant states of a Quantum Markov Semigroup (QMS) to be diagonal. We apply our result to complete the characterization of the set of invariant states for the asymmetric exclusion Quantum Markov Semigroup introduced in Ref.[10]

Keywords: Asymmetric exclusion quantum Markov semigroup; Fixed point algebra; Subharmonic projections.

1. Introduction

As far as we know, it does not exist a general criterion, or even a sufficient condition, to ensure that all invariant states of a given QMS are diagonal with respect to the same orthonormal basis. This note aims to fulfil this gap by taking advantage of the isometric isomorphism between the algebra of all fixed points $\mathcal{F}(\mathcal{T}) := \{x \in \mathcal{B}(\mathsf{h}) : \mathcal{T}_t(x) = x\}$, of a given quantum Markov semigroup (QMS) $\mathcal{T} = (\mathcal{T}_t)_{t\geq 0}$ on $\mathcal{B}(\mathsf{h})$, and the dual of the fixed points of the predual semigroup $(\mathcal{F}(\mathcal{T}_*))^*$. This isometric isomorphism holds in the case when the semigroup has a faithful invariant state, as observed by A. Frigerio and M. Verri in Ref.[6] It turns out that when $\mathcal{F}(\mathcal{T})$ is isometrically isomorphic to some $l_\infty(\Omega)$ with Ω a denumerable subset, as a von Neumann algebra, then there exists an orthonormal basis of h with respect to which every invariant state of the semigroup is diagonal and one can easily characterize its attraction domain.

The paper is organized as follows, in Section 2 we prove our main result Theorem 2.1, that we apply in Section 3 to the n-photon absorption and emission QMS and to complete the characterization of the set of invariant states for the asymmetric exclusion QMS initiated in Ref.[8,10]

2. Main results

Along this work h is a separable complex Hilbert space, $\mathcal{B}(\mathsf{h})$ will denote the von Neumann algebra of all bounded operators on h and $L_1(\mathsf{h})$, $L_2(\mathsf{h})$ are the spaces of finite trace and Hilbert-Schmidt operators, respectively.

Let $\mathcal{T}_t$ a quantum dynamical semigroup on a W*-algebra $\mathcal{M}$ and denote by $\mathcal{T}_{*t}$ the predual semigroup of $\mathcal{T}_t$ of positive contractions on $\mathcal{M}_*$. Let $\mathcal{F}(\mathcal{T})$ and $\mathcal{F}(\mathcal{T}_*)$ denote the fixed point set of $\mathcal{T}_t$ and $\mathcal{T}_{*t}$, respectively. If there exists a faithful family of normal invariant states, then $\mathcal{F}(\mathcal{T})$ is a W*-subalgebra of $\mathcal{M}$. The support projection p of an invariant state ρ reduces the semigroup $\mathcal{T}_{*t}$, i.e., $p\mathcal{M}_*p$ is invariant under the semigroup $\mathcal{T}_*$, and the same is true for the *recurrent subspace projection*

$$R = \sup\{p : p \text{ is the support projection of an invariant state } \rho\}.$$

$R = 0$ if the semigroup has no normal invariant states and $R = I$ when there exists a faithful family of normal invariant states, see Ref.[6]

Our main result, Theorem 2.1, is based on the following well known result of Frigerio and Verri in Ref.,[6] that we include here for the sake of completeness.

Proposition 2.1. *If a given quantum Markov semigroup $\mathcal{T}$ on a W*-algebra has a faithful invariant state, then $(\mathcal{F}(\mathcal{T}_*))^*$ and $\mathcal{F}(\mathcal{T})$ are isometrically isomorphic, the duality being given by*

$$(\rho, x) \to tr(\rho x), \quad \rho \in \mathcal{F}(\mathcal{T}_*), x \in \mathcal{F}(\mathcal{T}).$$

Proof. By the Hahn-Banach and Schatten Theorems, given $\xi \in (\mathcal{F}(\mathcal{T}_*))^*$, it can be extended to $\Lambda : L_1(\mathsf{h}) \to \mathbb{C}$ and there exists $a \in \mathcal{B}(\mathsf{h})$ such that $\Lambda(x) = \mathrm{tr}(xa)$, $x \in L_1(\mathsf{h})$, with $||\xi|| = ||\Lambda|| = ||a||$. Then if $\rho \in \mathcal{F}(\mathcal{T}_*)$, $\xi(\rho) = \mathrm{tr}(\rho a) = \mathrm{tr}(\rho \mathcal{T}_t a)$, and from here we have,

$$\xi(\rho) = \lim_{t\to\infty} \mathrm{tr}\left(\rho \frac{1}{t} \int_0^t \mathcal{T}_s(a)ds\right) = \mathrm{tr}(\rho F(a)),$$

where $w^* - \lim_{t\to\infty} \frac{1}{t} \int_0^t \mathcal{T}_s(a)ds := F(a) \in \mathcal{F}(\mathcal{T})$ and $||F(a)|| = ||a||$; the existence of the ergodic limit follows from Theorem 2.1 in Ref.[6] So there exists an element $a_\xi := F(a) \in \mathcal{F}(\mathcal{T})$ such that $\xi(\rho) = \mathrm{tr}(\rho a_\xi)$, for every $\rho \in \mathcal{F}(\mathcal{T}_*)$; if there would exists another element $y \in \mathcal{F}(\mathcal{T})$ with the same properties, we would have $\mathrm{tr}\rho(y - a_\xi) = 0$ for all $\rho \in \mathcal{F}(\mathcal{T}_*)$, contradicting the well known fact, proven in Theorem 2.1 in Ref.,[6] that $\mathcal{F}(\mathcal{T}_*)$ separates $\mathcal{F}(\mathcal{T})$. Thus we have a well defined map $(\mathcal{F}(\mathcal{T}_*))^* \to \mathcal{F}(\mathcal{T})$, $\xi \mapsto a_\xi$ with $\xi(\rho) = \mathrm{tr}(\rho a_\xi)$, for every $\rho \in \mathcal{F}(\mathcal{T}_*)$; which is an isometry, since $||a_\xi|| = ||F(a)|| = ||a|| = ||\xi||$. The surjectivity follows by using once again the existence of the ergodic limit and the fact that $\mathcal{F}(\mathcal{T}_*)$ separates $\mathcal{F}(\mathcal{T})$. □

Theorem 2.1. *Suppose that there exists a faithful invariant state for a quantum Markov semigroup $\mathcal{T}$ on $\mathcal{B}(\mathsf{h})$, a measure space $(\Omega, 2^\Omega, \mu)$, where $\Omega \subset \mathbb{N}$ and μ is the counting measure on 2^Ω, and a von Neumann algebras isomorphism,*

$$s : \ell_\infty(\Omega) \to \mathcal{F}(\mathcal{T}).$$

Then there exists an isometric isomorphism s_*,

$$s_* : \mathcal{F}(\mathcal{T}_*) \to \ell_1(\Omega),$$

such that:

(a) for $n \in \Omega$, let $e_n = \mathbf{1}_{\{n\}}$ be the indicator function of the singleton $\{n\}$. Then the $\mathcal{T}_{*t}$-invariant elements $\rho_n := s_*^{-1}(e_n)$ are states.

(b) the support projection of each invariant state ρ_n is $p_n = s(e_n)$, and we have the decomposition $\mathsf{h} = \oplus_{n\in\Omega} V_n$, with $V_n = Ran\, p_n = \overline{span\, \beta_n}$ where $\beta_n = \{u_l^{(n)} : l \geq 1\}$ is the ortho-normal set of eigenvectors of ρ_n, i.e.,

$$\rho_n = \sum_{j=1}^{\infty} \lambda_j^{(n)} |u_j^{(n)}\rangle\langle u_j^{(n)}|; \quad \text{and} \quad \lambda_j^{(n)} \geq 0, \sum_{j=1}^{\infty} \lambda_j^{(n)} = 1.$$

(c)

$$\mathcal{F}(\mathcal{T}_*) = \left\{ \sum_{n\in\Omega} \alpha_n \rho_n \mid (\alpha_n)_{n\geq 0} \in \ell_1(\Omega) \right\}.$$

Hence, all invariant states are diagonal w.r.t. the ortho-normal basis $\beta = \cup_{n\in\Omega} \beta_n$.

Proof. Let us define $p_n = s(e_n)$. Since s is a w^*−algebra isomorphism, one can easily see that

$$p_n p_m = s(e_n e_m) = s(\delta_{n,m} e_n) = \delta_{n,m} p_n.$$

And from the w^*-continuity of s we get

$$\sum_{n\in\Omega} p_n = \sum_{n\in\Omega} s(e_n) = s(\sum_{n\in\Omega} e_n) = s(\mathbf{1}) = I \in \mathcal{F}(\mathcal{T}).$$

Hence the family of operators $\{p_n\}$ is an identity resolution. Since $s(\ell_\infty(\Omega)) = \mathcal{F}(\mathcal{T})$ we have

$$\mathcal{F}(\mathcal{T}) = \left\{ \sum_{n\in\Omega} \lambda_n p_n \mid (\lambda_n)_{n\in\Omega} \in \ell_\infty(\Omega) \right\}.$$

For any given $\rho \in \ell_1(\mathsf{h})$, $(\mathrm{tr}\rho p_i)_{i\geq 1} \in l_1(\Omega)$, so we can define a map $s_* : \mathcal{F}(\mathcal{T}_*) \to \ell_1(\Omega)$, by means of $\rho \mapsto s_*(\rho)$ where

$$s_*(\rho) = (\mathrm{tr}\rho p_i)_{i\geq 1} \in \ell_1(\Omega),$$

or equivalently

$$s_*(\rho)(\cdot) = \mathrm{tr}\rho s(\cdot) \in \ell_\infty(\Omega)^*,$$

from this, clearly we have $s_*(\rho) \geq 0$ if $\rho \geq 0$. Then using the canonical injection $j : \ell_1(\Omega) \hookrightarrow \ell_1(\Omega)^{**} = \ell_\infty(\Omega)^*$, we have weak*-continuous maps for each n, $j(e_n) : \ell_\infty(\Omega) \to \mathbb{C}$, $j(e_n)(a) = a_n$, $a = (a_n)_{n\in\Omega} \in \ell_\infty$, and $j(e_n) \circ s^{-1} : \mathcal{F}(\mathcal{T}) \to \mathbb{C}$. Hence there exists a unique positive element $\rho_n \in \mathcal{F}(\mathcal{T}_*)$ s.t.

$$\left(j(e_n) \circ s^{-1}\right)(x) = \mathrm{tr}\left(\rho_n x\right), \quad x \in \mathcal{F}(\mathcal{T}).$$

In particular for $x = I$ and $x = p_m = s(e_m)$ we have respectively, $1 = \mathrm{tr}(\rho_n)$ and

$$\mathrm{tr}\left(\rho_n p_m\right) = j(e_n)(e_m) = \delta_{n,m}. \tag{1}$$

Using the definition of s_* and (1) it is clear that $s_*(\rho_n) = e_n$.

Given $(\lambda_n)_{n\in\Omega} \in \ell_1(\Omega)$, let $\rho = \sum_i \lambda_i \rho_i$, then we have $s_*(\rho) = \left(\mathrm{tr}(\rho p_n)\right)_{n\geq 0} = (\lambda_n)_{n\in\Omega}$. Hence the map s_* is onto. Finally s_* is an isometry because, $||s_*(\rho)||_{\ell_1} = ||\,(\mathrm{tr}\rho p_i)_{i\geq 1}\,||_{\ell_1} =$

$$\sup_{\Lambda\in\ell_1^*, ||\Lambda||=1} \left|\Lambda\left((\mathrm{tr}\rho p_i)_{i\geq 1}\right)\right| = \sup_{||(\alpha_i)_{i\in\Omega}||_\infty=1} \left|\sum_{i\in\Omega} \alpha_i \mathrm{tr}(\rho p_i)\right| =$$

$$\sup_{||(\alpha_i)_{i\in\Omega}||_\infty=1} \left|\mathrm{tr}\left(\rho s((\alpha_i)_{i\in\Omega})\right)\right| = \sup_{a\in\mathcal{F}(\mathcal{T}), ||a||=1} |\mathrm{tr}(a\rho)| =$$

$$\sup_{a\in\mathcal{F}(\mathcal{T}_*)^*, ||a||=1} |\mathrm{tr}(a\rho)| = ||\rho||_1.$$

This proves the first part of the theorem and item (a).

Now, using that $1 = \mathrm{tr}\rho_n = \mathrm{tr}p_n\rho_n$, we have

$$0 = \sum_j \langle \rho_n u_j^{(n)}, u_j^{(n)} \rangle - \sum_j \langle p_n \rho_n u_j^{(n)}, u_j^{(n)} \rangle = \sum_j \lambda_j^{(n)} - \sum_j \lambda_j^{(n)} ||p_n u_j^{(n)}||^2,$$

which implies $1 - ||p_n u_j^{(n)}||^2 = 0$ for each $j \geq 1$ s.t. $\lambda_j^{(n)} \neq 0$. Hence $u_j^{(n)} \in Ran\, p_n$ for each $j \geq 1$ s.t. $\lambda_j^{(n)} \neq 0$. So if q_n is the support projection of ρ_n, i.e., if $Ran\, q_n = \overline{span\, \beta_n}$, we conclude that $q_n \leq p_n$. We know Ref.,[8] that $q_n \in \mathcal{F}(\mathcal{T})$. Hence $q_n = \sum_{j=0}^{\infty} \alpha_j^{(n)} p_j$, with each $\alpha_j^{(n)} \in \{0, 1\}$. Therefore

$$q_n = p_n q_n = \sum_{j=0}^{\infty} \alpha_j^{(n)} p_n p_j = \alpha_n^{(n)} p_n.$$

Thus $q_n = p_n$. The remaining part of (b) follows from the above mentioned fact that the family $\{p_n\}$ is an identity resolution.

Item (c) follows immediately from the fact that s_* is an isomorphism and $s_*(\rho_n) = e_n$, $n \in \Omega$. □

Proposition 2.2. *Let us suppose that a QMS $\mathcal{T}$ satisfies the assumptions in Theorem 2.1. Then a state σ is invariant for $\mathcal{T}$ if and only if it has the form*

$$\sigma = \sum_{n \in \Omega} tr(\sigma p_n) \rho_n. \tag{2}$$

Proof. By (c) in Theorem 2.1, any invariant state σ is represented in the form

$$\sigma = \sum_{n \in \Omega} \alpha_n \rho_n, \quad \text{with} \sum_{n \geq 0} \alpha_n = 1,$$

thus by (1) we have

$$\mathrm{tr}(\sigma p_n) = \mathrm{tr}\big((\sum_{m \in \Omega} \alpha_m \rho_m) p_n\big) = \sum_{m \in \Omega} \alpha_m \mathrm{tr}(\rho_m p_n) = \alpha_n.$$

□

Corollary 2.1. *Let us suppose that a QMS $\mathcal{T}$ satisfies the assumptions in Theorem 2.1. Then*

(i) For any initial state σ there exists an invariant state σ_∞ such that in the weak topology of $L_1(\mathsf{h})$,

$$\sigma_\infty = \lim_{t \to \infty} \frac{1}{t} \int_0^t \mathcal{T}_{*s}(\sigma) ds. \tag{3}$$

Moreover,

$$\sigma_\infty = \sum_{n\geq 0} tr(\sigma p_n)\rho_n.$$

(ii) If σ is an invariant state, then its ergodic attraction domain

$$\mathbb{A}(\sigma) := \{\nu \; state : \; \sigma = weak - \lim_{t\to\infty} \frac{1}{t}\int_0^t \mathcal{T}_{*s}(\nu)ds \; in \; L_1(\mathsf{h})\} \quad (4)$$

is the set of states

$$\{\nu \in L_1(\mathsf{h}) : \; tr(\nu p_n) = tr(\sigma p_n), \; \forall n \geq 0\}.$$

Proof. The existence of σ_∞ is guaranteed by Theorem 2.1 in Ref.[6] Since for each $n \in \Omega$, $p_n \in \mathcal{F}(\mathcal{T})$, by Proposition 2.17 in Ref.[2] we have for every element $x \in \mathcal{B}(\mathsf{h})$, and all $t \geq 0$ that $\mathcal{T}_t(p_n x p_n) = p_n \mathcal{T}_t(x) p_n$, which implies

$$\mathcal{T}_{*t}(p_n \sigma p_n) = p_n \mathcal{T}_{*t}(\sigma) p_n. \quad (5)$$

Then using the Markovianity of $\mathcal{T}$ and (5) we have

$$tr(\sigma p_n) = tr(p_n \sigma p_n) = tr\big(\mathcal{T}_{*t}(p_n \sigma p_n)\big) = tr\big(p_n \mathcal{T}_{*t}(\sigma) p_n\big) = tr\big(\mathcal{T}_{*t}(\sigma) p_n\big).$$

Hence

$$\begin{aligned} tr(\sigma p_n) &= \lim_{t\to\infty} \frac{1}{t}\int_0^t tr\big(\mathcal{T}_{*s}(\sigma)p_n\big)\,ds \\ &= \lim_{t\to\infty} tr\big((\frac{1}{t}\int_0^t \mathcal{T}_{*s}(\sigma)\,ds)p_n\big) = tr(\sigma_\infty p_n). \end{aligned}$$

Therefore Proposition 2.2 implies that

$$\sigma_\infty = \sum_{n\geq 0} tr(\sigma p_n)\rho_n,$$

since σ_∞ is an invariant state.

Part (ii) is an immediate consequence of (i). □

3. Applications

In this section we apply Theorem 2.1 to some quantum Markov semigroups. The following notions were introduced in Ref.[2,3]

Definition 3.1. Let $\mathcal{T} = (\mathcal{T}_t)_{t\geq 0}$ be a QMS acting on $\mathcal{B}(\mathsf{h})$. A positive operator a is called *subharmonic* (*superharmonic, harmonic*) for $\mathcal{T}$ if for all $t \geq 0$, $\mathcal{T}_t(a) \geq a$ ($\mathcal{T}_t(a) \leq a$, $\mathcal{T}_t(a) = a$, respectively).

Remark 3.1. For any QMS, I is harmonic. If $\mathcal{T}$ is conservative and p is an orthogonal projection then p is subharmonic if and only if $I-p$ is superharmonic. Hence, p is harmonic if and only p is sub and superharmonic.

Theorem 3.1. *(Fagnola- Rebolledo, Ref.[3] Th. II.1).*
The support projection of a normal stationary state for a quantum Markov semigroup is subharmonic.

Using Theorem 3.1 we proved in Ref.,[8] the following result.

Proposition 3.1. *Assume that there exists a faithful invariant state of $\mathcal{T}$. Then any subharmonic or superharmonic operator is harmonic.*

3.1. *The n-photon absorption emission process*

In this section we introduce the Lindbladian for the n-photon creation and annihilation process. We follow Ref.,[4] where the case $n=2$ was studied. The case $n=1$ corresponds with the quantum Ornstein-Uhlenbeck QMS studied in Ref.[1]

Let h be the Hilbert space $\mathsf{h}=\ell^2(\mathbb{N})$ and let a, a^+ and N be the annihilation and creation operators. Denote by $(e_k)_{k\geq 0}$ the canonical orthonormal basis of h. Let n be a fixed positive integer and G be the operator defined on the domain $\mathrm{Dom}(N^n)$ of the n-power of the number operator by

$$G=-\frac{\lambda^2}{2}a^n a^{+n}-\frac{\mu^2}{2}a^{+n}a^n-i\omega a^{+n}a^n$$

with $\lambda\geq 0, \mu>0$, $\omega\in\mathbb{R}$ and let L_1, L_2 be the operators defined on $\mathrm{Dom}(N^n)$ by

$$L_1=\mu a^n, \qquad L_2=\lambda a^{+n}.$$

Clearly G generates a strongly continuous semigroup of contractions $(P_t)_{t\geq 0}$ with

$$P_t=\mathrm{e}^{-\frac{t}{2}(\lambda^2(N+1)\cdots(N+n)+(\mu^2+2i\omega)N(N-1)\cdots(N-n+1))}.$$

For every $x\in\mathcal{B}(\mathsf{h})$ the Lindblad formal generator is a sesquilinear form defined by

$$\pounds(x)[u,v]=\langle Gu,xv\rangle+\sum_{\ell=1}^{2}\langle L_\ell u,xL_\ell v\rangle+\langle u,xGv\rangle, \tag{6}$$

for $u,v\in\mathrm{Dom}(G)=\mathrm{Dom}(N^n)$. One can easily check that conditions for constructing the minimal quantum dynamical semigroup (QDS) associated with the above G, L_1, L_2 ((H-min) in Ref.[5]) hold and this semigroup $\mathcal{T} = (\mathcal{T}_t)_{t\geq 0}$ satisfies the so called Lindblad equation

$$\langle v, \mathcal{T}_t(x)u\rangle = \langle v, P_t^* x P_t u\rangle + \sum_{\ell=1}^{2}\int_0^t \langle L_\ell P_{t-s}v, \mathcal{T}_s(x) L_\ell P_{t-s}u\rangle\, ds, \quad (7)$$

for all $u, v \in \mathrm{Dom}(G)$.

It follows that the action of $\mathcal{L}$ on the diagonal elements $x = \sum_{j\geq 0} x_j |e_j\rangle\langle e_j|$ of the linear manifold $\mathcal{M} = \mathrm{span}\{|e_j\rangle\langle e_k| : j, k \geq 0\}$ of finite range operators is given by

$$\begin{aligned}\mathcal{L}(x) = \sum_{j\geq 0} \Big(&\lambda^2 (j+n)(j+n-1)\cdots(j+1)(x_{j+n} - x_j) + \\ &\mu^2 (j(j-1)_+ \cdots (j-n+1)_+ (x_{j-n} - x_j)\Big)|e_j\rangle\langle e_j|\end{aligned} \quad (8)$$

where $(j-m+1)_+ = \max\{j-m+1, 0\}$, $1 \leq m \leq n$.

The above generator can be written in the form

$$\mathcal{L}(x) = \sum_{j\geq 0}\sum_{0\leq r\leq n-1} \Big(\mathcal{L}_r(x_r)\Big)_j |e_{nj+r}\rangle\langle e_{nj+r}|, \quad (9)$$

where $x_r \in \mathcal{M}$ is defined by $(x_r)_j = x_{nj+r}$ and for $y \in \mathcal{M}$,

$$\begin{aligned}\mathcal{L}_r(y) = \sum_{j\geq 0}\Big(&\lambda^2 \prod_{\ell=1}^{n}(nj+r+\ell)(y_{j+1} - y_j) + \\ &\mu^2 \prod_{\ell=0}^{n-1}(nj+r-\ell)_+ (y_{j-1} - y_j)\Big)|e_j\rangle\langle e_j|.\end{aligned} \quad (10)$$

Notice that $(nj+r-0)_+ = nj+r$ for all j, $(nj+r-\ell)_+ = nj+r-\ell$, for all $j \geq 1$ and each $\mathcal{L}_r$ coincides with the generator of a birth and death process with birth intensities $\lambda_j = \lambda^2 \prod_{\ell=1}^{n}(nj+r+\ell)$ and death intensities $\mu_j = \mu^2 \prod_{\ell=0}^{n-1}(nj+r-\ell)_+$. Hence, for $\nu = \lambda/\mu < 1$, we can easily find n extremal invariant states. Indeed, a straightforward computation yields the states

$$\rho_r = (1-\nu^2)\sum_{j\geq 0} \nu^{2j} |e_{nj+r}\rangle\langle e_{nj+r}|, \quad 0 \leq r < n. \quad (11)$$

Proposition 3.2. *The states ρ_r, $0 \leq r < n$, are invariant.*

Proof. Let $\mathcal{L}_*$ be the generator of the predual semigroup $\mathcal{T}_* = (\mathcal{T}_{*t})_{t\geq 0}$, acting on the Banach space $L_1(\mathsf{h})$ of trace class operators on h. Consider the approximations $\rho_{r,m} = (1-\nu^2)\sum_{j=0}^{m} \nu^{2j}|e_{nj+r}\rangle\langle e_{nj+r}|$, of ρ_r by finite range operators.

The operators $\rho_{r,m}$ belong to the domain of $\mathcal{L}_*$. For $r = 0$ we have $\mathcal{L}_*(\rho_{0,m}) = \lambda^2(1-\nu^2)\prod_{\ell=1}^{n}(nm+\ell)\nu^{2m}(|e_{n(m+1)}\rangle\langle e_{n(m+1)}| - |e_{nm}\rangle\langle e_{nm}|)$. Thus, for all $m < k$, we have

$$\|\mathcal{L}_*(\rho_{0,k} - \rho_{0,m})\|_1 \leq 2\lambda^2(1-\nu^2)\Big(\prod_{\ell=1}^{n}(nm+\ell)\nu^{2m} + \prod_{\ell=1}^{n}(nk+\ell)\nu^{2k}\Big) \quad (12)$$

which converges to zero as $m, k \to \infty$ since $\nu < 1$. Similar computations and conclusion hold for $\rho_{r,n}$, $0 \leq r < n$. Since $\mathcal{L}_*$ is closed, this proves that $\rho_r \in \mathrm{Dom}(\mathcal{L}_*)$, $1 \leq r < n$, and $\mathcal{L}_*(\rho_r) = 0$, $0 \leq r < n$. □

We shall prove that, when $\lambda > 0$, any invariant state is a convex combination of ρ_r, $0 \leq r < 0$. This corresponds with the case $\Omega = \{0, 1, 2, \cdots n-1\}$ in Theorem 2.1. In particular one can easily see that any state of the form

$$\rho = \sum_r \alpha_r \rho_r, \quad 0 < \alpha_r \leq 1, \quad \text{and} \quad \sum_r \alpha_r = 1, \quad (13)$$

is invariant and faithful. Therefore, by Proposition 4.2 in Ref.,[10] the minimal semigroup associated with the GKS-L generator (6) is conservative and consequently it is the unique solution of the Lindblad equation (7).

Let p_r be the support projection of the invariant state ρ_r. By Theorem 3.1 and Proposition 3.1, each p_r is harmonic for the n-photon semigroup and consequently $p_r \in \mathcal{F}(\mathcal{T})$. Moreover, by Proposition 2.17 in Ref.[2] we have for every element $x \in \mathcal{B}(\mathsf{h})$,

$$\mathcal{T}(p_r x p_r) = p_r \mathcal{T}(x) p_r.$$

Hence for every $0 \leq r < n$, $\mathcal{T}(p_r x p_r) \in \mathcal{A}_r := p_r \mathcal{B}(\mathsf{h}) p_r$. This proves that the hereditary subalgebra $\mathcal{A}_r$, $0 \leq r < n$, is invariant under the action of the n-photon QMS.

In order to characterize all the invariant states, we recall the following results which, apart from introducing some important objects, also help in establishing the asymptotic behavior of the reduced semigroup $\mathcal{T}^r =$

$(\mathcal{T}_t^r)_{t\geq 0}$, defined on $\mathcal{A}_r$ by means of $\mathcal{T}_t^r(p_r x p_r) := p_r \mathcal{T}_t p_r$, for each $r = 0, \ldots, n-1$. The following result is due to Frigerio and Verri (see Refs.[6,7]).

Theorem 3.2. *Let $\mathcal{S}$ be a QMS on a von Neumann algebra $\mathcal{A}$ with a faithful normal invariant state ω and let $\mathcal{F}(\mathcal{S})$, $\mathcal{N}(\mathcal{S})$ be the von Neumann subalgebras of $\mathcal{A}$*

$$\mathcal{F}(\mathcal{S}) = \{ x \in \mathcal{A} \mid \mathcal{S}_t(x) = x, \ \forall t \geq 0 \},$$

$$\mathcal{N}(\mathcal{S}) = \{ x \in \mathcal{A} \mid \mathcal{S}_t(x^* x) = \mathcal{S}_t(x^*)\mathcal{S}_t(x), \ \mathcal{S}_t(x x^*) = \mathcal{S}_t(x)\mathcal{S}_t(x^*), \forall t \geq 0 \}.$$

Then: $\mathcal{N}(\mathcal{S}) = \mathcal{F}(\mathcal{S}) = \mathbb{C}\mathbf{1}$, implies that $\lim_{t\to\infty} \mathcal{S}_{*t}(\sigma) = \omega$ for all normal state σ on $\mathcal{A}$.

The computation of $\mathcal{F}(\mathcal{T}^r)$ and $\mathcal{N}(\mathcal{T}^r)$, can be performed by means of an application of the following result by Fagnola and Rebolledo, Ref.[2]

Theorem 3.3. *Suppose that both the minimal QDS $\mathcal{T}$ associated with the operators G, L_ℓ and $\widetilde{\mathcal{T}}$ associated with and the operators G^*, L_ℓ are Markov. Moreover suppose that there exists $D \subset \mathsf{h}$ dense which is a common core for G and G^* such that the sequence $(nG^*(n-G)^{-1})u)_{n\geq 1}$ converges for all $u \in D$. Then $\mathcal{N}(\mathcal{T}) \subset \{L_k, L_k^* : k \geq 1\}'$ and $\mathcal{F}(\mathcal{T}) = \{H, L_k, L_k^* : k \geq 1\}'$.*

Here the $\{X_1, X_2 \ldots\}'$ denotes the *generalised commutator* of the (possibly unbounded) operators $X_1, X_2, \ldots$. This is the subalgebra of $\mathcal{B}(\mathsf{h})$ of all the operators y such that $yX_k \subseteq X_k y$ (i.e. $\mathrm{Dom}(X_k) \subseteq \mathrm{Dom}(X_k y)$ and $yX_k u = X_k y u$ for all $u \in \mathrm{Dom}(X_k)$) for all $k \geq 1$.

We can now prove the following result

Theorem 3.4. *Let $r = 0, 1 \ldots, n-1$ and $\mathcal{T}^r$ be the reduced semigroup on the subalgebra $\mathcal{A}_r$. Then $\mathcal{F}(\mathcal{T}^r) = \mathcal{N}(\mathcal{T}^r) = \mathbb{C}\mathbf{1}$. It follows that, for any initial state σ_r on $\mathcal{A}_r$, we have*

$$\lim_{t\to\infty} \mathcal{T}_{*t}^r(\sigma_r) = \rho_r.$$

Proof. Let us prove that $\mathcal{N}(\mathcal{T}^r) = \mathbb{C}\mathbf{1}$. If $x \in \mathcal{N}(\mathcal{T}^r)$ is a self-adjoint operator we have that $x_{mk} = 0$ for all $m, k \geq 0$ such that $m, k \not\equiv r$, $(\mathrm{mod}\, n)$. Let us consider the case $r = 0$; the other cases are similar. If $xa^n \subset a^n x$ and $xa^{+n} \subset a^{+n} x$, after simple computations we obtain

$$\begin{aligned}((nm)!)^{\frac{1}{2}} \langle e_{nm}, x e_{nk} \rangle &= \langle x\, a^{+nm} e_0, e_{nk} \rangle \\ &= \langle a^{+nm} x e_0, e_{nk} \rangle = \langle e_0, x\, a^{nm} e_{nk} \rangle = 0\end{aligned}$$

for all $k < m$. Being self-adjoint, this proves that x is diagonal. Now for any $m, k \geq 0$ after direct computations we obtain

$$(nm+n)^{\frac{1}{2}}\cdots(nm+1)^{\frac{1}{2}}x_{nm+n,nk} = \langle xa^{+n}e_{nm}, e_{nk}\rangle =$$
$$\langle e_{nm}, a^{n}xe_{nk}\rangle = \langle e_{nm}, xa^{n}e_{nk}\rangle = (nk-n+1)^{\frac{1}{2}}\cdots(nk)^{\frac{1}{2}}x_{nm,nk-n},$$

hence with $k = m+1$ we obtain that $x_{nk,nk} = x_{nk-n,nk-n}$ for all $k \geq 1$. This proves that x is a multiple of the identity operator. For a general element $x \in \mathcal{N}(\mathcal{T}^r)$ we can use its decomposition as a linear combination of self-adjoint operators both in $\mathcal{N}(\mathcal{T}^r)$. Then we have that $\mathcal{F}(\mathcal{T}^r) = \mathcal{N}(\mathcal{T}^r) = \mathbb{C}\mathbb{1}$. The conclusion follows then from Theorem 3.2 (iv). □

The subharmonic projections are characterized by the following

Theorem 3.5. *A projection p is $\mathcal{T}$-subharmonic if and only if its range $R(p)$ is an invariant subspace for the operators P_t $(t \geq 0)$ and $L_\ell pu = pL_\ell u$ for all $u \in Dom(G) \cap R(p)$ and all $\ell \geq 1$.*

As an application of the above theorem we have the following

Theorem 3.6. *If $\lambda > 0$:*

(i) *Subharmonic projections for the n-photon absorption emission QMS $\mathcal{T}$ are 0, $\mathbb{1}$*

$$p_r = \sum_{k\geq 0} |e_{nk+r}\rangle\langle e_{nk+r}|, \quad 0 \leq r < n$$

and sums of orthogonal projections of the former class.

(ii) $\mathcal{F}(\mathcal{T}) = \{p_r : 0 \leq r < n\}'' = \{\sum_{0\leq r<n} \lambda_r p_r : \lambda_r \in \mathbb{C},\ 0 \leq r < n\}$.

Proof. Any invariant subspace of a normal compact operator is generated by eigenvectors (e.g. by Theorem 4 p. 272 in Ref.[11]). The operators $(P_t)_{t\geq 0}$ belong to $L_2(\mathsf{h})$ since they are limit of finite-range operators and the series of eigenvalues is square summable. All the e_k's are eigenvectors. Therefore any invariant closed subspace $\mathcal{I}_K$ has the form

$$\mathcal{I}_K = \overline{\text{span}}\{e_l : l \in K\}, \quad K \subset \mathbb{N}.$$

Assume $\lambda > 0$. Since $\mathcal{I}_K$ is invariant under the action of L_1, L_2, hence $L_1 e_j = \mu^2(\prod_{\ell=0}^{n-1}(j-\ell)_+)^{\frac{1}{2}} e_{j-n} \in \mathcal{I}_K$ and $L_2 e_j = \lambda^2(\prod_{\ell=1}^{n}(j+\ell))^{\frac{1}{2}} e_{l+n} \in \mathcal{I}_K$. Therefore if some $0 \leq r < n$ belongs to K then the whole coset $[r] = \{l \geq 0 :\ l \equiv r \pmod n\}$ is a subset of K and, consequently, $K = \cup_{l_r}[l_r]$ for some $0 \leq l_r < n$. This proves (i).

Since each projection p_r is harmonic we have $\{p_r : 0 \leq r < n\} \subset \mathcal{F}(\mathcal{T})$. Now, if $p \in \mathcal{F}(\mathcal{T})$ is a projection, then p is harmonic; therefore it is one of the projections in item (i). This proves the Theorem. □

By Corollary 2.1, the attraction domain $\mathbb{A}(\rho)$ of an invariant state ρ is given by

$$\mathbb{A}(\rho) = \{\nu \in L_1(\mathsf{h}) : \; tr(\nu p_r) = tr(\rho p_r), \; \forall \, 0 \leq r < n\}.$$

3.2. *The asymmetric exclusion QMS*

The GKSL generator of the asymmetric exclusion QMS acts on the von Neumann algebra of all bounded operators on $\mathsf{h} = \bigotimes^{\varphi}_{l \in \mathbb{Z}^d} \mathsf{h}_l$, the stabilized tensor product of $\mathsf{h}_l = \mathbb{C}^2$, $l \in \mathbb{Z}^d$, with respect to the stabilizing sequence $\varphi = (|0\rangle)_{l \in \mathbb{Z}^d}$.

Let $r \in \mathbb{Z}^d, r = (r_1, \ldots, r_d)$, $|r| := \sum_i |r_i|$, $S := \{\eta : \mathbb{Z}^d \to \{0,1\} \mid \exists \, n \in \mathbb{Z}_+ := \mathbb{N} \cup \{0\}$ so that $\eta(r) = 0$ if $|r| \geq n\}$. We shall denote by $\eta(r)$ or η_r the r-th coordinate of η. For any $\eta \in S$ let $supp(\eta) := \{r \in \mathbb{Z}^d | \eta(r) = 1\}$ and $|\eta| := \#supp(\eta)$. Therefore, $S = \{\eta : \mathbb{Z}^d \to \{0,1\} \mid |\eta| < \infty\}$ and $|\eta| = |\xi|$ if and only if $\#(supp(\eta) \backslash supp(\xi)) = \#(supp(\xi) \backslash supp(\eta))$. The elements of $\mathbb{Z}^d$ will be called *sites*; $|\eta|$ will be *the size of the configuration* and for $n \in \mathbb{Z}_+$, let S_n the set of configurations of size n. The subset $\beta = \{|\eta\rangle : \eta \in S\}$ is an orthonormal basis of h. For each $n \geq 1$, let $\beta_n = \{|\eta\rangle : \eta \in S_n\}$, $V_n := \overline{\text{span}\,\beta_n}$ and let us denote by p_n the orthogonal projection on V_n.

Let $r, s \in \mathbb{Z}^d$, $r \neq s$, and $\eta \in S$. Then we define η_{rs} as $\eta_{rs} = \eta + (-1)^{\eta(r)} \mathbb{1}_r + (-1)^{\eta(s)} \mathbb{1}_s$, where $\mathbb{1}_r$ is the indicator function of the site $\{r\}$. For every pair $r \neq s \in \mathbb{Z}^d$, we define the operator $C_{rs} : \mathsf{h} \to \mathsf{h}$ as,

$$C_{rs}|\eta\rangle := (1 - \eta(s))\eta(r)|\eta_{rs}\rangle \;\; \text{for} \;\; \eta \in S,$$

and it is extended by linearity and continuity to the whole of h.

The GKSL formal generator of asymmetric exclusion QMS is represented as

$$\mathcal{L}(x)[\eta, \xi] = \Phi(x)[\eta, \xi] + \langle G\eta, x\xi\rangle + \langle \eta, xG\xi\rangle, \tag{14}$$

where

$$\Phi(x) = \sum_{\{(r,s) \in \mathbb{Z}^d \times \mathbb{Z}^d : r \neq s\}} (2a^+_{rs} C^*_{rs} x C_{rs} + 2a^-_{rs} C_{rs} x C^*_{rs}), \;\; x \in \mathcal{B}(\mathsf{h}), \tag{15}$$

with a^+_{rs}, $a^-_{rs} \in \mathbb{R}_+ \setminus \{0\}$ for all $(r,s) \in \mathbb{Z}^d \times \mathbb{Z}^d$, $r \neq s$. The operator G is defined by $G = -\frac{1}{2}\Phi(I) - iH$ with H the self-adjoint operator

$$H = \sum_{\{(r,s)\in\mathbb{Z}^d\times\mathbb{Z}^d : r\neq s\}} \left(b^+_{rs} C^*_{rs} C_{rs} - b^-_{rs} C_{rs} C^*_{rs}\right), \quad b^+_{rs}, b^-_{rs} \in \mathbb{R}. \tag{16}$$

Definition 3.2. A family of strictly positive numbers $\{a^+_{rs}, a^-_{rs} : r, s \in \mathbb{Z}^d\}$ satisfies a *Kolmogorov reversibility condition* if for all $n \geq 2$, and any cycle in $\mathbb{Z}^d$, i.e., a finite sequence $r_0 \neq r_1, \ldots, r_n = r_0$, we have

$$\prod_{j=1}^{n} a^+_{r_{j-1}r_j} = \prod_{j=1}^{n} a^-_{r_{j-1}r_j} \quad \text{and} \tag{17}$$

$$\sum_{s\in\mathbb{Z}^d} \frac{a^+_{r_0 s}}{a^-_{r_0 s}} < \infty, \quad \forall r_0 \in \mathbb{Z}^d. \tag{18}$$

The proof of the following theorem is in Ref.[8]

Theorem 3.7. *Let $\{a^+_{rs}, a^-_{rs} : r, s \in \mathbb{Z}^d\}$ be a family of positive numbers. Then following conditions are equivalent:*

(a) *Kolmogorov reversibility condition holds,*
(b) *There exists a positive function q on $\mathbb{Z}^d$ such that $\frac{a^+_{rs}}{a^-_{rs}} = \frac{q(r)}{q(s)}$, for all $r \neq s \in \mathbb{Z}^d$, and $\sum_{r\in\mathbb{Z}^d} \frac{1}{q(r)} < \infty$*
(c) *There exists a faithful state $\rho = \sum_{\eta\in S} \rho(\eta)|\eta\rangle\langle\eta|$ invariant for the semigroup $(\mathcal{T}_t)_{t\geq 0}$ that satisfies the condition*

$$\rho(\eta_{rs}) = \frac{a^+_{rs}}{a^-_{rs}} \rho(\eta), \quad \textit{whenever } (1-\eta_s)\eta_r = 1. \tag{19}$$

Remark 3.2. Condition (19), that we call *infinitesimal detailed balance*, is fundamental to prove that the asymmetric exclusion quantum Markov semigroup satisfies the quantum detailed balance condition, see Ref.[10]

Let ρ be the invariant state in Theorem 3.7, then, see Ref.,[8] we have the decomposition

$$\rho = \sum_{j=0}^{\infty} \mathrm{tr}(\rho p_j) \rho_j, \tag{20}$$

where

$$\rho_j = \frac{1}{\mathrm{tr}(\rho p_j)} \sum_{\eta\in S_j} \rho_\eta |\eta\rangle\langle\eta|$$

with $\rho_\eta = Z\frac{1}{q(r_1)\cdots q(r_k)}$, $\{r_1,\ldots,r_k\} = \mathrm{supp}(\eta)$, $Z = \Pi_{r\in\mathbb{Z}^d}\frac{q(r)}{1+q(r)}$ and q is a positive function on $\mathbb{Z}^d$ as in part (*b*) of Theorem 3.7. For each $j \geq 0$, ρ_j is an invariant state and p_j is its support projection.

The algebra of fixed points and the subharmonic projections of this semigroup were characterized in Ref.[8] as follows.

Proposition 3.3. *For the asymmetric exclusion QMS $\mathcal{T}$,*

(i)

$$\mathcal{F}(\mathcal{T}) = \{p_n \mid n \in \mathbb{N}\cup 0\}'' = \left\{\sum_n \lambda_n p_n \mid (\lambda_n)_n \in \ell_\infty(\mathbb{N})\right\},$$

where $''$ means the double commutant. In other words, the algebra of fixed points is the von Neumann algebra generated by all the projections p_n.

(ii) The subharmonic projections are of the form $\sum_{n\in A} p_n$, for some subset A of $\mathbb{N}$. In particular, the only nonzero minimal subharmonic projections are the p_n's. Moreover, for every $n \geq 0$ the hereditary subalgebra $\mathcal{A}_n = p_n\mathcal{B}(\mathsf{h})p_n$ is invariant under the action of the semigroup.

As a consequence of the above proposition and item (c) of Theorem 2.1 we have that for the asymmetric exclusion semigroup

$$\mathcal{F}(\mathcal{T}_*) = \left\{\sum_{n\in\Omega} \alpha_n\rho_n \mid (\alpha_n)_{n\geq 0} \in \ell_1(\Omega)\right\},$$

with $\Omega = \mathbb{N}$. Hence, all its invariant states are diagonal w.r.t. the orthonormal basis $\beta = \cup_{n\in\Omega}\beta_n$. Moreover by Corollary 2.1, the attraction domain $\mathbb{A}(\sigma)$ of any invariant state σ is given by

$$\mathbb{A}(\sigma) = \{\nu \in L_1(\mathsf{h}) : \ tr(\nu p_n) = tr(\sigma p_n), \ \forall\, n \geq 0\}.$$

Acknowledgement: The authors are grateful to the organizers of the *30th Conference on Quantum Probability and Related Topics*, Santiago, Chile, for warm hospitality during the conference. We also thank Guido Raggio for driving our attention to the result of A. Frigerio in Proposition 2.1.

References

1. F. Cipriani, F. Fagnola and J.M. Lindsay, *Spectral analysis and Feller property for Ornstein-Uhlembeck semigroups.*, Comm. Math. Phys. **210**, 85 (2000).
2. F. Fagnola, and R. Rebolledo, *Lectures on Qualitative Analysis of Quantum Markov Semigroups*, in: Interacting Particle Systems, QP-PQ Quantum Probability and White Noise Analyisis **XIV** pp. 137-240, Word Scientific (2002).
3. F. Fagnola and R. Rebolledo, *Subharmonic Projections for a Quantum Markov Semigroup*, Journal of Mathematical Physics **43**, 1074–1082, (2002).
4. F. Fagnola and R. Quezada, *Two-photons absortion and emission process*, Infinite Dimensional Analysis, Quantum probability and Related Topics **8, No. 4** 573–591, (2005).
5. F. Fagnola, *Quantum Markov Semigroups and Quantum Markov Flows*, Proyecciones **18**, no.3, 1–144, (1999).
6. A. Frigerio, *Quantum, Dynamical Semigroups and Approach to Equilibrium*, Lett. in Math. Phys. **2**, 79-87, (1977).
7. A. Frigerio and M. Verri, *Long-Time Asymptotic Properties of Dynamical Semigroups on W^*-Algebras*, Math. Zeitschrift, **180**, 275–286, (1982).
8. J.C. García, L. Pantaleón-Martínez and R. Quezada, *Invariant states For the Asymmetric Exclusion Quantum Markov Semigroup*, Communications on Stochastic Analysis **Vol. 3, No. 3** 419–431, (2009).
9. L. Pantaleón-Martínez, *El Semigrupo Cuántico de Exclusión Asimétrica*, Tesis de Doctorado, UAM-I, México (2008).
10. L. Pantaleón-Martínez and R. Quezada, *The asymmetric exclusion quantum Markov semigroup*, Infinite Dimensional Analysis, Quantum Probability and Related Topics **12**, 367–385, (2009).
11. J. Wermer, *On invariant subspaces of normal operators.* Proc. Amer. Math. Soc. **3**, 270–277, (1952).

STATE ESTIMATION METHODS USING INDIRECT MEASUREMENTS

K. M. HANGOS*

Process Control Research Group, Computer and Automation Research Institute Budapest, Hungary
E-mail: hangos@scl.sztaki.hu
http://daedalus.scl.sztaki.hu/PCRG/

L. RUPPERT

Department of Analysis, Budapest University of Technology and Economics, Budapest, Hungary
E-mail: ruppertl@math.bme.hu

This paper presents mathematically well grounded statistical methods for state estimation in the indirect measurement setting, when the measurement is performed on an ancilla system that is put into interaction with the unknown one. Both the unknown and the ancilla quantum systems are assumed to be quantum bits. The measurements applied on the ancilla qubit are the classical von Neumann measurements using the Pauli matrices as observables. The repeated measurements performed on the ancilla enables us to construct estimators of the initial state of the unknown system. Based on the statistical properties of the considered indirect measurement scheme,[16] three related but different approaches are proposed and investigated: (i) a direct estimation procedure that is based on the estimated relative frequencies of the characterizing conditional probability densities, (ii) Bayesian recursive approach for state estimation, and (iii) a martingale approach that bases the estimator on the stopping times of the state evolution as a martingale driven by the repeated measurements.

The statistical properties, i.e. the unbiasedness and the efficiency of the proposed procedures are investigated both analytically and experimentally using simulation.

Keywords: Quantum state estimation, Indirect measurements, Martingales, Bayes estimate.

1. Introduction

It is well known that a projective measurement applied to a quantum system will change the state of the measured system in an irreversible way depending on the measurement outcome.[12] In addition, the probabilistic nature of the measurement result calls for applying a state estimation (or state tomography) approach if one wants to have information about the state of the quantum system.

Similarly to any realistic physical measurement, a quantum measurement is almost always realized by taking a measurement device that is put in interaction with the system to be measured, and then to "read" the meter on the measurement device. In the macroscopic measurement situation the measurement device is "small" compared to the system to be measured, thus the measurement back-action, i.e. the disturbance caused by the measurement is negligible, but that is not the case in the quantum setting. In quantum state estimation the above measurement configuration, when the 'unknown' quantum system is coupled with a 'measurement' (also called *'ancilla'*) system, and the measurements are only applied on the ancilla system [6] is termed an *indirect measurement scheme.* In the field of solid-state quantum bits a not fully realistic but conceptually simple model of indirect projective measurement is when the measured qubit interacts with another (ancillary) qubit, which is later measured in the "orthodox" projective way.[10] A cavity quantum electrodynamics experiment, in which a single photon could be detected non-destructively has also been reported,[11] where atomic interferometry was used to measure the phase shift in an atomic wavefunction, caused by a cycle of photon absorption and emission.

The notion of *weak measurements* is related to the notion of indirect measurements, but approaches the measurement back-action problem in a different way. A weak measurement[4,18] is designed not to demolish the system state completely, i.e. the post-measurement state still contains information about the original one, but on the price of a decreased information gained from the measurement. In the weak measurement setting the measured variable has an effective interaction with the unknown one "in the limit of weak coupling" thus minimizing the disturbance caused (and the information gained) by the measurement.

Besides of state estimation, weak measurements are used for other related tasks, such as state purification or noise reduction combined with suitable feedback, see e.g.,[15] In a particularly interesting paper Korotkov and Jordan[9] have shown that "it is possible to fully restore any unknown, pre-measured state, though with probability less than unity" for solid-state qubits and continuous time measurements. Recently, a similar approach for reversing the weak quantum measurement for a photonic qubit has also been reported.[8]

However, it is intuitively clear, that one must make a compromise between the information gained in a measurement and the disturbance or demolition caused by it. The general impossibility of determining the state of a single quantum system is proved[2] whatever measurement scheme is used. This indicates that the efficiency or precision provided by an indirect measurement scheme is necessarily smaller than that of a scheme that uses von Neumann measurements.

The aim of this work is to propose mathematically well grounded statistical methods for for state estimation for the indirect measurement setting and compare their efficiency to the usual direct approaches.

2. The investigated indirect measurement scheme

Indirect measurement means that the projective measurements are performed on the ancilla system (being in state θ_M) attached to the one we are interested in (θ_S). The *Bloch-vector* representation of the states of quantum bits is used[12] here in the form of

$$\rho_S(k) = \frac{1}{2}(I + \theta_S(k)\sigma^S) \ , \ \rho_M(k) = \frac{1}{2}(I + \theta_M(k)\sigma^M), \tag{1}$$

where θ_S and θ_M are 3 dimensional real vectors.

In the composite system (in state ρ_{S+M}) an indirect measurement corresponds to the observables of the form $I \otimes A_M$, where A_M is a self adjoint operator on the Hilbert space of system M. For the sake of simplicity, it is assumed, that A_M is a Pauli spin operator.

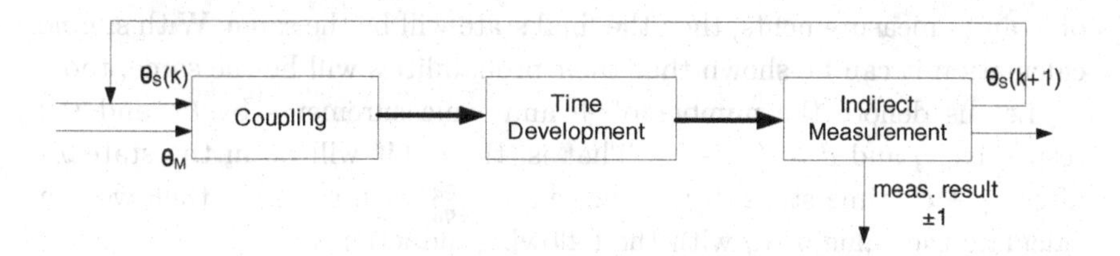

Fig. 1. Signal flow diagram of indirect measurement

The above described *measurement strategy* is shown schematically in Fig. 1. At each time instant of the discrete time set, an ancilla qubit prepared in a known state is coupled to the unknown system S. They evolve according to the bipartite dynamics for the sampling time h, and at the end of the sampling interval, a von Neumann measurement is performed on the ancilla qubit. At the next time instant, the previous steps are repeated.

It has been shown,[16] that the parameters of the strategy can be chosen such that a selective indirect scheme for estimating the second coordinate of the unknown qubit state vector θ_{S2}. The effect of the interaction parameters and that of the ancilla qubit state can then be collected in a constant $0 \leq c \leq 1$.

2.1. *Mathematical problem statement*

In the following let $x(k)$ be the second coordinate of the Bloch vector of the unknown qubit in the kth time step and c be a parameter that characterizes the state of the ancilla qubit and that of the interaction parameters. Then[16]

$$x(k+1) = \begin{cases} \frac{x(k)+c}{1+cx(k)}, \text{ with } \frac{1+cx(k)}{2} \text{ probability: + measurement} \\ \\ \frac{x(k)-c}{1-cx(k)}, \text{ with } \frac{1-cx(k)}{2} \text{ probability: - measurement} \end{cases} \quad (2)$$

Note that if $c = 1$ then we obtain the standard measurement scheme for the direct state estimation of a qubit (see below in sub-section 2.2), where

$$x(1) = \pm 1 \quad , \quad Prob(\pm 1) = \frac{1}{2}(1 \pm x(0)).$$

All of the possible cases of states can be ordered in a line such a way, that after each measurement we jump in the neighboring state on the left or right side:[15]

$$\cdots \rightleftarrows x_{-3} \rightleftarrows x_{-2} \rightleftarrows x_{-1} \rightleftarrows x_0 \rightleftarrows x_1 \rightleftarrows x_2 \rightleftarrows x_3 \rightleftarrows \ldots$$

where $x_0{=}x(0)$. Hence if two outcome sequences contain the same number of + and - measurements, then the final state will be the same. With simple calculation it can be shown that their probabilities will be the same, too.

Let us denote the number of + and - measurements by ℓ_+ and ℓ_-, respectively, and $d = \ell_+ - \ell_-$. That is, the qubit will be in the state x_d after $\ell_+ + \ell_-$ time steps. If we denote $x_d = \frac{z_d}{q_d}$ with $\ell_+ > \ell_-$, then we can calculate the value of x_d with the following induction

$$(z_d, q_d) \rightarrow (z_d + cq_d, q_d + cz_d) = (z_{d+1}, q_{d+1}) \quad (3)$$

where $z_0 = x_0$ and $q_0 = 1$. It is easy to see that z_d and q_d are linear in x_0. Moreover if $z_d = a_d + b_d x_0$, then $q_d = b_d + a_d x_0$. Thus

$$x_d = \frac{a_d + b_d x_0}{b_d + a_d x_0} \quad (4)$$

Similarly we get:

$$x_{-d} = \frac{-a_d + b_d x_0}{b_d - a_d x_0} \quad (5)$$

where a_d and b_d are the same as in x_d. They can be calculated using the recursion, and they are polynomials of c.

It follows from the above properties that if we have the *measurement record*, that is, the observed measurement outcomes (+1s and -1s), and know the *parameter* c, then we can calculate the state of the unknown qubit $x(k)$ $(= x_{\ell_+ - \ell_-})$ at any time instance from the initial state x_0. Therefore, **the only meaningful problem statement is to estimate the initial state x_0 from the measurement record and from c.**

2.2. *The standard measurement scheme for direct state estimation of qubits*

The performance of the proposed methods in the indirect setting will be compared to that of the direct state estimation methods using the same observables and projective measurements.

The most widespread and convenient method for estimating the state of a qubit is to use direct measurements on multiple copies of identically prepared qubits. The so called standard scheme applies the Pauli matrices σ_i, $i = 1, 2, 3$ as observables. Then the estimator for the Bloch vector θ of the qubit is

$$\Phi_{\mathbf{n}} = \begin{bmatrix} 2\nu(n_1, \sigma_1, +1) - 1 \\ 2\nu(n_2, \sigma_2, +1) - 1 \\ 2\nu(n_3, \sigma_3, +1) - 1 \end{bmatrix} \tag{6}$$

when σ_i is measured n_i times with the relative frequency of its $+1$ outcome $\nu(n_i, \sigma_i, +1)$, $1 \leq i \leq 3$. It can be shown that this estimate is unbiased and efficient.

The choice $n_1 = n_2 = n_3 = r$ constitutes the so-called *standard measurement scheme* for qubits whose mean quadratic error matrix is

$$V_{3r}^{st}(\theta) = \frac{1}{r} \begin{bmatrix} 1 - \theta_1^2 & 0 & 0 \\ 0 & 1 - \theta_2^2 & 0 \\ 0 & 0 & 1 - \theta_3^2 \end{bmatrix} . \tag{7}$$

3. Conditional probability density function approaches

Traditionally, maximum-likelihood (ML) or Bayesian estimation procedures are very popular for quantum state estimation with compatible parametrization and distance.[3] Therefore, it is straightforward to use them in the indirect setting, too.

3.1. *A conditional histogram approach*

A simple method that is based on the direct estimation of the conditional probabilities in Eq. (2) is proposed in this subsection.

Let us assume to have a single copy of the unknown and ancilla qubit pair, and we have collected the outcomes of a projective measurement on the ancilla qubit $\{y(k)|k = 1, ..., N\}$. Let us fix the number of measurements N and the number of investigated systems states M, such that

$$k = 1, 2, ..., N; \quad N >> M$$

The systems starts from an unknown initial state x_0 to be estimated.

Data collection The first step of the estimation procedure is to collect the conditional relative frequencies ν_d of the "+1" measurement outcome to each considered relative systems states, i.e. to compute the pairs

$$(\nu_d, x_d), \quad d \in \{-M, ..., -1, 0, 1, ...M\} \tag{8}$$

Estimation The estimation is performed in three substeps.

(a) First we can construct an estimate for the relative states x_d by using Eq. (2)

$$\hat{x}_d = \frac{2\nu_d - 1}{c} \tag{9}$$

(b) Then we can use Eqs. (4) and (5) for $d \geq 0$ and $d < 0$, respectively to derive estimators for the unknown initial state x_0

$$\hat{x}_0^{(d)} = \frac{s \cdot a_d - b_d \hat{x}_d}{s \cdot a_d \hat{x}_d - b_d} = \frac{s \cdot a_d c - 2b_d \nu_d - b_d}{2s \cdot a_d \nu_d - s \cdot a_d - b_d c} \tag{10}$$

where $s = \text{sign}(d)$.

(c) Finally, we can construct an overall estimate of x_0 from the above estimates $\hat{x}_0^{(d)}$ by averaging them, for example.

Properties of the estimator The estimators from (10) are biased, because they are nonlinear functions of the relative frequencies ν_d. The magnitude of the bias depends on the parameters of the problem, i.e. on x_0 and c.

Tuning the parameters of the algorithm The parameters of the algorithm can be chosen based on the parameters of the problem (x_0 and c).

- *The number of measurements N.* The information is concentrated in the beginning of the trajectory, therefore we need to have it so large that we do not get too close to ± 1. N is chosen reasonably if it is in the order of $1/c^2$.
- *The number of state points M.* It cannot be too large, but it should be large enough to get meaningful estimates. As x_d becomes more distant from x_0 the estimation (10) is less efficient. Therefore, x_d should not be too close to ± 1. A meaningful approach is to choose $M = \alpha \cdot \sqrt{N}$ with $0 < \alpha \leq 1$.
- The optimal way of combining the estimates $\hat{x}_0^{(d)}$ to form an overall estimate $\hat{x}_0^*$ is to be determined from statistical considerations. If d is close to 0 then (10) will have similar variance like (9). It is known that if we have independent measurements then the optimal linear measurement is if we have weights in ratio of reciprocal of variances. In this case the variance of $\hat{x}_0^{(d)}$ is constant times $\frac{1}{N_d}$ (with N_d being the data points belonging to the relative state x_d), so the weights will be N_d in the weighted average.

Simulation investigations The values $c = 0.1$, $x_0 = 0.4$ and $N = 1000$ were used in the simulations. Two different M values were investigated: $M = 10$ and $M = 20$.

If the estimator (9) resulted in a value with absolute value greater than 1, then $\hat{x}_0^{(d)} = \pm 1$ was used depending on the positivity of (9). The estimate was constructed from the individual estimators (10) by taking the weighted average.

In order to improve the efficiency of the estimate, we have repeated the simulation experiments using multiple copies of the same unknown and ancilla qubit pairs, and have averaged the estimates. The figures illustrating the simulation results depict the empirical mean values and the empirical mean square errors (MSEs) as a function of the used qubit pairs.

In the $M = 10$ case a large bias was observed because of the low number of relative frequency data, that is seen in the upper sub-figure of Figure 2. For bigger numbers of the used qubit pairs this biasness vanishes, hence the mean square error becomes quite accurate as shown in the lower sub-figure.

In the $M = 20$ case one could expect to have better estimates but this is not the case. Because of the low number of relative frequency data the additional estimates that are far away form the initial state are so bad that they make the estimate even worse. In Figure 2 we can see that the convergence of the estimate to the true value is much slower, it is not

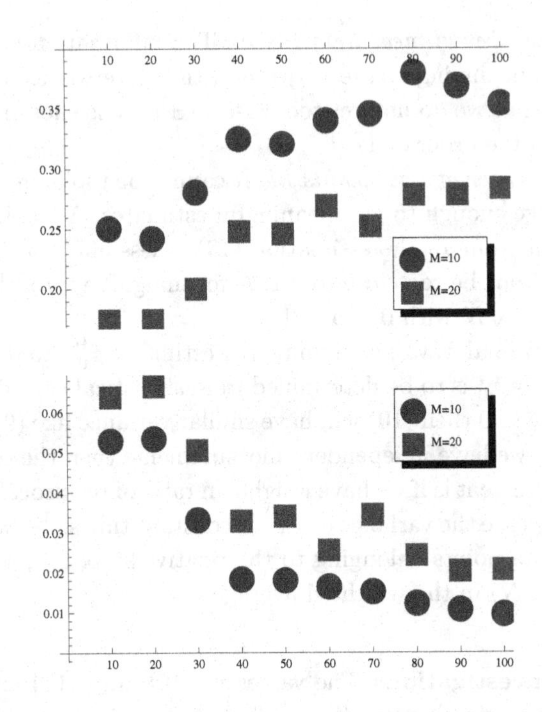

Fig. 2. The mean and the MSE as a function of the number of used qubits

surprising that mean square error is greater, too.

3.2. *Bayesian state filtering*

In this section a Bayesian parameter estimation method is proposed to estimate the initial state x_0. Let us suppose that $f_0(x)$ is an apriori density function of x_0. A sequence of indirect measurements is performed as before, and after each measurement the estimate of x_0 and the state evolution $x(k)$ are updated.

If we denote the outcome of the kth measurement by $y(k) \in \{-1, 1\}$, then we can calculate the actual state $x(k)$ recursively from Eq. (2) as

$$x(k) = \frac{x(k-1) + y(k)c}{1 + y(k) \cdot c \cdot x(k-1)}, \tag{11}$$

where $x(0) = x_0$. Of course, as we seen before, after each time step

$$x(k) = \frac{a_d x_0 + b_d}{a_d + b_d x_0},$$

with some constant a_d and b_d that only depend on c and on the actual difference between the positive and negative measurement outcomes d. We will calculate the actual constants together with the simple recursion (11) in each step, because we update the probabilities also recursively.

Let the posterior probability distribution of the estimate x_0 be $f_k(x)$ after k step. Then the probabilities evaluate in the following way

$$f_{k+1}(x) = Prob(x_0 = x | y(1), \ldots, y(k+1)) =$$
$$= Prob(y(k+1)|x_0 = x, y(1), \ldots, y(k)) \cdot Prob(x_0 = x, y(1), \ldots, y(k)) \propto$$
$$\propto (1 + y(k+1) \cdot c \cdot x(k)) \cdot f_k(x) \tag{12}$$

where $\propto$ means proportionality up to a normalizing constant.

Recursive estimation method Let us fix the length of the measurement sequence to be N. If we do not know anything about the initial state, then the standard procedure is to use the uniform distribution on the state space as prior, i.e. let $f_0(x) = 1/2$. As the calculations are not analytically feasible, we only calculate the values of $f_k(x)$ on a grid of discrete values of $x_j \in [-1, 1]$.

Then, for each step (from $k = 0$ to $N - 1$) the following substeps are performed:

(1) Perform a measurement and record $y(k+1)$
(2) Update the density function $f_{k+1}(x) = (1 + y(k+1) \cdot c \cdot x(k)) \cdot f_k(x)$
(3) Update the new state distribution $x(k+1) = \frac{x(k)+y(k+1)\cdot c}{1+y(k+1)\cdot c\cdot x(k)}$

Having completed the above substeps, the whole sequence can be repeated on another qubit pair, etc. The obtained posterior $f_N(x)$ of the previous copy is used as the new apriori distribution on x_0 for the next copy of the unknown and ancilla qubit pairs:

$$f_0(x) := \frac{f_N(x)}{\int_{-1}^{1} f_N(x)dx}$$

Simulation results The above described method was applied on multiple copies of qubit pairs with $x_0 = 0.4$ and $c = 0.1$. Altogether 10000 copies were used, i.e. the above recursion was performed 10000 times. The length of each sequence was $N = 100$.

Figures 3 show the evolution of the probability density function of the estimate of x_0 after 10 and 1000 steps. The convergence seems to be smooth, but the last sub-figure reveals a numerical issue that need to be solved, namely, we should somehow change the grid of calculation dynamically, to obtain accurate results.

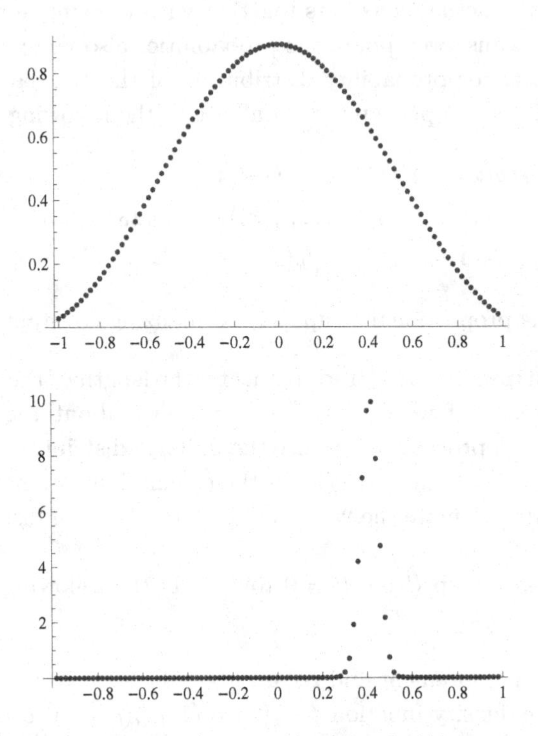

Fig. 3. Density function after 10 and 1000 steps

From the probability distribution we can easily get a point estimate for x_0. The easiest way is to calculate the mean of the density function. Another possibility is to fit a normal distribution to the density function using the least square method. Both of the above methods give results with similar properties.

The empirical statistics of the estimate have been analyzed as functions of the used qubit pairs. In order to obtain more precise result we repeated the whole procedure 100 times, and averaged the results to calculate the empirical mean value and empirical mean square error (MSE). The empirical mean of the estimator can be shown in the upper sub-figure of Figure 4, while the MSE is in the lower sub-figure. We can see in Figure 4 that the estimate is biased; that is natural because we started from the uniform distribution on the $[-1, 1]$ interval. But it can also be seen that the esti-

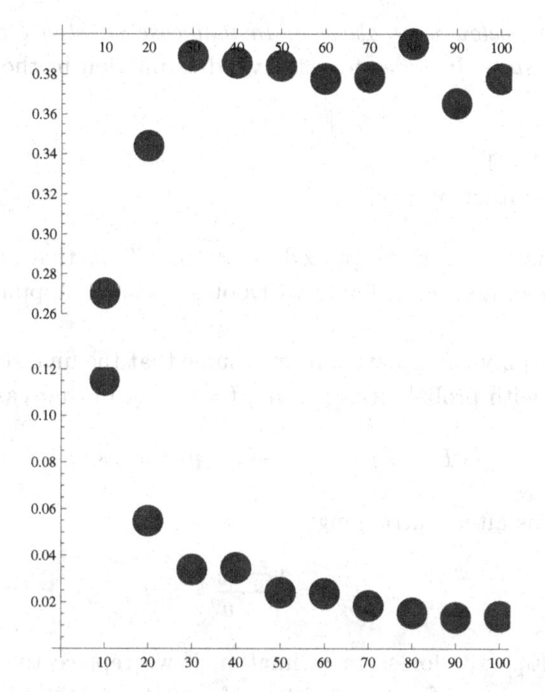

Fig. 4. The mean and the MSE of the estimator as a function of the number of used qubits

mate is asymptotically unbiased. The lower sub-figure of Figure 4 shows a convergence of MSE to zero in the order of $1/N$.

4. The martingale approach

The above two approaches result in a biased estimate of the initial state, therefore a novel estimation method based on the martingale property of the state evolution process is proposed in this section.

4.1. *The martingale generated by the state evolution*

The process described in (2) is a martingale, because it is easy to check that

$$E(x(k+1)) = x(k).$$

We can make use of this property if *we fix the values u, v $(u < x_0 < v)$, and then we start the process from x_0 by performing indirect measurements until* *we reach either below u or above v. In that case we stop the process and note the final state.* In order to get a valid estimation method we assume that

- $|u| < 1$, $|v| < 1$
- x_0 is in the interval $[u, v]$.

Assume that we stop the procedure at time T, in that case T will be a stopping time, and accordingly to Doob's optional stopping theorem[19] $E(x(T)) = x_0$.

We can simplify the situation if we assume that the final state is exactly either u or v, with probabilities p and q $(= 1 - p)$. In that case

$$E(x(T)) = p\,u + (1 - p)\,v = x_0 \tag{13}$$

and one obtains after rearranging:

$$p = \frac{v - x_0}{v - u}$$

We can use Eq. (13) for state estimation if we replace the probabilities $(p, 1 - p)$ with relative frequencies after N repeated settings using different unknown and ancilla qubit pairs. Let us denote $\nu_u = N_u/N, \nu_v = N_v/N$, where N_i $(i \in u, v)$ is the number of experiments with the final state i. Then

$$\hat{x}_0 = \nu_u\,u + \nu_v\,v = \nu_u\,u + (1 - \nu_u)\,v = v + (u - v)\nu_u$$

The variance of this estimate is:

$$Var(\hat{x}_0) = (u - v)^2\,Var(\nu_u) = (u - v)^2\,\frac{p(1 - p)}{N}$$

because N_u has a binomial distribution with parameters (N, p). After substitution we obtain:

$$Var(\hat{x}_0) = \frac{1}{N}(v - x_0)(x_0 - u)$$

Note if we have $u = -1$ and $v = 1$ then we get an equivalent method to the direct standard method (see sub-section 2.2) with variance:

$$Var(\hat{x}_0) = \frac{1 - x_0^2}{N}$$

4.2. *Stopping relative to the initial state*

The above described procedure is not feasible because we do not know neither the state $x(k)$ nor x_0, so we can not know when we reach the stopping time. Therefore, an *initial state relative* method of detecting a stopping time is proposed where the final states are defined using a given distance from the unknown initial state.

Data collection Let be $d(k)$ the number of $+1$ measurements minus the number of -1 measurements after the first k measurements. Let us repeat the measurements until $d(k)$ reaches $\pm D$, where D is a given integer, let be in this case $\tau = k$.

In this case τ is a stopping time, and the final state will be x_+ or x_-, accordingly if $d(\tau)$ is equal to $+D$ or $-D$. We can calculate x_+ and x_- from the recursion (3) to have $x_+ = \frac{y_D}{q_D}$, where $y_D = a_D + b_D x_0$ and $q_D = b_D + a_D x_0$. Therefore

$$x_+ = \frac{a_D + b_D x_0}{b_D + a_D x_0}$$

Similarly we get:

$$x_- = \frac{-a_D + b_D x_0}{b_D - a_D x_0}$$

where a_D and b_D are the same as in x_+.

The estimator We can use the above results to estimate the initial state x_0 using the $[x_-, x_+]$ interval.

The probability that the process ends up in x_+ is

$$p_+ = \frac{x_0 - x_-}{x_+ - x_-} = \frac{b_D + a_D x_0}{2b_D} = \frac{1}{2}\left(1 + \frac{a_D}{b_D} x_0\right) \tag{14}$$

that only depends on the ratio of a_D and b_D, hence let us introduce the notation $\gamma_D = \frac{b_D}{a_D}$. From the previous probability we can easily get an estimate for x_0. Denote ν_+ the relative frequency of the x_+ outcome, then

$$\hat{x}_0 = \gamma_D(2\nu_+ - 1) \tag{15}$$

Note that the estimate (15) is unbiased, because (14) is linear in x_0. Let us also remark that if we choose D big enough, then x_+ and x_- will be close to ± 1, so we use the whole trajectory and the estimation will be equivalent to the direct standard estimation method. Easy to see that as D goes to infinity γ_D decreasingly converges to 1, and so (15) converges to the direct standard estimator.

The variance of (15) is

$$Var(\hat{x}_0) = 4\gamma_D^2 Var(\nu_+) = \gamma_D^2 - x_0^2 \tag{16}$$

because ν_+ has a binomial distribution.

4.3. *Simulation investigations*

The properties of the estimator given in Eq. (15) were investigated by simulation, too. Here again we investigated the case when $c = 0.1$ and $x_0 = 0.4$.

The simulations were performed for two different D values: first for $D = 10$ and then for $D = 100$. For these numbers we can easily calculate the a_D and b_D values that give $\gamma_{10} \approx 1.3106178$ and $\gamma_{100} \approx 1$. For each number of used qubit pairs we repeated the whole estimation procedure 100 times and calculated the empirical mean values and empirical mean square errors (MSEs).

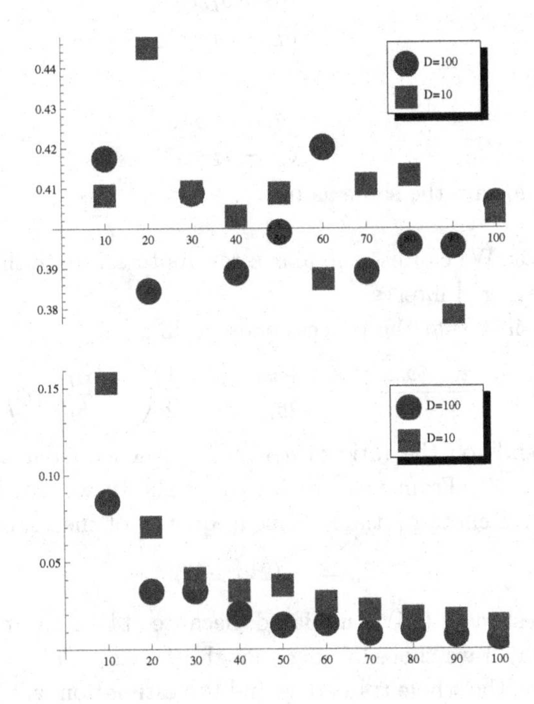

Fig. 5. The mean and the MSE as a function of the number of used qubits

For $D = 10$ we observed that the mean is unbiased, the empirical values are close to the real initial state $x_0 = 0.4$ (the upper sub-figure of Figure 5). In the lower sub-figure it is seen that the variance converges to zero. The order of the convergence is $1/N$, where N is the number of used qubit pairs.

For $D = 100$ the mean is also unbiased (Figure 5), and the variance is converging to zero with order of $1/N$, but a smaller coefficient is obtained than previously, which is in good agreement with (16). Note that this case is practically equivalent to the direct standard measurement, since $\gamma_{100} \approx 1$.

5. Comparison and discussion

The three developed estimation method, the conditional histogram, the Bayes and the martingale approaches are compared in this section from the viewpoint of their efficiency.

The efficiency of the indirect estimation methods is evaluated by comparing the mean square error (MSE) of their estimates with the theoretical variance of the direct standard estimation method

$$\eta = \frac{MSE_{method}}{Var_{dstandard}} \tag{17}$$

where $Var_{dstandard} = \frac{1-x_0^2}{N}$ with N being the used qubit pairs.

The efficiencies of the investigated methods can be seen in Figure 6. If we use only one indirect measurement, then the efficiency will be $1/c^2 = 100$, so using repeated measurements we are having a huge improvement in estimation quality.

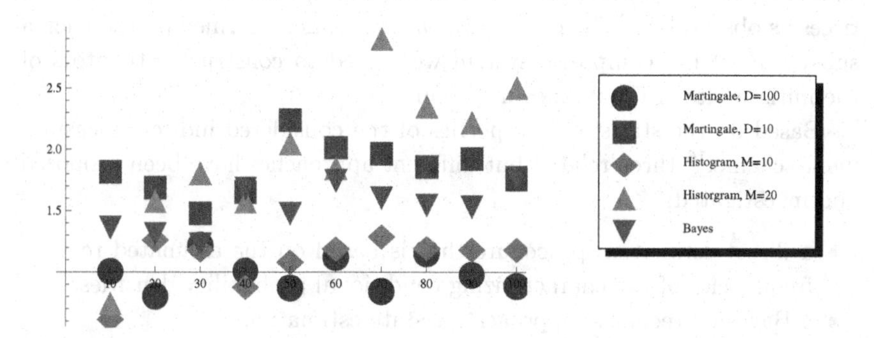

Fig. 6. The efficiency of different methods

Conditional histogram approach The $M = 10$ case is represented by diamonds and the $M = 20$ case is denoted by upper triangle in Figure 6. As we have seen before, this approach provides biased estimates, therefore its relatively good efficiency should be taken with reservation. In the better case ($M = 10$) we can achieve quite efficient estimation, although this seem to be a tough problem without any information on the initial state.

Bayes approach In Figure 6 this is represented by lower triangles. The estimate in this case is also biased but the efficiency is not very good either. On the other hand, the method does not need any parametrical setting, that makes the estimation robust. Nevertheless, the computation demand is high, and it only produces medium efficiency.

Martingale approach The $D = 10$ case is represented by squares, while the $D = 100$ case is denoted by circles in Figure 6. As we have seen before, the estimate based on the martingale approach is unbiased, and its efficiency with $D = 100$ is close to the optimal value 1. The optimal choice of D is crucial, but it is enough to choose D big enough.

The above discussion shows that *the martingale approach gives the best quality estimate from the statistical viewpoint, and it is unbiased and efficient at the same time.* In addition, we can easily tune its parameter D.

6. Conclusions

The simplest possible discrete time indirect measurement scheme has been investigated here, where both the unknown and the ancilla quantum systems are quantum bits. The measurements applied on the ancilla qubit were chosen to be the classical von Neumann measurements using the Pauli matrices as observables. The repeated measurements performed on the ancilla subsystem of the composite system were used to construct estimators of the initial state of the unknown system.

Based on the statistical properties of the considered indirect measurement scheme,[16] three related but different approaches have been proposed and investigated:

- a direct estimation procedure that is based on the estimated relative frequencies of the characterizing conditional probability densities,
- a Bayesian recursive approach for state estimation,
- a martingale approach that bases the estimator on the stopping times of the state evolution as a martingale driven by the repeated measure-

ments.

The unbiasedness and the mean square error of the estimate has been analyzed by analytical computations and computer simulation. It has been shown that the martingale approach gives an unbiased estimate, while the other two approaches result in an asymptotically unbiased estimate.

From our investigations we can conclude that the martingale approach gives the best quality estimate from the statistical viewpoint, that is unbiased, and efficient at the same time. The other methods can also provide acceptable results, but either its parameters are more difficult to select properly (the relative frequency method) or it is computationally demanding (the Bayes method). Therefore, the martingale approach is the best candidate to use in more complex settings.

The detailed derivations of our results can be found in a recent research report.[17]

Acknowledgement

This work is supported by the Hungarian Research Grant OTKA K67625, and by the Control Engineering Research Group of the Budapest University of Technology and Economics.

References

1. A. M. Branczyk, P. E. M. F. Mendonca, A. Gilchrist, A. C. Doherty, and S. D. Bartlett. Quantum Control of a Single Qubit, `arXiv:quant-ph/0608037v2`, 2006.
2. G. M. D'Ariano and H. P. Yuen. Impossibility of measuring the wave function of a single quantum system. *Physical Review Letters*, **76**, 2832-2835, 1996.
3. G. M. D'Ariano, M. G. A. Paris, and M. F. Sacchi, Quantum tomographic methods, in *Quantum State Estimation*, eds. M. Paris and J. Rehácek, Lect. Notes Phys. **649**, 7-58, 2004.
4. L. Diósi. Weak measurement in quantum mechanics. *Encyclopedia of Mathematical Physics*, 276-282, Oxford, 2006.
5. K. Jacobs. Feedback Control Using Only Quantum Back-Action. `arXiv:quant-ph/0904.3754v1`, 2009.
6. A.N. Jordan, B. Trauzettel, and G. Burkard. Weak measurement of quantum dot spin qubits. *Physical Review B*, 76, 155324, 2007.
7. N. Khaneja and S. J. Glaser. Cartan decomposition of su(2^n), constructive controllability of spin systems and universal quantum computing. `arXiv:quant-ph/0010100v1`, 2000.
8. Y.-S. Kim, Y.-W. Cho, Y.-S. Ra and Y.-H. Kim. Reversing the weak quantum measurement for a photonic qubit. `arXiv:quant-ph/0903.3077v1`, 2009.

9. A. N. Korotkov and A. N. Jordan. Undoing a weak quantum measurement of a solid-state qubit. `arXiv:cond-mat/0606713v1`, 2006.
10. A. N. Korotkov. Quantum efficiency of binary-outcome detectors of solid-state qubits. *Physical Review B*, 78, 174512, 2008.
11. G. Nogues, A. Rauschenbeutel, S. Osnaghi, M. Brune, J. M. Raimond and S. Haroche. Seeing a single photon without destroying it. *Nature*, 400, 239-242, 1999.
12. D. Petz. *Quantum Information Theory and Quantum Statistics.* Springer, Berlin and Heidelberg, 2008.
13. D. Petz, K.M. Hangos and A. Magyar, Point estimation of states of finite quantum systems, *J. Phys. A: Math. Theor.*, 40, 7955-7969, 2007.
14. R. Romano and D. D'Alessandro. Incoherent control and entanglement for two-dimensional coupled systems. `arXiv:quant-ph/0510020v1`, 2005.
15. L. Ruppert, A. Magyar and K. Hangos. Compromising non-demolition and information gaining for qubit state estimation. *Quantum Probability and Related Topics*, World Scientific, p. 212-224, 2008
16. L. Ruppert, A. Magyar, K. M. Hangos and D. Petz. Towards optimal quantum state estimation of a qubit by using indirect measurements. *Technical report of the Systems and Control Laboratory* **SCL-002/2008**, `http://daedalus.scl.sztaki.hu/`, 2008.
17. L. Ruppert and K. M. Hangos. State estimation methods using indirect measurements. *Technical report of the Systems and Control Laboratory* **SCL-001/2010**, `http://daedalus.scl.sztaki.hu/`, 2010.
18. G.A. Smith, A. Silberfarb, I.H. Deutsch, and P.S. Jessen. Efficient quantum state estimation by continuous weak measurement and dynamical control. *Phys. Rev. Lett.*, 97, 180403, 2006.
19. D. Williams, *Probability with Martingales.* Cambridge University Press, 1991.

ON THE CLASSIFICATION OF INVARIANT STATE OF GENERIC QUANTUM MARKOV SEMIGROUPS: THE GAUSSIAN GAUGE INVARIANT CASE

SKANDER HACHICHA

Université de sousse

Institut supérieur des sciences appliquées et Technologie de Sousse, TUNISIA

E-mail: skander_hachicha@yahoo.fr

We study the invariant state of generic quantum Markov semigroups using the Fagnola-Rebolledo criterion. This class arising from the stochastic limit of a discrete system with generic Hamiltonian H_S, acting on h, interacting with a Gaussian, gauge invariant, reservoir.

Keywords: Generic quantum Markov semigroup, invariant state, convergence to equilibrium.

1. Introduction

A quantum Markov semigroup (QMS) on the algebra $B(h)$ of all bounded operators on a complex separable Hilbert space h is a semigroup $(\mathcal{T}_t)_{t\geq 0}$ of completely positive, identity preserving, normal maps on $B(h)$. QMS arose in several contexts in the physical literature to describe the irreversible evolution of a small system coupled with a bigger environment (see [6] and the references therein).

The formal generators of these semigroups were introduced by Accardi and Kozyrev[2] who considered both zero and positive temperature Gaussian fields.

The system Hamiltonian H_S determines a privileged orthonormal basis of h. The action of generic QMS can be written separately on diagonal and off-diagonal (with respect to this basis) operators. The action on diagonal operators describes a classical Markov jump process; the action on off-diagonal operators is given by the conjugation with a contraction semigroup and its adjoint semigroup on h (see,[14]).

Here we are concerned with the positive temperature case. The complex separable Hilbert space h is the space of the discrete system S. We denote by V a finite or countable index set, such that $(|\sigma\rangle)_{\sigma\in V}$ is an orthonormal

basis of h, and write $\mathcal{V}$ to denote the linear manifold generated by finite combinations of vectors $|\sigma\rangle$.
Let H_S be the system Hamiltonian

$$H_S = \sum_{\sigma \in V} \epsilon_\sigma |\sigma\rangle\langle\sigma|$$

with eigenvalues $(\epsilon_\sigma)_{\sigma \in V}$. Following Accardi and Kozyrev,[2] we call the Hamiltonian H_S generic if the eigenspace associated with each eigenvalue ϵ_σ in one dimensional and one has $\epsilon_\sigma - \epsilon_\tau = \epsilon_{\sigma'} - \epsilon_{\tau'}$ for $\sigma \neq \tau$ if and only if $\sigma = \sigma$ and $\tau = \tau'$.
The environment is described by a Boson Fock space over the Hilbert space $L^2(\mathbb{R}^d; \mathbb{C})$ with creation and annihilation operators $A(g)$ and $A^+(g)$.
The interaction between the system and the environment has the dipole type form

$$H_I = D \otimes A^+(g) + D^+ \otimes A(g),$$

where D, D^+ are system operators, i.e. operators on h, with domain containing $\mathcal{V}$, such that $\langle v, Du\rangle = \langle D^+ v, u\rangle$ for $u, v \in \mathcal{V}$and satisfying the analyticity condition (Γ is the Euler Gamma function)

$$\sum_{n \geq 1} |\langle\sigma', D^n \sigma\rangle \Gamma(n\theta)^{-1} < +\infty$$

for all σ, σ' and some $\theta \in]0,1[$(a bounded operator, for instance). The function g is called the form factor (or cutoff) of the interaction and will be supposed to belong to the Schwarz space of smooth functions rapidly decreasing at infinity. The form generator of the generic QMS is (see[2] and[3])

$$\pounds\,(x) = \frac{1}{2} \sum_{\substack{\sigma,\sigma' \in V \\ \sigma \neq \sigma'}} \left(\gamma_{\sigma\sigma'} \left(2\,|\sigma\rangle\,\langle\sigma'\,|x|\,\sigma'\rangle\,\langle\sigma| - \{|\sigma\rangle\,\langle\sigma|\,, x\}\right)\right.$$
$$\left. + i\xi_{\sigma\sigma'}\,[x, |\sigma\rangle\,\langle\sigma|]\right) \tag{1}$$

for all x in $\mathcal{V}$, where, for all x and y bounded operators on h, $\{x, y\} = xy + yx$ and $[x, y] = xy - yx$.
The real constants $\gamma_{\sigma\sigma'}$, called the generalized susceptivities, are positive and can be interpreted as rates of jumps from σ to σ'. The real constants $\xi_{\sigma\sigma'}$ are called generalized Lamb shifts.
In [,[2] p. 27 (1.1.85)] this arises from the limit of matrix elements of the evolution of the system coupled with the reservoir.
Clearly (1) makes sense as the bounded generator of a norm-continuous QMS when V is finite. Here we consider countable sets V, therefore $\pounds\,(x)$

is well defined as a quadratic form on $\mathcal{V} \times \mathcal{V}$ because the sums are always weakly converging on this domain.
Sometimes it will be convenient to distinguish these coefficients in terms of the order of the eigenvalues (energies) associated with the indexes σ, σ' and we will denote, for $\epsilon_\sigma > \epsilon_{\sigma'}$

$$\gamma^-_{\sigma\sigma'} = \gamma_{\sigma\sigma'}, \quad \xi^-_{\sigma\sigma'} = \xi_{\sigma\sigma'}, \quad \gamma^+_{\sigma'\sigma} = \gamma_{\sigma'\sigma}, \quad \xi^+_{\sigma'\sigma} = \xi_{\sigma'\sigma}.$$

In other words we add a $-$ (resp $+$) to $\gamma_{\sigma\sigma'}$and $\xi_{\sigma\sigma'}$ to stress that it is the rate of a transition to a lower (resp. upper) level.
In our model the constants $\gamma^-_{\sigma\sigma'}$, $\xi^-_{\sigma\sigma'}$, $\gamma^+_{\sigma'\sigma}$, $\xi^+_{\sigma'\sigma}$ are given by (see,[2] and Th.1.6 p.19 (1.1.58), (1.1.59))

$$\gamma^-_{\sigma\sigma'} = 2\mathfrak{Re}\,(g \mid g)^-_\omega \left|\langle \sigma', D\sigma\rangle\right|^2, \; \xi^-_{\sigma\sigma'} = \mathfrak{Im}\,(g \mid g)^-_\omega \left|\langle \sigma', D\sigma\rangle\right|^2$$

$$\gamma^+_{\sigma'\sigma} = 2\mathfrak{Re}\,(g \mid g)^+_\omega \left|\langle \sigma', D\sigma\rangle\right|^2, \; \xi^+_{\sigma'\sigma} = \mathfrak{Im}\,(g \mid g)^+_\omega \left|\langle \sigma', D\sigma\rangle\right|^2, \qquad (2)$$

$\omega = \varepsilon_\sigma - \varepsilon_{\sigma'}$, $\epsilon_\sigma > \epsilon_{\sigma'}$

The constants $(g \mid g)^-_\omega$, $(g \mid g)^+_\omega$ are given by

$$(g \mid g)^-_\omega = \pi \int_{S(\omega)} \frac{|g(k)|^2 e^{\beta\omega(k)}}{e^{\beta\omega(k)} - 1} d_S k - i.P.P \int_{\mathbb{R}^d} \frac{|g(k)|^2 e^{\beta\omega(k)}}{(e^{\beta\omega(k)} - 1)(\omega(k) - \omega)} dk$$

$$(g \mid g)^+_\omega = \pi \int_{S(\omega)} \frac{|g(k)|^2}{e^{\beta\omega(k)} - 1} d_S k - i.P.P \int_{\mathbb{R}^d} \frac{|g(k)|^2}{(e^{\beta\omega(k)} - 1)(\omega(k) - \omega)} dk$$

where (see [,[2] p.15]) $\beta > 0$ is the inverse temperature, $S(\omega)$ denotes the surface in $\mathbb{R}^d$,
$S(\omega) = \{k \in |\omega(k) = \omega\}$, $d_S k$ the surface integral and p.p. the principal part of the integral. Moreover, the functions ω and g satisfy the following physical assumptions:
a) the dispersion function $\omega : \mathbb{R}^d \longrightarrow [0, +\infty[$ is radial and strictly increasing along each half-line starting form 0,
b) the cutoff g belongs to the Schwarz space of smooth functions rapidly decreasing at infinity,
c) the cutoff g and the dispersion function ω are related by the analytical condition

$$\int_{\mathbb{R}} \left| \int_{\mathbb{R}^d} e^{it\omega(k)} |g(k)|^2 dk \right| dt < +\infty.$$

Frequent choices of the dispersion function are $\omega(k) = |k|$ and $\omega(k) = |k|^2 + m$ $(m \geq 0)$ and we shall write also $\omega(r)$ for $\omega(rk/|k|)$ $(r \geq 0,\ k \in \mathbb{R}^d)$.

With these choices of the dispersion function the analytical condition c) holds for $d \geq 2$ and $d \geq 3$ respectively. Therefore we shall also suppose that d) $d \geq 2$.
Therefore, throughout this paper, we shall assume that the following summability condition

$$\mu_\sigma := \sum_{\sigma' \in V, \epsilon_{\sigma'} < \epsilon_\sigma} \gamma^-_{\sigma\sigma'} < \infty \tag{3}$$

and

$$\lambda_\sigma := \sum_{\sigma' \in V, \epsilon_\sigma < \epsilon_{\sigma'}} \gamma^+_{\sigma'\sigma} < \infty \tag{4}$$

$$\sum_{\sigma' \in V, \epsilon_{\sigma'} < \epsilon_\sigma} |\xi^-_{\sigma\sigma'}| + |\xi^+_{\sigma'\sigma}| < \infty \tag{5}$$

holds for all $\sigma \in V$ and define

$$\kappa_\sigma := \sum_{\sigma' \in V, \epsilon_{\sigma'} < \epsilon_\sigma} \xi^-_{\sigma\sigma'} + \xi^+_{\sigma'\sigma}$$

Under this summability condition, the form generator (1) can be represented in the generalized canonical GKSL form

$$\pounds\,(x) = G^* x + \sum_{\substack{\sigma,\sigma' \in V \\ \sigma \pm \sigma'}} L^*_{\sigma\sigma'} x L_{\sigma\sigma'} + xG$$

where

$$G = -\sum_{\sigma \in V} \left(\frac{\mu_\sigma + \lambda_\sigma}{2} + i\kappa_\sigma \right) |\sigma\rangle\langle\sigma|$$

$$L_{\sigma\sigma'} = \sqrt{\gamma^-_{\sigma\sigma'}} |\sigma'\rangle\langle\sigma|, \ \ if \epsilon_{\sigma'} < \epsilon_\sigma \qquad L_{\sigma\sigma'} = \sqrt{\gamma^+_{\sigma\sigma'}} |\sigma'\rangle\langle\sigma|, \ \ if \epsilon_\sigma < \epsilon_{\sigma'} \tag{6}$$

Since the operators $L_{\sigma\sigma'}$ are bounded, if G is also bounded, it follows that the form generator $\pounds$ is bounded and generates a norm continuous quantum Markov semigroup on $B(h)$ (see [,[910]]). The case when G is unbounded will be studied in.[4]

In the present paper we give a classification of the invariant states of the above class of semigroups under the condition: some of $\gamma_{\sigma\sigma'}$ are equal to zero. In the remainder of this Section, as example, all invariant states in different situations of the generic QMS in $B(\mathbb{C}^3)$ are worked out. In the next Section, by using Fagnola-Rebolledo's criterion,[7] we introduce the definition of A−chain in order to classify the invariant states of the generic QMS.

Example 1.1. The form generator of the generic QMS in $B(\mathbb{C}^3)$ is given, for all $x \in B(\mathbb{C}^3)$, by

$$\pounds(x) = \frac{1}{2} \sum_{\substack{1\leq j,k\leq 3 \\ j\neq k}} (\gamma_{jk} (2 |e_j\rangle \langle e_k |x| e_k\rangle \langle e_j| - \{|e_j\rangle \langle e_j|, x\}) \\ + i\xi_{jk} [x, |e_j\rangle \langle e_j|]) \tag{7}$$

where $(|e_j\rangle)_{1\leq j\leq 3}$ is an orthonormal basis for $\mathbb{C}^3$. In this case, we have

$$\begin{aligned} \mu_1 &= \sum_{m<1} \gamma_{1m} = 0, \\ \mu_2 &= \sum_{m<2} \gamma_{2m} = \gamma_{21}, \\ \mu_3 &= \sum_{m<3} \gamma_{3m} = \gamma_{31} + \gamma_{32}. \end{aligned} \tag{8}$$

Moreover,

$$\begin{aligned} \lambda_1 &= \sum_{1<m} \gamma_{1m} = \gamma_{12} + \gamma_{13}, \\ \lambda_2 &= \sum_{2<m} \gamma_{2m} = \gamma_{23}, \\ \lambda_3 &= \sum_{3<m} \gamma_{3m} = 0. \end{aligned} \tag{9}$$

The form generator (7) can be represented in the generalized canonical GKSL form as follows

$$\pounds(x) = G^* x + \sum_{1\leq j\neq k\leq 3} L^*_{jk} x L_{jk} + xG,$$

where

$$G = - \sum_{1\leq j\leq 3} \left(\frac{\mu_j + \lambda_j}{2} + i\kappa_j \right) |e_j\rangle\langle e_j|$$

and

$$L_{jk} = \begin{cases} \sqrt{\gamma_{jk}} |e_k\rangle\langle e_j| \text{ if } k < j \\ \sqrt{\gamma_{kj}} |e_k\rangle\langle e_j| \text{ if } k > j \end{cases} \tag{10}$$

Observe that, if $0 < \mu_1+\lambda_1$, $0 < \mu_2+\lambda_2$ and $0 < \mu_3+\lambda_3$, the semigroup has a unique invariant state (see e.g.,[4]). Furthermore, the predual generator is given by

$$\mathcal{L}_*(\rho) = \frac{1}{2} \sum_{\substack{1\le j,k\le 3 \\ j\neq k}} \left(\gamma_{jk}\left(2e_j(\rho)\left|e_k\right\rangle\left\langle e_k\right| - \left\{\left|e_j\right\rangle\left\langle e_j\right|, \rho\right\}\right)\right.$$
$$\left. + i\xi_{jk}\left[\rho, \left|e_j\right\rangle\left\langle e_j\right|\right]\right), \tag{11}$$

where ρ is a nonnegative element in $\mathcal{F}_1(\mathbb{C})$ such that $tr(\rho) = 1$ and $e_j(\rho) = \langle e_j, \rho e_j\rangle$.

With the notation $\rho_j \equiv \rho_{jj} \equiv \langle e_j, \rho e_j\rangle$, the invariant state of the GQMS in $B(\mathbb{C}^3)$ is characterized by:

- $\rho_1 = Z^{-1}(\gamma_{21}\gamma_{31} + \gamma_{32}\gamma_{21} + \gamma_{23}\gamma_{31})$
- $\rho_2 = Z^{-1}(\gamma_{12}\gamma_{32} + \gamma_{31}\gamma_{21} + \gamma_{13}\gamma_{32})$
- $\rho_3 = Z^{-1}(\gamma_{13}\gamma_{23} + \gamma_{12}\gamma_{23} + \gamma_{21}\gamma_{13})$

where Z is a normalization constant and, for $n \neq m$,

$$(\lambda_m + \mu_m + \lambda_n + \mu_n + i(\kappa_m - \kappa_n))\langle e_m, \rho e_n\rangle = 0.$$

Remark 1.1.

(1) Non-classical invariant states may exist if and only if there is a pair of indices $n \neq m$ such that $\lambda_m = \mu_m + \lambda_n = \mu_n = 0$ and $\kappa_m = \kappa_n = 0$.
(2) When all ρ_j's are strictly positive, the state (ρ_1, ρ_2, ρ_3) is the unique (classical) faithful invariant state.

Let ρ be a $\mathcal{T}-$invariant state. For $\rho = \sum_{i,j} \rho_{ij}\left|e_i\right\rangle\left\langle e_j\right|$, direct computation gives $\mathcal{L}_*(\rho) = 0$ and therefore

- $\langle e_1, \mathcal{L}_*(\rho)\, e_1\rangle = -\lambda_1\rho_1 + \gamma_{21}\rho_2 + \gamma_{31}\rho_3 = 0$
- $\langle e_2, \mathcal{L}_*(\rho)\, e_2\rangle = \gamma_{12}\rho_1 - (\lambda_2 + \mu_2)\rho_2 + +\gamma_{32}\rho_3 = 0$
- $\langle e_3, \mathcal{L}_*(\rho)\, e_3\rangle = \gamma_{13}\rho_1 + \gamma_{23}\rho_2 - \mu_3\rho_3 = 0$

replacing the second equation, that must be linearly dependant from the first and second one in order to find non-trivial solutions, by the normalization condition $\rho_1 + \rho_2 + \rho_3 = 1$ one find

- $\rho_1 = Z^{-1}(\gamma_{21}\gamma_{31} + \gamma_{32}\gamma_{21} + \gamma_{23}\gamma_{31})$
- $\rho_2 = Z^{-1}(\gamma_{12}\gamma_{32} + \gamma_{31}\gamma_{21} + \gamma_{13}\gamma_{32})$
- $\rho_3 = Z^{-1}(\gamma_{13}\gamma_{23} + \gamma_{12}\gamma_{23} + \gamma_{21}\gamma_{13})$

with Z being a normalization constant. Thus, for $n \neq m$, we obtain

$$\langle e_m, \mathcal{L}_*(\rho)\, e_n\rangle = (\lambda_m + \mu_m + \lambda_n + \mu_n + i(\kappa_m - \kappa_n))\langle e_m, \rho e_n\rangle = 0.$$

2. The Fagnola-Rebolledo Criterion

This criterion appeared in[8] and was generalized in,[76]
Under the summability condition (3) (4) (5), the form generator (1) can be represented in the generalized canonical GKSL form

$$\pounds(x) = \sum_k (L_k^* x L_k - \frac{1}{2}\{L_k^* L_k, x\}) + i[H, x] \tag{12}$$

where

$$H = \sum_{\sigma \in V} \kappa_\sigma |\sigma\rangle\langle\sigma|$$

and $k = (\sigma, \sigma')$

$$L_k = \sqrt{\gamma_{\sigma\sigma'}^-}|\sigma'\rangle\langle\sigma|, \;\; if \epsilon_{\sigma'} < \epsilon_\sigma \qquad L_k = \sqrt{\gamma_{\sigma\sigma'}^+}|\sigma'\rangle\langle\sigma|, \;\; if \epsilon_\sigma < \epsilon_{\sigma'} \tag{13}$$

Lemma 2.1. *Suppose that*

$$\gamma_{\sigma\sigma'} \neq 0, \;\; \epsilon_\sigma \neq \epsilon_{\sigma'},$$

Then the generator (12) is such that

$$\{L_k, L_k^*, H\}' = \{L_k, L_k^*\}' = \mathbb{C}1$$

and therefore, $\exp(t\pounds)$ *converges towards the equilibrium.*

Proof. Since

$$\{L_k, L_k^*, H\}' \subseteq \{L_k, L_k^*\}'$$

and

$$\{L_k, L_k^*\}' = \{|\sigma'\rangle\langle\sigma|, \epsilon_\sigma \neq \epsilon_{\sigma'}\}'$$

Therefore,if $x \in \{|\sigma'\rangle\langle\sigma|, \epsilon_\sigma \neq \epsilon_{\sigma'}\}'$ then $\forall \epsilon_\sigma \neq \epsilon_{\sigma'}$

$$x|\sigma'\rangle\langle\sigma| = |\sigma'\rangle\langle\sigma|x \tag{14}$$

Therefore $\forall\sigma''$

$$x|\sigma'\rangle\delta_{\sigma,\sigma''} = |\sigma'\rangle\langle\sigma, x\sigma''\rangle$$

hence if we choose $\sigma \neq \sigma''$ then

$$0 = |\sigma'\rangle\langle\sigma, x\sigma''\rangle$$

and this is possible iff

$$\langle\sigma, x\sigma''\rangle = 0; \;\; \forall\sigma \neq \sigma''$$

If $x = x^*$ this implies that x has the form

$$x = \sum \lambda_\sigma |\sigma\rangle\langle\sigma|$$

in this case the identity (14) becomes

$$\lambda_{\sigma'}|\sigma'\rangle\langle\sigma| = \lambda_\sigma|\sigma\rangle\langle\sigma'|; \ \ \forall\sigma \neq \sigma'$$

Therefore $x = \lambda.1$, i.e. x is a multiple of the identity.

□

Condition $\gamma_{\sigma\sigma'} \neq 0$, $\epsilon_\sigma \neq \epsilon_{\sigma'}$, is very strong. A weaker condition can be obtained with the following arguments.

Definition 2.1. We say that $\sigma_1, \sigma_n \in V$ belong to $\mathrm{A}-$*chain* if there exist $\sigma_2, \cdots, \sigma_{n-1} \in V$ such that

$$(\sigma_j, \sigma_{j+1}) \in \mathrm{A}; \ \ \forall j = 1, \cdots, n-1$$

Lemma 2.2. *Define*

$$\mathrm{A} := \{(\sigma', \sigma) : \gamma_{\sigma\sigma'} \neq 0\}$$

If, for any $\sigma_1, \sigma_n \in V$ *such that*

$$\epsilon_{\sigma_n} < \epsilon_{\sigma_1}$$

σ_1, σ_n *belong to an* $\mathrm{A}-$*chain, then the conclusion of Lemma 3-1 is true.*

Proof. Notice that

$$\{|\sigma'\rangle\langle\sigma| : (\sigma', \sigma) \in \mathrm{A}\}$$

This algebra contains all the elements of the form

$$|\sigma_1\rangle\langle\sigma_2||\sigma_2\rangle\langle\sigma_3||\sigma_3\rangle\langle\sigma_4| \cdots |\sigma_{n-1}\rangle\langle\sigma_n| = |\sigma_1\rangle\langle\sigma_n|$$

We assume that any pair $\sigma_1, \sigma_n \in V$ belongs to such a chain. It contains all operators of the form

$$|\sigma'\rangle\langle\sigma|; \ \ \epsilon_\sigma \neq \epsilon_{\sigma'}$$

hence,by the argument of the preceeding Lemma,

$$\mathcal{A}_V' = \{L_k, L_k^*\}' = \mathbb{C}1$$

and this ends the proof. □

Acknowledgments

The author is grateful to Professors L. Accardi and F. Fagnola for fruitful discussions and many stimulating comments.

References

1. L. Accardi, F. Fagnola and S. Hachicha, *Generic q-Markov semigroups and speed of convergence of q-algorithms*, Inf. Dimen. Anal. Quant. Probab. Rel. Top. Vol. **9**, No.4 (2006), 567-594.
2. L. Accardi and S. Kozyrev, *Quantum interacting particle systems (Trento, 2000)*, 1-195, QP–PQ: Quantum Probab. White Noise Anal. **14**, World Sci. Publishing, River Edge, NJ, 2002.
3. L. Accardi, Y. G.Lu and I. Volovich, *Quantum theory and its stochastic limit*, Springer-Verlag, Berlin, (2002).
4. R. Carbone, F. Fagnola and S. Hachicha, *Generic Quantum Markov Semigroups: the Gaussian Gauge Invariant Case*, Open Sys. Information Dyn. **14** (2007), 425444.
5. A. M. Chebotarev and F. Fagnola, *Sufficient conditions for conservativity of minimal quantum dynamical semigroups*, J. Funct. Anal. **153**, No. 2 (1998), 382-404.
6. A. Dhahri, F. Fagnola and R. Rebolledo, *The decoherence-free subalgebra of a quantum Markov semigroup with unbounded generator*, Inf. Dimen. Anal. Quant. Probab. Rel. Top. Vol. **13** (2010), 413-433.
7. F. Fagnola and R. Rebolledo, *Algebraic conditions for convergence of a quantum Markov semigroup to a steady state*, Inf. Dimen. Anal. Quant. Probab. Rel. Top. Vol. **11** (2008), 467-474.
8. A. Frigerio and M. Verri, *Long-time asymptotic properties of dynamical semigroups on W^*-algebras*, Math.Z. **180** (1982), 275-286.
9. V. Gorini, A. Kossakowski and E.C.G. Sudarshan, *Completely positive dynamical semigroups of N-level systems*, J. Math. Phys. **17** (1976), 8-21.
10. G. Lindblad, *On the generators of Quantum Dynamical Semigroups*, Commun. Math. Phys. **48** (1976), 1-19.

INDEPENDENCE GENERALIZING MONOTONE AND BOOLEAN INDEPENDENCES

TAKAHIRO HASEBE*

Graduate School of Science, Kyoto University,
Kyoto 606-8502, Japan
**E-mail: hsb@kurims.kyoto-u.ac.jp*

We define conditionally monotone independence in two states which interpolates monotone and Boolean ones. This independence is associative, and therefore leads to a natural probability theory in a non-commutative algebra.

Keywords: Monotone independence; Boolean independence; conditionally free independence; associative law

1. Introduction

1.1. *Independence in algebraic probability space*

Non-commutative probability theory provides quantum mechanics with an algebraic framework. The basis is a pair $(\mathcal{A}, \varphi)$ of a unital $*$-algebra $\mathcal{A}$ and a state φ. We call this pair *an algebraic probability space.* Many probabilistic concepts can be extended to this setting. For instance, a $*$-subalgebra of $\mathcal{A}$ corresponds to a σ-field in probability theory. Among them, independence is a fundamental concept. The usual independence in probability theory can be naturally extended to algebraic probability spaces and is called *tensor independence.* Independence is not unique: for instance, *free independence* was introduced to solve problems in operator algebras.[17] This is defined as follows. Let Λ be an index set and $\{\mathcal{A}_\lambda\}_{\lambda\in\Lambda}$ be $*$-subalgebras of $\mathcal{A}$.

Definition 1.1. (Free independence[17]). We assume that all $\mathcal{A}_\lambda$ contain the unit of $\mathcal{A}$. Then $\{\mathcal{A}_\lambda\}$ is said to be *free independent* if

$$\varphi(X_1 \cdots X_n) = 0$$

holds whenever $\varphi(X_1) = \cdots = \varphi(X_n) = 0$, $X_k \in \mathcal{A}_{\lambda_k}$, $\lambda_1 \neq \cdots \neq \lambda_n$. $\lambda_1 \neq \cdots \neq \lambda_n$ means that $\lambda_i \neq \lambda_{i+1}$ for $1 \leq i \leq n-1$.

We can also define free independence for $*$-subalgebras without the unit of $\mathcal{A}$; however we omit the definition. There are other kinds of independence. Boolean independence and monotone independence are fundamental ones.

Definition 1.2. We assume that for every λ, $\mathcal{A}_\lambda$ does not contain the unit of $\mathcal{A}$.
(1) (Boolean independence[16]). $\{\mathcal{A}_\lambda\}$ is said to be *Boolean independent* if

$$\varphi(X_1 \cdots X_n) = \varphi(X_1) \cdots \varphi(X_n)$$

for $X_k \in \mathcal{A}_{\lambda_k}$, $\lambda_1 \neq \cdots \neq \lambda_n$.
(2) (Monotone independence[12]). We assume moreover that Λ is equipped with a linear order $<$. Then $\{\mathcal{A}_\lambda\}$ is said to be *monotone independent* if

$$\varphi(X_1 \cdots X_i \cdots X_n) = \varphi(X_i)\varphi(X_1 \cdots X_{i-1}X_{i+1} \cdots X_n)$$

holds whenever i satisfies $\lambda_{i-1} < \lambda_i$ and $\lambda_i > \lambda_{i+1}$ (one of the inequalities is eliminated if $i = 1$ or $i = n$).

Remark 1.1. (1) It is easy to prove that φ becomes a $*$-homomorphism on $\bigcup_{\lambda \in \Lambda} \mathcal{A}_\lambda$ if the subalgebras $\mathcal{A}_\lambda$ are Boolean independent (or monotone independent) and they contain the unit of $\mathcal{A}$. Therefore, it is unavoidable to assume that $\mathcal{A}_\lambda$ do not contain the unit of $\mathcal{A}$.
(2) Only monotone independence is not symmetric. More precisely, Y and X are not necessarily monotone independent if X and Y are monotone independent. This property requires us to generalize the axioms for monotone cumulants,[8] which leads to a new viewpoint on cumulants.

There were several attempts to interpolate independences. For instance, a q-deformation of the CCR interpolates tensor and free independences (and moreover, independence for Fermions); the reader is referred to Ref. 5. Another one is *bm-independence* in Ref. 18 which interpolates monotone and Boolean independences. In this approach monotone independence is generalized by extending a linearly ordered set, denoted as Λ in Definition 1.2(2), to a partially ordered set.

An interpolation called *conditionally (or c- for short) free independence* was introduced in Ref. 6. This generalizes free and Boolean independences by using two states in an algebra. C-free independence effectively uses the fact that Boolean independence becomes trivial for $*$-subalgebras containing the unit of the whole algebra.

Definition 1.3. (Conditionally free independence[6]). Let $(\mathcal{A}, \varphi, \psi)$ be a tuple of a unital $*$-algebra $\mathcal{A}$ and states φ, ψ. Let $\mathcal{A}_\lambda$ be $*$-subalgebras of $\mathcal{A}$. $\mathcal{A}_\lambda$

are assumed to contain the unit of $\mathcal{A}$. $\mathcal{A}_\lambda$ are said to be *c-free independent* if

$$\varphi(X_1 \cdots X_n) = \prod_{k=1}^{n} \varphi(X_k) \tag{1}$$

whenever $\psi(X_1) = \cdots = \psi(X_n) = 0$, $X_k \in \mathcal{A}_{\lambda_k}$, $\lambda_1 \neq \cdots \neq \lambda_n$.

If $\varphi = \psi$, c-free independence is no other than free independence, and if $\psi = 0$ on $\bigcup_{\lambda \in \Lambda} \mathcal{A}_\lambda$, c-free independence is Boolean independence. In addition, c-free independence generalizes monotone independence in a sense.[10] The reader is also referred to Fig. 1. We will explain these in Subsection 1.2 and in Section 2.

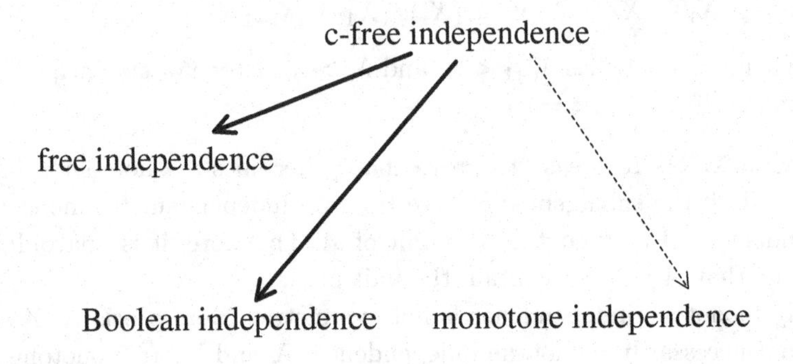

Fig. 1. Every arrow indicates that the initial independence generalizes the terminal. An arrow without dots means that it preserves the associative law; a dotted arrow means that it does not preserve the associative law.

1.2. *Free product of algebras and independence*

In probability theory, independent random variables can be realized by taking products of probability spaces. Algebraically, the product of base spaces corresponds to the tensor product of the commutative algebras of functions. To realize the independences defined in the previous subsection, a free product of algebras is used instead of the tensor product. For (possibly non-unital) $*$-algebras $\mathcal{A}_i$, let $\mathcal{A}_1 *_n \mathcal{A}_2$ be the algebraic free product without identification of units:

$$\mathcal{A}_1 *_n \mathcal{A}_2 = \bigoplus_{n=1}^{\infty} \bigoplus_{(\varepsilon_1,\cdots,\varepsilon_n)=(1,2,1,2,\cdots),(2,1,2,1,\cdots)} \mathcal{A}_{\varepsilon_1} \otimes \cdots \otimes \mathcal{A}_{\varepsilon_n}. \tag{2}$$

For unital $*$-algebras $\mathcal{A}_i$, let $\mathcal{A}_1 * \mathcal{A}_2$ be the algebraic free product with identification of units. This can be realized by decomposing the algebras as $\mathcal{A}_i = \mathbb{C}1_{\mathcal{A}_i} \oplus \mathcal{A}_i^0$ as vector spaces and by defining

$$\mathcal{A}_1 * \mathcal{A}_2 = \mathbb{C}1_{\mathcal{A}_1 * \mathcal{A}_2} \oplus \bigoplus_{n=1}^{\infty} \bigoplus_{(\varepsilon_1, \cdots, \varepsilon_n) = (1,2,1,2,\cdots),(2,1,2,1,\cdots)} \mathcal{A}_{\varepsilon_1}^0 \otimes \cdots \otimes \mathcal{A}_{\varepsilon_n}^0. \quad (3)$$

The reader is referred to Refs. 11,15 for details.

Free, Boolean and monotone independences can be realized by taking suitable products of states on a free product of algebras. In the following definitions, we assume that $\mathcal{A}_i$ is a $*$-algebra, φ_i is a state on $\mathcal{A}_i$ for $i = 1, 2$ and $(\varepsilon_1, \cdots, \varepsilon_n) = (1, 2, 1, \cdots)$ or $(2, 1, 2, \cdots)$.

Definition 1.4. (1) (Free product[2,17]). We assume that $\mathcal{A}_i$ are unital. The *free product* $\varphi_1 * \varphi_2$ on $\mathcal{A}_1 * \mathcal{A}_2$ is defined by the following rule:

$$\varphi_1 * \varphi_2(X_1 \cdots X_n) = 0$$

holds whenever $\varphi_{\varepsilon_i}(X_i) = 0$, $X_i \in \mathcal{A}_{\varepsilon_i}$ for $i = 1, \cdots, n$.
(2) (Boolean product[4,16]). The *Boolean product* $\varphi_1 \diamond \varphi_2$ on $\mathcal{A}_1 *_n \mathcal{A}_2$ is defined by

$$\varphi_1 \diamond \varphi_2(X_1 \cdots X_n) = \varphi_{\varepsilon_1}(X_1) \cdots \varphi_{\varepsilon_n}(X_n)$$

for $X_i \in \mathcal{A}_{\varepsilon_i}$.
(3) (Monotone product[9,12]). The *monotone product* $\varphi_1 \triangleright \varphi_2$ on $\mathcal{A}_1 *_n \mathcal{A}_2$ is defined by

$$\varphi_1 \triangleright \varphi_2(X_1 \cdots X_n) = \varphi_1\Big(\overrightarrow{\prod_{i: X_i \in \mathcal{A}_1}} X_i\Big)\Big(\prod_{i: X_i \in \mathcal{A}_2} \varphi_2(X_i)\Big),$$

where $\overrightarrow{\prod}_{i \in V} X_i$ for a subset $V \subset \{1, \cdots, n\}$ is the product of X_i, $i \in V$ in the same order as they appear in $X_1 \cdots X_n$.

A product of states defined on a free product of algebras with (resp. without) identification of units is said to be *associative* if for any algebraic probability spaces $(\mathcal{A}_i, \varphi_i)$ $(i = 1, 2, 3)$, $(\varphi_1 \cdot \varphi_2) \cdot \varphi_3 = \varphi_1 \cdot (\varphi_2 \cdot \varphi_3)$ under the natural isomorphism $(\mathcal{A}_1 * \mathcal{A}_2) * \mathcal{A}_3 \cong \mathcal{A}_1 * (\mathcal{A}_2 * \mathcal{A}_3)$ (resp. $(\mathcal{A}_1 *_n \mathcal{A}_2) *_n \mathcal{A}_3 \cong \mathcal{A}_1 *_n (\mathcal{A}_2 *_n \mathcal{A}_3)$).

The above four independences are all associative. The associative laws of free, Boolean and c-free products are not difficult to prove. The associative law of the monotone product was proved in Ref. 9. Associative products of states on a free product of algebras are classified in papers Refs. 3,13–15 under some natural conditions.

Definition 1.5. (C-free product[6]). Let $\mathcal{A}_i$ be a unital $*$-algebra. Let φ_i, ψ_i be states on $\mathcal{A}_i$ for $i = 1, 2$. The *c-free product* of (φ_1, ψ_1) and (φ_2, ψ_2) is a pair of states $(\varphi_1 {}_{\psi_1}\!*_{\psi_2} \varphi_2, \psi_1 * \psi_2) = (\varphi_1, \psi_1) * (\varphi_2, \psi_2)$ on $\mathcal{A}_1 * \mathcal{A}_2$ defined by the following rules.

(1) $\psi_1 * \psi_2$ is the free product.
(2) $\varphi_1 {}_{\psi_1}\!*_{\psi_2} \varphi_2$ is defined by the condition that

$$\varphi_1 {}_{\psi_1}\!*_{\psi_2} \varphi_2(X_1 \cdots X_n) = \prod_{i=1}^{n} \varphi_{\varepsilon_i}(X_i) \tag{4}$$

holds whenever $\psi_{\varepsilon_i}(X_i) = 0$, $X_i \in \mathcal{A}_{\varepsilon_i}$ for $i = 1, \cdots, n$.

We can naturally define the concept of associativity for two states and then the c-free product is associative:

$$(\varphi_1 {}_{\psi_1}\!*_{\psi_2} \varphi_2)_{(\psi_1 * \psi_2)}\!*_{\psi_3} \varphi_3 = \varphi_1 {}_{\psi_1}\!*_{(\psi_2 * \psi_3)} (\varphi_2 {}_{\psi_2}\!*_{\psi_3} \varphi_3).$$

1.3. *The additive convolution associated with a product of states*

Once an associative product of states is given, we can define an associative convolution of probability measures. Let $\mathbb{C}[z]$ (resp. $\mathbb{C}[z]_0$) be the unital (resp. non-unital) $*$-algebra generated from an indeterminate z equipped with the involution $z^* = z$. $\mathbb{C}[z]$ is used for the free and tensor products, and $\mathbb{C}[z]_0$ is used for the monotone and Boolean products. The following argument is, however, the same for both $\mathbb{C}[z]_0$ and $\mathbb{C}[z]$, and therefore we do not consider $\mathbb{C}[z]_0$. A state φ on $\mathbb{C}[z]$ can be expressed by a probability measure μ defined by

$$\int x^n \mu(dx) = \varphi(z^n).$$

Moreover, if the moment sequence $\{\varphi(z^n)\}_{n=0}^{\infty}$ is determinate,[1] the probability measure is unique. Using a product of states $\varphi_1 \cdot \varphi_2$ on $\mathbb{C}[z_1, z_2] = \mathbb{C}[z_1] * \mathbb{C}[z_2]$, we can define the additive convolution of probability measures as follows. Let $\mathcal{A}_i$ be $\mathbb{C}[z_i]$ and μ_i be the probability measure corresponding to the moments $\varphi_i(z_i^n)$. Then the convolution $\mu_1 \cdot \mu_2$ is defined by the moments $\varphi_1 \cdot \varphi_2((z_1 + z_2)^n)$, if the resulting moments are determinate. A similar argument holds for a product of pairs of states, in such a case as the c-free product: one can define a convolution $(\mu_1, \nu_1) \cdot (\mu_2, \nu_2)$ for pairs of probability measures.

We denote respectively by $\mu \boxplus \nu$, $\mu \uplus \nu$ and $\mu \triangleright \nu$ the convolutions of probability measures μ, ν for the free, Boolean and monotone products. We also denote by $(\mu_1, \nu_1) \boxplus (\mu_2, \nu_2) = (\mu_1 {}_{\nu_1}\boxplus_{\nu_2} \mu_2, \nu_1 \boxplus \nu_2)$ the c-free convolution of (μ_1, ν_1) and (μ_2, ν_2).

2. Free, Boolean and monotone products in terms of conditionally free product

The left component of the c-free product generalizes free, Boolean and monotone (and anti-monotone) products.[6,10] Let $\mathcal{A}$ be a $*$-algebra with a decomposition $\mathcal{A} = \mathbb{C}1_{\mathcal{A}} \oplus \mathcal{A}^0$, where $\mathcal{A}^0$ is a $*$-subalgebra. In this case, we can define a delta state δ by $\delta(\lambda 1_{\mathcal{A}} + a^0) = \lambda$ for $\lambda \in \mathbb{C}$ and $a^0 \in \mathcal{A}^0$. When we use a delta state, we always assume the decomposition of an algebra. Let $(\mathcal{A}_i, \varphi_i)$ be algebraic probability spaces. Then the free, Boolean and monotone products appear as

$$\varphi_1 {}_{\varphi_1}\!*_{\varphi_2} \varphi_2 = \varphi_1 * \varphi_2 \text{ on } \mathcal{A}_1 * \mathcal{A}_2, \tag{5}$$

$$\varphi_1 {}_{\delta_1}\!*_{\delta_2} \varphi_2 = \varphi_1 \diamond \varphi_2 \text{ on } \mathcal{A}_1^0 *_n \mathcal{A}_2^0, \tag{6}$$

$$\varphi_1 {}_{\delta_1}\!*_{\varphi_2} \varphi_2 = \varphi_1 \triangleright \varphi_2 \text{ on } \mathcal{A}_1^0 *_n \mathcal{A}_2^0, \tag{7}$$

where δ_i is a delta state on $\mathcal{A}_i$ w.r.t. a decomposition $\mathcal{A}_i = \mathbb{C}1_{\mathcal{A}_i} \oplus \mathcal{A}_i^0$ as $*$-algebras.

As a result, the convolutions of (compactly supported) probability measures satisfy

$$(\mu, \mu) \boxplus (\nu, \nu) = (\mu \boxplus \nu, \mu \boxplus \nu), \tag{8}$$

$$(\mu, \delta_0) \boxplus (\nu, \delta_0) = (\mu \uplus \nu, \delta_0), \tag{9}$$

$$(\mu, \delta_0) \boxplus (\nu, \nu) = (\mu \triangleright \nu, \nu), \tag{10}$$

where δ_0 is a delta measure at 0.

The motivation for the article Ref. 7 is the problem concerning the associative laws of the c-free, free, Boolean and monotone products. Associative laws of the free and Boolean products naturally follow from that of the c-free product. For instance, we look at the associative law of the Boolean product. We observe that

$$\Big((\varphi_1, \delta_1)*(\varphi_2, \delta_2)\Big)*(\varphi_3, \delta_3) = (\varphi_1 \diamond \varphi_2, \delta_1 * \delta_2)*(\varphi_3, \delta_3) = ((\varphi_1 \diamond \varphi_2) \diamond \varphi_3, (\delta_1 * \delta_2) * \delta_3)$$

and

$$(\varphi_1, \delta_1)*\Big((\varphi_2, \delta_2)*(\varphi_3, \delta_3)\Big) = (\varphi_1, \delta_1)*(\varphi_2 \diamond \varphi_3, \delta_2 * \delta_3) = (\varphi_1 \diamond (\varphi_2 \diamond \varphi_3), \delta_1 * (\delta_2 * \delta_3)).$$

Since the c-free product is associative, $(\varphi_1 \diamond \varphi_2) \diamond \varphi_3 = \varphi_1 \diamond (\varphi_2 \diamond \varphi_3)$. Therefore, the Boolean product is associative. This situation is, however, different in the case of the monotone product:

$$\Big((\varphi_1, \delta_1) * (\varphi_2, \varphi_2)\Big) * (\varphi_3, \varphi_3) = (\varphi_1 \triangleright \varphi_2, \varphi_2) * (\varphi_3, \varphi_3).$$

We cannot repeat the calculation anymore. Therefore, the associative law of the monotone product does not follow from that of the c-free product. This difficulty also appears in cumulants which will be explained in details in Ref. 7. These situations are summarized in Fig. 1.

To solve this difficulty, we would like to define a new product of pairs of states which generalizes not only the monotone and Boolean products but the associative laws of them. Such a product is called a c-monotone product (see Fig. 2). First we consider probability measures to look for such a product of states.

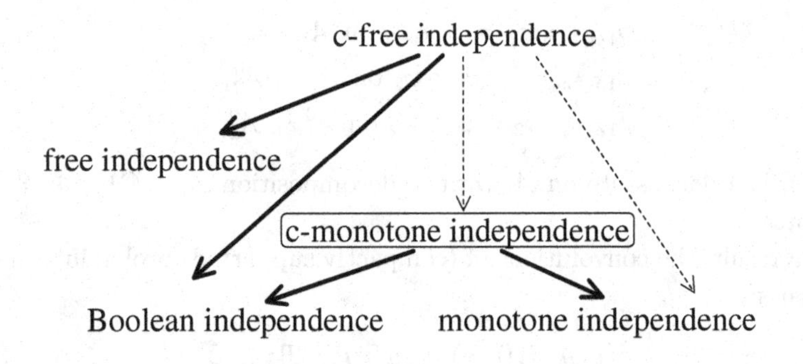

Fig. 2. The meaning of the arrows is the same as in Fig. 1.

We define the reciprocal Cauchy transform of a probability measure μ by

$$H_\mu(z) := \Big(\int_{\mathbb{R}} \frac{1}{z-x}\mu(dx)\Big)^{-1}, \quad z \in \mathbb{C}\backslash\mathbb{R}. \tag{11}$$

Since $\mu_{\delta_0}\boxplus_\nu \nu = \mu \triangleright \nu$, it is natural to calculate $\mu_1{}_{\delta_0}\boxplus_{\nu_2}\mu_2$ by using three probability measures. This operation is characterized simply as follows.

Proposition 2.1. *For compactly supported probability measures μ_i, ν_i, $i = 1, 2$,*

$$H_{\mu_1{}_{\delta_0}\boxplus_{\nu_2}\mu_2} = H_{\mu_1} \circ H_{\nu_2} + H_{\mu_2} - H_{\nu_2}. \tag{12}$$

The above result follows from a characterization of the c-free convolution; we refer the reader to Ref. 7 for the proof. We note that this includes the characterizations of the monotone and Boolean convolutions:[12,16] $H_{\mu\rhd\nu}(z) = H_\mu(H_\nu(z))$ and $H_{\mu\uplus\nu}(z) = H_\mu(z) + H_\nu(z) - z$.

Now we consider the convolution

$$(\mu_1, \nu_1) \rhd (\mu_2, \nu_2) := (\mu_1 {}_{\delta_0}\boxplus_{\nu_2} \mu_2, \nu_1 \rhd \nu_2). \tag{13}$$

If this convolution is associative, this generalizes not only the Boolean convolution but also the monotone convolution, including their associative laws, since

$$\Big((\mu_1, \mu_1)\rhd(\mu_2, \mu_2)\Big)\rhd(\mu_3, \mu_3) = (\mu_1\rhd\mu_2, \mu_1\rhd\mu_2)\rhd(\mu_3, \mu_3) = ((\mu_1\rhd\mu_2)\rhd\mu_3, (\mu_1\rhd\mu_2)\rhd\mu_3)$$

and

$$(\mu_1, \mu_1)\rhd\Big((\mu_2, \mu_2)\rhd(\mu_3, \mu_3)\Big) = (\mu_1, \mu_1)\rhd(\mu_2\rhd\mu_3, \mu_3\rhd\mu_3) = (\mu_1\rhd(\mu_2\rhd\mu_3), \mu_1\rhd(\mu_2\rhd\mu_3)).$$

Therefore, the monotone convolution appears in the left component naturally and this calculation can be repeated more than twice.

We now prove that the above convolution is indeed associative.

Proposition 2.2. *The convolution* $\rhd$ *defined for pairs of states is associative.*

Proof. It suffices to show that

$$(\mu_1 {}_{\delta_0}\boxplus_{\nu_2} \mu_2) {}_{\delta_0}\boxplus_{\nu_3} \mu_3 = \mu_1 {}_{\delta_0}\boxplus_{(\nu_2\rhd\nu_3)} (\mu_2 {}_{\delta_0}\boxplus_{\nu_3} \mu_3).$$

By Proposition 2.1,

$$\begin{aligned} H_{(\mu_1 {}_{\delta_0}\boxplus_{\nu_2}\mu_2) {}_{\delta_0}\boxplus_{\nu_3}\mu_3} &= H_{\mu_1 {}_{\delta_0}\boxplus_{\nu_2}\mu_2} \circ H_{\nu_3} + H_{\mu_3} - H_{\nu_3} \\ &= H_{\mu_1} \circ H_{\nu_2} \circ H_{\nu_3} + H_{\mu_2} \circ H_{\nu_3} - H_{\nu_2} \circ H_{\nu_3} + H_{\mu_3} - H_{\nu_3} \end{aligned}$$

and

$$\begin{aligned} H_{\mu_1 {}_{\delta_0}\boxplus_{(\nu_2\rhd\nu_3)}(\mu_2 {}_{\delta_0}\boxplus_{\nu_3}\mu_3)} &= H_{\mu_1} \circ H_{\nu_2\rhd\nu_3} + H_{\mu_2 {}_{\delta_0}\boxplus_{\nu_3}\mu_3} - H_{\nu_2\rhd\nu_3} \\ &= H_{\mu_1} \circ H_{\nu_2} \circ H_{\nu_3} + H_{\mu_2} \circ H_{\nu_3} - H_{\nu_2} \circ H_{\nu_3} + H_{\mu_3} - H_{\nu_3}. \end{aligned}$$

Therefore, the two probability measures coincide. □

We call the convolution $\rhd$ *a c-monotone convolution.*

3. C-monotone product and c-monotone independence

Since the c-monotone convolution $\triangleright$ is associative, one expects that the corresponding product of states is also associative. This is indeed the case. Roughly, we define a product $(\varphi_1, \psi_1) \triangleright (\varphi_2, \psi_2) := (\varphi_1 {}_{\delta_1}\!*_{\psi_2} \varphi_2, \psi_1 \triangleright \psi_2)$. The free product of algebras however needs to be treated more carefully.

Definition 3.1. Let $(\mathcal{A}_i, \varphi_i, \psi_i)$ $(i = 1, 2)$ be algebraic probability spaces. We define a c-monotone product on $\mathcal{A}_1 *_n \mathcal{A}_2$ by

$$(\varphi_1, \psi_1) \triangleright (\varphi_2, \psi_2) := (\varphi_1 \triangleright_{\psi_2} \varphi_2, \psi_1 \triangleright \psi_2), \tag{14}$$

where $\varphi_1 \triangleright_{\psi_2} \varphi_2$ is defined as follows. We denote the unitization of each $\mathcal{A}_i$ by $\widetilde{\mathcal{A}_i} := \mathbb{C}1_{\widetilde{\mathcal{A}_i}} \oplus \mathcal{A}_i$. Then the states φ_i and ψ_i naturally extend to $\widetilde{\varphi}_i$ and $\widetilde{\psi}_i$ on $\widetilde{\mathcal{A}_i}$ defined by $\widetilde{\varphi}_i(1_{\widetilde{\mathcal{A}_i}}) = 1$ and $\widetilde{\psi}_i(1_{\widetilde{\mathcal{A}_i}}) = 1$. See Ref. 11 for details. Let $\widetilde{\delta}_i$ denote the delta state associated to the decomposition $\widetilde{\mathcal{A}_i} = \mathbb{C}1_{\widetilde{\mathcal{A}_i}} \oplus \mathcal{A}_i$. $\varphi_1 \triangleright_{\psi_2} \varphi_2$ is defined as the restriction of $\widetilde{\varphi_1}_{\widetilde{\delta}_1}\!*_{\widetilde{\psi_2}} \widetilde{\varphi_2}$ on $\mathcal{A}_1 *_n \mathcal{A}_2$.

We refer the reader to Ref. 7 for the proof of the following fact.

Theorem 3.1. *Let $(\mathcal{A}_i, \varphi_i, \psi_i)$ $(i = 1, 2)$ be algebraic probability spaces. The calculation rules for a c-monotone product are what follows.*

(1) $\varphi_1 \triangleright_{\psi_2} \varphi_2(bax) = \varphi_2(b)\varphi_1 \triangleright_{\psi_2} \varphi_2(ax)$ *for* $a \in \mathcal{A}_1$, $b \in \mathcal{A}_2$ *and* $bax \in \mathcal{A}_1 *_n \mathcal{A}_2$ *(this is also the case when x is absent).*
(2) $\varphi_1 \triangleright_{\psi_2} \varphi_2(xab) = \varphi_1 \triangleright_{\psi_2} \varphi_2(xa)\varphi_2(b)$ *for* $a \in \mathcal{A}_1$, $b \in \mathcal{A}_2$ *and* $xab \in \mathcal{A}_1 *_n \mathcal{A}_2$ *(this is also the case when x is absent).*
(3) $\varphi_1 \triangleright_{\psi_2} \varphi_2(a_1 b_1 \cdots b_{n-1} a_n) = (\varphi_2(b_j) - \psi_2(b_j))\varphi_1 \triangleright_{\psi_2} \varphi_2(a_1 b_1 a_2 \cdots b_{j-1} a_j)\varphi_1 \triangleright_{\psi_2} \varphi_2(a_{j+1} b_{j+1} \cdots b_{n-1} a_n)$
$+ \psi_2(b_j)\varphi_1 \triangleright_{\psi_2} \varphi_2(a_1 b_1 a_2 \cdots b_{j-1} a_j a_{j+1} b_{j+1} \cdots b_{n-1} a_n)$ *for* $a_k \in \mathcal{A}_1$, $b_k \in \mathcal{A}_2$, $1 \le j \le n-1$, $n \in \mathbb{N}$.

By using the calculation rules, we can prove the associative law. See Ref. 7 for the proof.

Theorem 3.2. *The c-monotone product is associative.*

Since we have an associative product of states, independence can be formulated for $*$-subalgebras.

Definition 3.2. Let $(\mathcal{A}, \varphi, \psi)$ be an algebraic probability space with two states and I be a linearly ordered set. $*$-subalgebras $\{\mathcal{A}_i\}_{i \in I}$ which do not contain the unit of $\mathcal{A}$ are said to be *c-monotone independent* if

(1) $\varphi(a_1a_2\cdots a_n) = \varphi(a_1)\varphi(a_2\cdots a_n)$ for $a_k \in \mathcal{A}_{i_k}$ with $i_1, i_2, \cdots, i_n \in I$ and $i_1 > i_2$;
(2) $\varphi(a_1a_2\cdots a_n) = \varphi(a_n)\varphi(a_1\cdots a_{n-1})$ for $a_k \in \mathcal{A}_{i_k}$ with $i_1, i_2, \cdots, i_n \in I$ and $i_n > i_{n-1}$;
(3) $\varphi(a_1a_2\cdots a_n) = (\varphi(a_j) - \psi(a_j))\varphi(a_1\cdots a_{j-1})\varphi(a_{j+1}\cdots a_n) + \psi(a_j)\varphi(a_1a_2\cdots a_{j-1}a_{j+1}\cdots a_n)$
for $a_k \in \mathcal{A}_{i_k}$ with $i_1, i_2, \cdots, i_n \in I$, $i_{j-1} < i_j > i_{j+1}$ and $2 \leq j \leq n-1$;
(4) $\mathcal{A}_i$ are monotone independent w.r.t. ψ.

Remark 3.1. (1) This independence interpolates Boolean and monotone independences. That is, if $\psi = 0$ on $\bigcup_{i\in I}\mathcal{A}_i$, we recover the Boolean independence; if $\psi = \varphi$ on $\bigcup_{i\in I}\mathcal{A}_i$, we recover the monotone independence.
(2) C-monotone independence is not symmetric, i.e., c-monotone independence of $\mathcal{A}_1$ and $\mathcal{A}_2$ does not imply the c-monotone independence of $\mathcal{A}_2$ and $\mathcal{A}_1$. In other words, the concept of "mutual independence" fails to hold.
(3) As in the cases of Boolean and monotone independences, if the subalgebras $\mathcal{A}_i$ contain the unit of the whole algebra $\mathcal{A}$, c-monotone independence becomes trivial.
(4) The c-monotone product was defined by using the c-free product; however, the associative law of the c-monotone product does not follow from that of the c-free product. Therefore, we completed Fig. 2.

Example 3.1. Let $(\mathcal{A}, \varphi, \psi)$ be an algebraic probability space with two states; let $\mathcal{A}_1$ and $\mathcal{A}_2$ be c-monotone independent $*$-subalgebras. For $a, a' \in \mathcal{A}_1$ and $b, b' \in \mathcal{A}_2$, joint moments are calculated as follows.

$$\varphi(ab) = \varphi(ba) = \varphi(a)\varphi(b),$$
$$\varphi(aba') = \varphi(a)\big(\varphi(b) - \psi(b)\big)\varphi(a') + \psi(b)\varphi(aa'),$$
$$\varphi(bab') = \varphi(b)\varphi(a)\varphi(b').$$

It is worthy to compare the above with the c-free case.[6] Let $\mathcal{A}_1$ and $\mathcal{A}_2$ be c-free independent $*$-subalgebras. Then

$$\varphi(ab) = \varphi(ba) = \varphi(a)\varphi(b),$$
$$\varphi(aba') = \varphi(a)\big(\varphi(b) - \psi(b)\big)\varphi(a') + \psi(b)\varphi(aa'),$$
$$\varphi(bab') = \varphi(b)\big(\varphi(a) - \psi(a)\big)\varphi(b') + \psi(a)\varphi(bb').$$

Thus difference arises from $\varphi(bab')$.

Acknowledgement

The author would like to thank Mr. Hayato Saigo for many discussions about independence and cumulants. He is grateful to Professor Izumi Ojima for discussions about independence, interacting Fock spaces and other subjects. He thanks Professor Marek Bożejko for guiding him to the notion of conditionally free independence and the important reference.[10] He also thanks Professor Uwe Franz for fruitful discussions and for giving a seminar on the categorical treatment of independence. He also thanks Professor Shogo Tanimura, Mr. Ryo Harada, Mr. Hiroshi Ando and Mr. Kazuya Okamura for their comments and encouragements. This work was supported by Japan Society for the Promotion of Science, KAKENHI 21-5106. Finally the author thanks the financial support by Global COE Program at Kyoto University.

References

1. N. I. Akhiezer, *The Classical Moment Problem* (English transl.), Oliver and Boyd, 1965.
2. D. Avitzour, Free products of C^*-algebras, Trans. Amer. Math. Soc. **271**, no. 2 (1982), 423-435.
3. A. Ben Ghorbal and M. Schürmann, Non-commutative notions of stochastic independence, Math. Proc. Comb. Phil. Soc. **133** (2002), 531-561.
4. M. Bożejko, Positive definite functions on the free group and the noncommutative Riesz product, Bull. Un. Mat. Ital. (6) **5-A** (1986), 13-21.
5. M. Bożejko and R. Speicher, An example of a generalized Brownian motion, Commun. Math. Phys. **137** (1991), 519-531.
6. M. Bożejko and R. Speicher, ψ-independent and symmetrized white noises, Quantum Probability and Related Topics (L. Accardi, ed.), World Scientific, Singapore **VI** (1991), 219-236.
7. T. Hasebe, Conditionally monotone independence I: Independence, additive convolutions and related convolutions, submitted.
8. T. Hasebe and H. Saigo, The monotone cumulants, to appear in Ann. Inst. Henri Poincaré Probab. Stat. arXiv:0907.4896v3.
9. U. Franz, Monotone independence is associative, Infin. Dim. Anal. Quantum Probab. Rel. Topics **4**, no. 3 (2001), 401-407.
10. U. Franz, Multiplicative monotone convolutions, Banach Center Publ., **73** (2006), 153-166.
11. U. Franz, Lévy processes on quantum groups and dual groups, in *Quantum independent increment processes II*, Lecture Notes in Math., vol. 1866, Springer-Verlag, 2006.
12. N. Muraki, Monotonic convolution and monotonic Lévy-Hinčin formula, preprint, 2000.
13. N. Muraki, The five independences as quasi-universal products, Infin. Dim. Anal. Quantum Probab. Rel. Topics **5**, no. 1 (2002), 113-134.

14. N. Muraki, The five independences as natural products, Infin. Dim. Anal. Quantum Probab. Rel. Topics **6**, no. 3 (2003), 337-371.
15. R. Speicher, On universal products, in Free Probability Theory, papers from a Workshop on Random Matrices and Operator Algebra Free Products, Toronto, Canada 1995, ed. D. V. Voiculescu, Fields Inst. Commun. 12 (Amer. Math. Soc., 1997), 257-266.
16. R. Speicher and R. Woroudi, Boolean convolution, in Free Probability Theory, papers from a Workshop on Random Matrices and Operator Algebra Free Products, Toronto, Canada 1995, ed. D. V. Voiculescu, Fields Inst. Commun. 12 (Amer. Math. Soc., 1997), 267-280.
17. D. Voiculescu, Symmetries of some reduced free product algebras, Operator algebras and their connections with topology and ergodic theory, Lect. Notes in Math. **1132**, Springer (1985), 556-588.
18. J. Wysoczański, Bm-central limit theorems for positive definite real symmetric matrices, Infin. Dim. Anal. Quantum Probab. Rel. Topics **11**, no. 1 (2008), 33-51.

ROLES OF WHITE NOISE IN STOCHASTIC ANALYSIS AND SOME OF FUTURE DIRECTIONS

TAKEYUKI HIDA

Nagoya, Japan

White Noise analysis may be thought of a well-established theory. This is true in a sense, however we are surprised to find that there are so many profound properties still remain undiscovered. In this report, we shall have a quick review of white noise theory, then we shall propose some of future directions to be investigated, from our viewpoint. Further, we shall discuss a new noise which is of Poisson type.

2000 AMS Subject Classification 60H40

1. Preliminaries

The first section will be devoted to a short note which gives a summary of the white noise theory with some new interpretations on the idea of this theory.

We are, in general, interested in mathematical approach to random complex, developing phenomena systematically. Such phenomena, hopefully, would be expressed as functionals of independent elemental random variables indexed by the time variable t. We take, as a typical example of such variables, *white noise* $\dot{B}(t), t \in R$, obtained by taking the time derivatives of a Brownian motion $B(t)$. Since $\dot{B}(t)$ is no more an ordinary random variable, but a generalized random variable. So, we may say that a white noise $\dot{B}(t), t \in R^1$ is actually a system of *idealized elemental random variables.*

We shall take them to be basic objects that are rigorously defined, so that we shall discuss the functionals expressing the random phenomena in question in the form

$$\varphi(\dot{B}(s), s \in R^1, t).$$

The white noise analysis deals with the calculus of such functionals in line with the infinite dimensional analysis.

Before we come to the contents of the white noise analysis, its advantages or characteristics will be explained.

We follow the following steps.

1) Each $\dot{B}(t)$ has the identity, so that the collection of nonlinear functionals should be taken care of systematically, and the collection of reasonable class of functionals forms a large topological vector space involving *generalized white noise functionals.* The space, thus obtained, plays significant and new roles in stochastic analysis.

2) The new space satisfies invariance and duality. These should be *aesthetic* nature of white noise analysis.

The infinite dimensional rotation group gives invariance of white noise measure, and the space of generalized white noise functionals allows us to find various kinds of duality. Often, optimality is involved there.

3) The theory has good applications. Perhaps, it is better to say the white noise theory have good collaboration with other fields of science; quantum dynamics, molecular biology, information sciences and others,

4) We can think of a new noise, i.e. a system of i.e.r.v.'s. We now propose a new noise of Poisson type, parametrized by a space variable, not by time t as in the case of $\dot{B}(t)$. It is interesting to see dissimilarities between the white noise (which is Gaussian) and a new noise of Poisson type. We shall briefly mention on this subject.

2. Generalized white noise functionals and the rotation group

I. Brief summary of generalized functionals.

Let E be a nuclear space dense in $L^2(R^1)$. The characteristic functional $C(\xi), \xi \in E$, of white noise $\{\dot{B}(t)\}$ is given by

$$C(\xi) = E(\exp[i\langle \dot{B}, \xi\rangle]) = \exp[-\frac{1}{2}\|\xi\|^2].$$

There exists a probability measure μ, which is to be the probability distribution of white noise, on E^* the dual space of E, such that

$$C(\xi) = \int_{E^*} e^{i\langle x,\xi\rangle} d\mu(x).$$

The classical Hilbert space (L^2) of ordinary Brownian functionals is a collection of functions of smeared variables (which are the bilinear forms) $\langle x, \xi \rangle$ with $x \in E^*(\mu)$ and with ξ smooth enough. The space (L^2) admits a direct sum decomposition into the spaces H_n of homogeneous chaos (or multiple Wiener integrals) of degree n (a Fock space):

$$(L^2) = \bigoplus_0^\infty H_n.$$

Generalized white noise functionals

We extend the space H_n to the space $H_n^{(-n)}$ so that it involves all the Hermite polynomials in $\dot{B}(t)$'s of degree n. It is defined to be isomorphic to the symmetric Sobolev space of degree $-\frac{n+1}{2}$ over R^n up to constant $\sqrt{n!}$. To construct such a space we have to employ the *renormalization* technique (in reality, serious considerations are involved there).

Taking a positive decreasing sequence $\{c_n\}$, which should be chosen depending on the problem involved, we have the space $(L^2)^-$, that is called the space of **generalized white noise functionals**

$$(L^2)^- = \bigoplus_n c_n H_n^{(-n)}. \tag{2.1}$$

Note that $\dot{B}(t)$ is a member of the space $H_1^{(-1)}$, so that it has now a definite *identity.*

Depending on the problem, we may further extend $H_1^{(-1)}$ to be $H_1^{(-N)}, N > 1$, that is isomorphic to the symmetric Sobolev space of order $-N(N > 1)$ over R^1, e.g. [16].

One thing to be noted is that in the space $(L^2)^-$ we can discuss lots of dualities in connection with time development.

There is another way to introduce the space of generalized white noise functionals due to Kubo and Takenaka [12] which may be thought of as an infinite dimensional analogue of the Schwartz space of distributions, where a Gel'fand triple is used:

$$(S) \subset (L^2)^- \subset (S)^*.$$

This is a significant method of introducing generalized functionals, however we do not use this triple in the paper, so we shall not go into further.

II, Infinite dimensional rotation group.

We start with the definition of the infinite dimensional rotation group after H. Yoshizawa 1970.

Take a nuclear space E which is dense in $L^2(R^1)$.

Let $O(E)$ be a collection of members g such that

1) g is a linear isomorphism of E,

2) g is orthogonal:

$$\|g\xi\| = \|\xi\|,$$

where $\| \cdot \|$ is the $L^2(R^1)$-norm. The compact-open topology is introduce in the group $O(E)$.

For each $g \in O(E)$ is its adjoint g^* which leaves the white noise measure μ invariant:

$$g^* \mu = \mu.$$

This is one of the significant properties of $O(E)$.

Since $O(E)$ is very big (neither compact nor locally compact), we take suitable subgroups. The entire group is divided into two parts: Class I and Class II.

The Class I involves members that can be defined by using a base (or coordinate vectors), say $\{\xi_n\}$ of E:

There is an example of a class I subgroup.

$$G_\infty = ind.lim_n G_n,$$

where G_n is isomorphic to $SO(n)$.

While, any member of the class II should come from a diffeomorphism of the parameter space $\bar{R}$, the one-point compactification of R^1.

Let us give a concrete definition of the class II.

Definition 2.1. Let g be a member of $O(E)$ defined by the formula

$$(g\xi)(u) = \xi(f(u))\sqrt{|f'(u)|}, \tag{2.2}$$

where f is a diffeomorphism of $\bar{R}$. If such g belongs to $O(E)$, then g is said to be in the class II. If a subgroup G of $O(E)$ involves members in class II only, then G is a subgroup of class II.

Definition 2.2. A one parameter subgroup $g_t, t \in R^1$, of class II is called a **whisker** if it is expressed in the form

$$(g_t)(\xi)(u) = \xi(\psi_t(u))\sqrt{|\psi_t'(u)|} \tag{2.3}$$

where $\psi_t(u) = f^{-1}(f(u) + t)$, and if g_t is continuous in t.

III. Whiskers

First we shall have a brief review of the known results on whiskers so that some hints to find new good subgroups of $O(E)$ will be found.

Each whisker, say $\{g_t; t \text{real}\}$, should be defined by a system of parametrized diffeomorphisms $\{\psi_t(u)\}$ of $\bar{R} = R \cup \infty$, as in (2.3).

More practically, we restrict our attention to the case where $g_t, t \in R$ has the (infinitesimal) generator

$$\alpha = \frac{d}{dt} g_t|_{t=0}$$

Assume that a family $\{\psi_t(u), t \in R\}$ should be such that $\psi_t(u)$ is measurable in (t, u) and satisfies

$$\psi_t \cdot \psi_t = \psi_{t+s}$$
$$\psi_0(u) = u.$$

Then, we have an expression for $\psi_t(u)$:

$$\psi_t(u) = f(f^{-1}(u) + t) \tag{2.4}$$

where f is continuous and strictly monotone. Its (infinitesimal) generator α, if f is differentiable, can be expressed in the form

$$\alpha = a(u)\frac{d}{du} + \frac{1}{2}a'(u), \tag{2.5}$$

where

$$a(u) = f'(f^{-1}(u)). \tag{2.6}$$

See e.g. [5] and [6].

We have already established the results that there exists a three dimensional subgroup involving the whiskers with significant *probabilistic* meanings. The generators are expressed by $a(u) = 1, a(u) = u$ and $a(u) = u^2$, respectively.

We can show a list:

$$s = \frac{d}{du},$$

$$\tau = u\frac{d}{du} + \frac{1}{2},$$

$$\kappa = u^2\frac{d}{du} + u$$

They describe the *projective invariance* of Brownian motion.

They form a base of a three dimensional Lie algebra isomorphic to $sl(2, R)$:

$$[\tau, s] = -s$$

$$[\tau, \kappa] = \kappa$$

$$[\kappa, s] = 2\tau$$

There is a remark that the shift with generator s is sitting as a *key member* of the algebra. Even, it corresponds to the *flow of Brownian motion.*

Theorem 2.1. *The shift S_t, the generator of which is $-s$, has a unitary representation on the space $H_n^{(-n)}, n \geq 1$.*

Proof is easy, because each member of $H_n^{(-n)}$ has an integral representation in terms of a kernel function F_n belonging to the Sobolev space of order $\frac{-(n+1)}{2}$ over R^n. Under the shift, the Sobolev norm of F_n does not change. Continuity in t and other properties are proved easily.

Note. As we shall see later in IV that comes next, the *dilation* τ is the key member in the algebra defined by half whiskers.

We are now in a position to have general relationships among the generators of the form (2.5) with the condition (2.6).

Use the notation

$$\alpha_a = a(u)\frac{d}{du} + \frac{1}{2}a'(u),$$

with $a(u) \in \mathbf{C}^1$. The collection of such α_a's is denoted by $\mathbf{D}$. Then,

Proposition 2.1. *It holds that, with the notation* $\{a, b\} = ab' - a'b$, *for any* α_a *and* α_b

$$[\alpha_a, \alpha_b] = \alpha_{\{a,b\}},$$

$$[\alpha_a, \alpha_b] = -[\alpha_b, \alpha_a].$$

Proposition 2.2. *The collection* $\mathbf{D}$ *forms a base of a Lie algebra, which is denoted by* $\mathbf{A}$. *It has no identity.*

IV. Half whiskers.

We are now in search of *new* whiskers that show significant probabilistic properties. Now a whisker is changed to a half whisker so that we can discover under mild restrictions.

We have already discussed half whiskers in some details (see [8],[10]), so we shall not present full known results, but state why we stick to subgroups of class II and show some of our attempts.

Functionals discussed in white noise analysis may be expressed either in the forms like

$$f(\langle \dot{B}, \xi_n \rangle, n \in Z^+), \ Z^+ = \{1, 2, \cdots\}$$

where $\xi_n \in E$ is a base of $L^2(R^1)$ and

$$\varphi(\dot{B}(t), t \in R^1)$$

We say that the former is of digital type and the latter is of analogue type.

The latter is better in many ways; for one thing t stands for the time, so that developing phenomena can be expressed, However, we have to pay a price. Since $\dot{B}(t)$ is a generalized, indeed idealized, random variable, non-linear functions of $\dot{B}(t)$'s are not simply defined, namely we need *renormalizations.* Having had renormalizations we can define a larger space of generalized functionals.

One may think it is easy to let digital formula tend to analogue ones. This can not be done easily. If we discuss operators, like differential operators, we have to be extremely careful (see Section 3, (3)).

As for the *harmonic analysis* arising from the infinite dimensional rotation group, the digital case is related to the subgroups of class I, while in the analogue case subgroups of class II are the main tools. One can agree with the idea of finding whiskers or half whiskers.

Now we come back to class II subgroups of $O(E)$,

We recall the notes [14] p. 60, where a new class of whiskers has been proposed; in reality, most of them are half whiskers. Let us repeat the proposal.

$$\alpha^p = u^{p+1}\frac{d}{du} + \frac{p+1}{2}u^p, \ u \geq 0, \tag{2.7}$$

is suggested to be studied, where p is not necessarily integer. Since fractional power p is involved, we tacitly assume that u is non-negative, We, therefore, take a white noise with time-parameter $[0, \infty)$. The basic nuclear space E is chosen to be a space of functions on $[0, \infty)$ and is suitably topologized.

We can say a partial answer to the problem to discover new whiskers,.

As was remarked before, number 1, the power of u^1 is the key number and, in fact, it is exceptional in case where the variable u runs through R^1, that is, corresponds to a whisker with generator τ. Thus, a special care will be taken in the case $p = 0$, that is α^0.

We remind the relationship between f and $a(u)$ for the function $\psi_t(u)$ and α. Assuming differentiability of f we have the formula of the generator. For $a(u) = u^p$, the corresponding $f(u)$ is determined. Namely,

$$u^p = f'(f^{-1}(u)).$$

An additional requirement for f is that f should be a map from the entire $[0.\infty)$ onto itself. Hence, we have

$$f(u) = c_p u^{\frac{1}{1-p}}, \tag{2.8}$$

where $c_p = (1-p)^{1/(1-p)}$.

We, therefore, have

$$f^{-1}(u) = (1-p)^{-1} u^{1-p}. \tag{2.9}$$

We are ready to define a transformation g_t^p acting on D_{00} by

$$(g_t \xi)(u) = \xi(c_p(\frac{u^{1-p}}{1-p} + t)^{\frac{1}{1-p}}) \sqrt{\frac{c_p}{1-p}(\frac{u^{1-p}}{1-p} + t)^{\frac{p}{1-p}} u^{-p}}. \tag{2.10}$$

Note that f is always positive and maps $(0, \infty)$ onto itself in the ordinary order in the case $p < 1$; while in the case $p > 1$ the mapping is in the reversible order.

The exceptional case $p = 0$ is referred to the shift and has already been well discussed.

Then, we claim, still assuming $p \neq 1$, the following theorem.

Theorem 2.2.

i) *g_t^p is a rotation for every $t > 0$.*

ii) *The collection $\{g_t^p, t \geq 0\}$ forms a continuous semi-group with the product $g_t^p \cdot g_s^p = g_{t+s}^p$ for $t, s \geq 0$.*

iii) *The generator of g_t^p is α^p up to constant.*

Definition 2.3. A continuous semi-group $g_t, t \geq 0$, each member of which comes from $\psi_t(u)$ is called a *half whisker*.

Duality. We do not define the notion of "Duality", however it expresses very significant meaning in white noise theory. This has recently been recognized. See, e.g.[15,16].

With respect to α^0 we can see a duality

$$\alpha^p \Longleftrightarrow \alpha^{-p}$$

3. A proposal and some of future directions

A proposal.

We have found a new noise of Poisson type (in [11]). The author should like to mention the results briefly, since they are strongly in line with our aim to explain the roles of white noise in stochastic analysis.

Our approach to stochastic analysis is based on the system of *idealized elemental random variables.* If the system is parametrized by the time t, then the property "elemental" suggests us to consider the time derivative of elemental additive process; Brownian motion and Poisson process, where we tacitly assume separability, or the associated probability distribution (measure) forms an abstract Lebesgue space.

On the other hand, if we choose another parameter, instead of time, say space variable, then we may have another kind of noise. Concerning this question, a compound Poisson process has given us a hint.

Let us start out with a functional $C^P(\xi)$, where the variable ξ runs through a certain nuclear space E, say isomorphic to the Schwartz space $\mathcal{S}$;

$$C^P(\xi) = \exp[\int (e^{iu\xi(u)} - 1)dn(u)], \tag{3.1}$$

where $dn(u)$ is a measure on $(0, \infty)$.

To fix the idea, we assume that the measure $dn(u)$ is equivalent to the Lebesgue measure, i.e. it is of the form $dn(u) = \lambda(u)du$ with $\lambda(u) > 0$ a.e. In addition, we assume that $\lambda \in L^1((0,\infty)) \cap L^2((0,\infty))$.

Theorem 3.1. *Under these assumptions, the functional $C^P(\xi)$ is a characteristic functional.*

Proof. i) $C^P(\xi)$ is continuous in ξ,

ii) $C^P(0) = 1$.

iii) Positive definiteness is shown by noting the fact that $\exp[(e^{izu} - 1)\lambda]$ is a characteristic function.

Hence, by the Bochner-Minlos theorem, there exists a probability measure ν^P on E^* such that

$$C^P(\xi) = \int_{E^*} e^{i<x,\xi>} \nu^P(dx). \tag{3.2}$$

We introduce a notation $P'(u, \lambda(u))$ or write it simply $P'(u)$. We understand that ν^P-almost all $x \in E^*$ is a sample function of $P'(u)$.

Theorem 3.2. *The $P'(u)$ has independent value at every point u.*

Proof is easy if the integral in (3.1) is understood to be extended over the support of ξ.

The bilinear form $\langle P', \xi \rangle$ is a random variable with mean $\int \xi(u)\lambda(u)du$ and the variance is $\int \xi(u)^2\lambda(u)du$. Hence $\langle P', \xi \rangle$ extends to $\langle P', f \rangle$ with $f \in L^2((0, \infty), \lambda du)$. If f and g are orthogonal in $L^2(\lambda du)$, then $\langle P', f \rangle$ and $\langle P', g \rangle$ are uncorrelated. Thus, we can form a random measure and hence, we can define the space $H_1(P)$ like H_1 in the case of Gaussian white noise. The space $H_1(P)$ can also be extended to a space $H_1(P)^{(-1)}$ of generalized linear functionals of $P'(u)$'s. Note that there we can give an identity to $P'(u)$ for any u.

Our conclusion is that single noise $P'(u)$ of Poisson type with the parameter u can be found. cf.[18]

The author wishes to discuss the following topics as future directions.

1) More whiskers and half whisker. Finding new subgroups of class II.

2) Duality in $(L^2)^-$ like,[15,16] and that in the algebras coming from the infinite dimensional rotation group.

3) The passage from digital to analogue with special emphasis on the passage of operators.

4) Path integrals. This topic was one of the motivation of white noise analysis. Still now we are given lots of interesting problems. For example to have more singular potentials, or time depending potentials, and so on. See e.g.[19] and references there. White Noise approach to the dynamics with the Coulomb potential is another interesting question (proposed by S. Oryu).

Acknowledgements The author is grateful to Professor R. Rebolledo, the organizer of the Conference "Probabilidades Cuánticas y Temas Relacionados", who has given the author the opportunity to talk at this 30th meeting.

References

1. T. Hida, Stationary stochastic processes. Princeton Univ. Press, 1970.
2. T. Hida Brownian motion. Iwanami Pub. Co. 1975, english ed. 1980, Springer-Verlag.
3. T. Hida. Analysis of Brownian functionals. Carleton Math. Notes. no.13 , 1975. Carleton Univ.
4. I.M. Gel'fand, M.I. Graev and N.Ya. Vilenkin, Generalized functions. vol. 5, 1962 (Russian original), Academic Press, 1966.
5. T. Hida, Stationary stochastic processes. Princeton Univ. Press, 1970.
6. T. Hida Brownian motion. Iwanami Pub. Co. 1975, english ed. 1980, Springer-Verlag.
7. T. Hida, Analysis of Brownian functionals. Carleton Math. Notes 13, Carleton Univ. 1975.
8. T. Hida, White noise and class II subgroups of the infinite dimensional rotation group. Proceedings of Stochastic Analysis and Applications. Tunis, 2009.
9. T. Hida and Si Si, Lectures on white noise functionals. World Sci. Pub. Co. 2008.
10. T. Hida and Si Si, Class II semi-subgroups of the infinite dimensional rotation group and associated Lie algebras. Prof. Z. Huang volume, 2010. World Sci. Pub. Co.
11. T. Hida, Si Si and Win Win Htay, A note on new noise of Poisson type. to appear.
12. I. Kubo and S. Takenaka, Calculus on Gaussian white noise, I-IV. Proc. Japan Academy 56 (1980), 376-380, 57 (1981) 433-437, 58 (1982) 186-189.
13. P. Lévy, Problèmes concrets d'analyse fonctionnelle. Gauthier-Villars, 1951.
14. T. Shimizu and Si Si, Professor Takeyuki Hida's mathematical notes. informal publication 2004.
15. Si Si, An aspect of quadratic Hida distributions in the realization of a duality between Gaussian and Poisson noises. Infinite Dim. Analysis, Quantum Probability and Related Topics. 11 (2008), 109-118.
16. Si Si, Multiple Markov generalized Gaussian processes and their dualities. Infinite Dim. Analusis, Quantum Probability and Related Topics. 13 (2010) 99-110.
17. Si Si, Introduction to Hida distributions. World Sci. Pub. Co. to appear.
18. Si Si A.H. Tsoi and Win Win Htay, Jump finding of a stable process. Quantum Information. vol. V, World Sci. Pub. Co. (2000), 193-201.
19. L. Streit et al, The Feynman integral for time-dependent anharmonic oscillators. J. Math. Phys. 38 (6) (1997), 3278-3299.

ON DIFFICULTIES APPEARING IN THE STUDY OF STOCHASTIC VOLTERRA EQUATIONS

A. KARCZEWSKA

Faculty of Mathematics, Computer Science and Econometrics,
University of Zielona Góra, ul. Szafrana 4a, 65-516 Zielona Góra, Poland
E-mail: A.Karczewska@wmie.uz.zgora.pl
www.wmie.uz.zgora.pl

In a note we discuss several problems and difficulties arising during the study of stochastic Volterra equations of convolution type. The resolvent approach to such equations enables us to obtain results in an analogous way like in the semigroup approach used to stochastic differential equations. However, in the resolvent case new difficulties arise because the solution operator corresponding to the Volterra equation in general does not create a semigroup. In this paper we present some consequences and complications coming from deterministic and stochastic convolutions connected with the stochastic Volterra equations under consideration.

Keywords: Stochastic Volterra equation; Stochastic convolution: Resolvent operator.

1. Introduction

In the paper we discuss some main difficulties appearing during the study of stochastic Volterra equations of convolution type in a separable Hilbert space. The aim of the paper is to discuss the so-called resolvent approach to stochastic Volterra equations which are more complicated and more difficult for study than stochastic differential equations.

Many authors have studied stochastic differential equations in Hilbert spaces with semigroup methods. This started by R. Curtain and P. Falb[7] and D. Dawson.[11] Then this has been continued by R. Curtain and A. Pritchard,[8] G. Da Prato and J. Zabczyk,[10] W. Grecksch and C. Tudor[15] and their collaborators. The semigroup methods are known as a powerful tool for analyzing stochastic differential equations. The assumption made in the approach that the linear part of the equation is an infinitesimal generator of a linear semigroup, is equivalent to the requirement that the equation studied has a unique solution continuously depending on the initial condition.

Let $(\Omega, \mathcal{F}, (\mathcal{F}_t)_{t\geq 0}, P)$ be a probability space and let H be a separable Hilbert space. In the paper we shall study the following stochastic Volterra equation

$$X(t) = X_0 + \int_0^t a(t-\tau)AX(\tau)d\tau + \int_0^t \Psi(\tau)\,dW(\tau), \quad t \geq 0, \qquad (1)$$

where X_0 is a $\mathcal{F}_0$-measurable random variable with values in H, $a \in L^1_{\text{loc}}(\mathbb{R}_+; \mathbb{R})$ and A is a closed linear operator in H with a dense domain $D(A)$ equipped with the graph norm

$$|h|_{D(A)} := (|h|_H^2 + |Ah|_H^2)^{1/2}, \quad h \in D(A).$$

In the equation (1), W is a cylindrical Wiener process, and Ψ is an appropriate predictable process, both described below.

The resolvent approach to stochastic Volterra equation (1) is a natural generalization of the semigroup approach. Such approach to Volterra equations provides us results, obtained in an elegant way, analogous to those received in the semigroup approach. But in the resolvent case new difficulties appear because resolvent (or solution operator) corresponding to the Volterra equation under consideration in general does not create any semigroup.

The resolvent approach to the stochastic Volterra equations started with Ph. Clément and G. Da Prato.[2] This approach has been continued, among others, by S. Bonaccorsi and L. Tubaro[1] and their collaborators and A. Karczewska and C. Lizama.[18–21]

Let us note that Volterra equations discussed below, deterministic or stochastic, arise in a variety of applications as model problems, see *e.g.*[24] and references therein. Well-known techniques like localization, perturbation and coordinate transformation allow to transfer results for many problems to integro-differential equations. In these applications, the operator A is a differential operator acting in spatial variables, like the Laplacian, the Stokes operator, or the elasticity operator. The kernel function should be thought as a kernel like $a(t) = e^{-\eta t}t^{\beta-1}/\Gamma(\beta)$, where $\eta \geq 0$, $\beta \in (0,2)$.

Let us emphasize that stochastic Volterra equations of convolution type (the equations (1) or (2) below) contain a big class of equations and are an abstract version of several problems, see *e.g.*[24] For example, if $a(t) = 1$ and f is a function of C^1-class, the equation (2) is equivalent to the Cauchy problem

$$\dot{u}(t) = Au(t) + \dot{f}(t) \quad \text{with} \quad u(0) = f(0).$$

Analogously, in the case $a(t) = t$ and f of C^2-class, the equation is equivalent to

$$\ddot{u}(t) = Au(t) + \ddot{f}(t) \text{ with initial conditions } u(0) = f(0) \text{ and } \dot{u}(0) = \dot{f}(0).$$

The case $a(t) = t^{\alpha-1}/\Gamma(\alpha), \alpha > 0, t > 0$, conducts us to a class of so-called fractional Volterra equations, studied among others by Y. Fujita,[14] A. Karczewska and C. Lizama[20] and M. Li and Q. Zheng.[22] Particularly, for $\alpha \in (1,2)$, we have very interesting class of integro-differential equations interpolating the heat and the wave equations.

The equation (1) is a stochastic version of the deterministic Volterra equation, see (2) below. The motivation for studying the equation (1) is the following. In order to take into account random fluctuations appearing in models governed by deterministic Volterra equation, it is natural to consider this equation with a very irregular exterior force, e.g., $\int_0^t \Psi(\tau)dW(\tau)$ or $W(t)$, $t \geq 0$, where Ψ, W are appropriate processes. There are several examples of stochastic Volterra equations arising in applications. One of the most interesting is that presented by Ph. Clement and G. Da Prato.[3] Here the problem of heat conduction in materials with memory and with random fluctuations has been conducted to the stochastic Volterra equations of convolution type with completely positive kernel functions. Another example may come from the paper of Ph. Clement, G. Da Prato and J. Prüss,[5] where a model of viscoelasticity with white noise perturbation has been studied in detail. The next good example is the use of stochastic integral equations in the rainfall-runoff models described in detail by Th.V. Hromadka and R.J. Whitley.[17] A rich source of models described by the equation (1) with the kernel function $a(t) = t^{\alpha-1}/\Gamma(\alpha)$, $\alpha > 0$, $t > 0$, should be, in the not far future, bioengineering. There are many examples of using fractional calculus in the biological modelling, see e.g. R.L. Magin.[23] Some of them could be studied taking into account random fluctuations.

The purpose of this note is to give a brief introduction to some of the basic tools for the study of stochastic Volterra equations of convolution type and to give an overview of some directions on research in this area.

2. Volterra equations – resolvent approach

The equation (1) is a stochastic counterpart of the following deterministic Volterra equation

$$X(t) = X_0 + \int_0^t a(t-\tau)AX(\tau)d\tau + f(t), \qquad t \geq 0, \tag{2}$$

where $X_0 \in H$, f is a continuous H-valued function and $a(t), A$ are the same like in (1).

In the paper we write $a(t)$ for the kernel function a. The notation $a(t)$ will mean the function and not the value of the function a at t. Such notation will allow to distinguish the function $a(t)$ and the article a.

The resolvent approach to the Volterra equation (2) is a natural generalization of the semigroup approach usually used to differential equations. It has been introduced many years ago, probably by A. Friedman and M. Shinbrot.[13] In eighties and nineties of twentieth century it was used *e.g.* by Ph. Clément, G. Da Prato, S.-O. Londen, J.A. Nohel and J. Prüss. And, as we have already written, Ph. Clément and G. Da Prato[2] were the first who applied the resolvent approach to stochastic Volterra equations of convolution type.

In the resolvent approach to (1) and (2) we assume that the deterministic Volterra equation (2) is well-posed. This is equivalent to the fact that the equation (2) admits the resolvent, that is, there exists a family of operators $S(t), t \geq 0$, as follows.

Definition 2.1. A family $(S(t))_{t\geq 0}$ of bounded linear operators in the space H is called `resolvent` for (2) if the following conditions are satisfied:

(1) $S(t)$ is strongly continuous on $\mathbb{R}_+$ and $S(0) = I$;
(2) $S(t)$ commutes with the operator A, that is, $S(t)(D(A)) \subset D(A)$ and $AS(t)x = S(t)Ax$ for all $x \in D(A)$ and $t \geq 0$;
(3) the following `resolvent equation` holds for all $x \in D(A)$, $t \geq 0$

$$S(t)x = x + \int_0^t a(t-\tau)AS(\tau)x d\tau \,. \tag{3}$$

There are the following useful connections between the operators $S(t)$, $t \geq 0$, and some scalar functions.

Suppose $S(t)$ is the resolvent for (2) and let $-\mu \in \sigma(A)$ be an eigenvalue of A with eigenvector $x \neq 0$. Then

$$S(t)\, x = s(t;\mu)\, x, \qquad t \geq 0, \tag{4}$$

where $s(t;\mu)$ is the solution of the one-dimensional Volterra equation

$$s(t;\mu) + \mu \int_0^t a(t-\tau)\, s(\tau;\mu)\, d\tau = 1, \qquad t \geq 0. \tag{5}$$

By $W^{1,p}_{\mathrm{loc}}(\mathbb{R}_+; H)$ we denote the Sobolev space of order $(1,p)$ of Bochner locally p-integrable functions mapping $\mathbb{R}_+$ into the space H.

Definition 2.2. A resolvent $S(t)$, for the equation (2), is called `differentiable` if $S(\cdot)x \in W^{1,1}_{\text{loc}}(\mathbb{R}_+; H)$ for any $x \in D(A)$ and there exists a function $\varphi \in L^1_{\text{loc}}(\mathbb{R}_+)$ such that $|\dot{S}(t)x| \leq \varphi(t)|x|_{D(A)}$ a.e. on $\mathbb{R}_+$, for every $x \in D(A)$.

Similarly, if $S(t)$ is differentiable then

$$\dot{S}(t)\, x = \mu\, r(t;\mu)\, x, \qquad t \geq 0,$$

where $r(t;\mu)$ is the solution of the one-dimensional equation

$$r(t;\mu) + \mu \int_0^t a(t-\tau)\, r(\tau;\mu)\, d\tau = a(t), \qquad t \geq 0. \tag{6}$$

In some special cases the functions $s(t;\mu)$ and $r(t;\mu)$ may be found explicitely. For example, for $a(t) = 1$, we have $s(t;\mu) = r(t;\mu) = e^{-\mu t}$, $t \geq 0$, $\mu \in \mathbb{C}$. For $a(t) = t$, we obtain $s(t;\mu) = \cos(\sqrt{\mu}t)$, $r(t;\mu) = \sin(\sqrt{\mu}t)/\sqrt{\mu}$, $t \geq 0$, $\mu \in \mathbb{C}$.

3. Consequences of convolution

The convolution which occurs in the deterministic Volterra equation (2) causes the following problems:

(1) The family of operators $S(t), t \geq 0$, in general, does not create any semigroup. So, in consequence we can not use powerful semigroup tools.
(2) The resolvent $S(t), t \geq 0$, is generated by the pair $(A, a(t)), t \geq 0$, that is the operator A and the kernel function $a(t)$, in contrary to semigroup which is generated only by the generator.
(3) The operators $S(t), t \geq 0$, are not exponentially bounded „automatically". W. Desch and J. Prüss[12] proved that the family $S(t), t \geq 0$, in contrary to the semigroup case, is not exponentially bounded in general, even if the kernel $a \in C^\infty(\mathbb{R}_+) \cap L^1(\mathbb{R}_+)$.
(4) The resolvent case is more complicated and more difficult to study than the semigroup case.

The main problem appearing in the resolvent approach to the equation (2) was to obtain sufficient conditions for the existence of „good" enough resolvent operators admited by (2). Particularly, the following convergence

$$S_n(t)x \to S(t)x, \qquad \text{when} \quad n \to \infty, \qquad x \in H, \;\; t \geq 0,$$

was needed.

Unfortunately, the class of locally integrable scalar kernel functions was too big for obtaining reasonable results. We had to restrict our considerations to the class of completely positive functions, introduced by Ph. Clément and J.A. Nohel[6] and recalled below.

Definition 3.1. Function $a \in L^1(0,T)$ is `completely positive` on the interval $[0,T]$, $T < \infty$, if for any $\gamma \geq 0$, the solutions to the equations (5) and (6) are nonnegative on $[0,T]$.

Examples The following functions are completely positive:

(1) $a(t) = t^{\alpha-1}/\Gamma(\alpha)$, $t > 0$, $\alpha \in (0,1]$, $\Gamma(\alpha)$ – gamma function;
(2) $a(t) = e^{-t}$, $t \geq 0$.

Let us note that the class of completely positive kernels contains the class of positive, decreasing, log-convex kernels and it is contained in the class of kernels which are positive pointwise and positive type. For the definitions we refer to the book by G. Gripenberg, S.-O. Londen and O. Staffans.[16] Moreover, the set of completely positive functions is meager or the first category in the space of the integrable functions. This is interesting and profitable that the class of completely positive kernels appears quite naturally in applications, particularly it is used in the theory of viscoelasticity, see *e.g.*[24]

The following theorem provides „good" resolvent operators to the equation (2) and the appropriate convergence of the resolvents.

Theorem 3.1 (Karczewska and Lizama, 2007/2009).
Assume that the operator A in (2) is the generator of a C_0-semigroup in H and suppose the kernel function $a(t)$ is completely positive. Then:

(1) the pair $(A, a(t))$ generates an exponentially bounded resolvent $S(t)$, $t \geq 0$;
(2) there exist bounded operators A_n such that the pairs (A_n, a) admit resolvent families $S_n(t)$ satisfying $||S_n(t)|| \leq Me^{w_0 t}(M \geq 1,\, w_0 \geq 0)$ for all $t \geq 0$, $n \in \mathbb{N}$;
(3) for all $x \in H$, $t \geq 0$ the following convergence holds

$$S_n(t)x \to S(t)x, \quad when \quad n \to \infty\,; \tag{7}$$

(4) the convergence (7) is uniform in t on every interval $[0,T]$.

Theorem 3.1 is analogous to Hille–Yosida's theorem and extends results due to Ph. Clément and J.A. Nohel.[6]

Let us note that other versions of Theorem 3.1, formulated in Banach space are available, see[20,21] and.[19] For instance, in the case when the kernel $a(t)$ is completely positive and A generates a cosine family or when the kernel $g_\alpha(t) = t^{\alpha-1}/\Gamma(\alpha)$, $\alpha \in (0,2)$ and A generates a cosine family.

4. Probabilistic complications

As we have already written, the resolvent case is more complicated than the semigroup one. First of all, the stochastic process $X(t)$, $t \geq 0$, given by the Volterra equation (1) is not, in general, Itô's process. So, in consequence, Itô's formula usually used in the studies of stochastic differential equations is not available directly in this case.

Let us define the stochastic terms in the equation (1). Assume that H and U are separable Hilbert spaces and $Q : U \to U$ is a linear symmetric and non-negative operator. Denote by $\{\lambda_i\}, \{e_i\}, i \in \mathbb{N}$, eigenvalues and eigenvectors of the operator Q, respectively. Then the family $\{e_i\}$, $i \in \mathbb{N}$, forms a basis in the space U.

In the paper, W is an U–valued Wiener process with the covariance operator Q, where Q is not necessarily nuclear. We use the fact that any Wiener process in a separable Hilbert space can be written in the form

$$W(t) = \sum_{j=1}^{+\infty} \sqrt{\lambda_j}\, e_j\, \beta_j(t)\,, \tag{8}$$

where $\{\lambda_j\}, \{e_j\}$ are eigenvalues and eigenvectors of the operator Q, respectively like above and $\beta_j(t)$ are independent real-valued Wiener processes.

We define $U_0 := Q^{\frac{1}{2}}(U)$ and by $L_2^0 := L_2(U_0, H)$ we denote the set of all Hilbert-Schmidt operators acting from U_0 into H.

By $\mathcal{N}^2(0,T;L_2^0)$ we denote all predictable processes Ψ with values in L_2^0 such that

$$||\Psi||_T := \left\{ \mathbb{E}\left(\int_0^T ||\Psi(\tau)||^2_{L_2^0} d\tau \right) \right\} < \infty .$$

The consequence of the form (8) of the Wiener process W is the below formula (9) of the definition and meaning of the stochastic integral in the equation (1):

$$\int_0^t \Psi(\tau)\, dW(\tau) = \lim_{m\to+\infty} \sum_{j=1}^{m} \int_0^t \Psi(\tau) \sqrt{\lambda_j}\, e_j\, \beta_j(\tau), \quad t \in [0,T] \tag{9}$$

in $L^2(\Omega; H)$. So, the stochastic Itô-type integral in (1) may be defined as the limit of the sum of one-dimensional integrals. For more details we refer to.[19]

In the resolvent approach, the following types of interpretation of the solutions to the equation (1) are possible.

Definition 4.1. Predictable process $X(t)$, $t \in [0,T]$, with values in H is called **strong solution** to (1), if X takes values in $D(A)$, P-a.s., $\int_0^T |a(T-\tau)AX(\tau)|_H \, d\tau < \infty$, P-a.s. and for all $t \in [0,T]$ the equation (1) is satisfied P-a.s.

Definition 4.2. An H-valued predictable process $X(t)$, $t \in [0,T]$, is said to be a **weak solution** to (1), if $P(\int_0^t |a(t-\tau)X(\tau)|_H d\tau < +\infty) = 1$ and if for all $\xi \in D(A^*)$ and all $t \in [0,T]$ the following equation holds P-a.s.

$$\langle X(t), \xi\rangle_H = \langle X_0, \xi\rangle_H + \langle \int_0^t a(t-\tau)X(\tau)\, d\tau, A^*\xi\rangle_H + \langle \int_0^t \Psi(\tau)\, dW(\tau), \xi\rangle_H.$$

Definition 4.3. Predictable process $X(t), t \in [0,T]$, with values in H is called **mild solution** to (1), if $\mathbb{E}(\int_0^t |S(t-\tau)\Psi(\tau)|^2_{L_2^0} \, d\tau) < \infty$ for $t \le T$ and for all $t \in [0,T]$, $X(t) = S(t)X_0 + \int_0^t S(t-\tau)\Psi(\tau)\, dW(\tau)$, P-a.s.

Let us note that a strong solution, if exists, is always a weak solution to the equation (1). Under some assumptions, see details in,[18,19] a weak solution to (1) is a mild solution, too. Conversly, a mild solution may be a weak solution to (1), as well.

Let us define the following stochastic convolutions

$$W_\Psi(t) := \int_0^t S(t-\tau)\, \Psi(\tau)\, dW(\tau) \tag{10}$$

and

$$W_S(t) := \int_0^t S(t-\tau)\, dW(\tau)\,. \tag{11}$$

Analogously like in the semigroup case, in the resolvent case both convolutions (10) and (11) play a prominent role. They are good candidates to be not only weak solutions but even strong solutions to stochastic Volterra equations. We shall write about this in the sequel.

One of the important tool that has been established for the study of the stochastic convolution (11) in the semigroup case, is the factorization method due to G. Da Prato, S. Kwapień and J. Zabczyk.[9] This powerful method gives sufficient, and in some cases necessary and sufficient conditions for the convolution (11) to be regular in time and in space, as well.

Unfortunately, the factorization method can not be applied to the convolutions (10) and (11) in the resolvent case because that method bases on the assumption that the operators $S(t), t \geq 0$, create a C_0-semigroup.

However, some regularity results for the convolution (11) are available. Below we recall the first and the most interesting results due to Ph. Clément and G. Da Prato[2] obtained under the following assumptions on the operator A, the function $a(t)$, $t \geq 0$, and the solution to the equation (4).

Hypothesis 4.1.

(i) A is a self-adjoint negative operator and $Ae_k = -\mu_k e_k$, for some positive numbers μ_k, $k \in \mathbb{N}$.
(ii) $-\mathrm{Tr}(A^{-1}) = \sum_{k=1}^{\infty}(1/\mu_k) < +\infty$.
(iii) The kernel function $a(t)$, $t \geq 0$, is completely positive.

Under these assumptions, from (4) we have

$$S(t)\, e_k = s(\mu_k; t)\, e_k, \qquad k \in \mathbb{N}.$$

Then, in this case, the stochastic convolution W_S defined by (11) is given formally by

$$W_S(t) = \sum_{k=1}^{\infty} \int_0^t s(\mu_k; t-\tau)\, e_k\, d\beta_k(\tau), \tag{12}$$

where $s(\mu_k; t)$, e_k, β_k are as above.

Now, we are able to recall the main regularity results for the stochastic convolution (11).

Theorem 4.1 (Clément and Da Prato 1996). *Assume that Hypothesis 4.1 holds. Then for any $t \geq 0$ the series*

$$\sum_{k=1}^{\infty} \int_0^t s(\mu_k, t-\tau) e_k d\beta_k(\tau)$$

is convergent in $L^2(\Omega)$ to a Gaussian random variable $W_S(t)$ of the form (12) with mean 0 and covariance operator Q_t determined by

$$Q_t e_k = \int_0^t s^2(\mu_k, \tau) d\tau e_k, \quad k \in \mathbb{N}.$$

In order to show hölderianity of the stochastic convolution (12), we need additional assumptions.

Hypothesis 4.2. There exists $\theta \in (0,1)$ and $C_\theta > 0$ such that, for all $0 < \tau < t$ we have

$$\int_\tau^t s^2(\mu,\sigma)d\sigma \le C_\theta \mu^{\theta-1}|t-\tau|^\theta, \qquad \sum_{k=1}^{\infty} \mu_k^{\theta-1} < +\infty,$$

$$\text{and} \qquad \int_0^\tau [s(\mu,\tau-\sigma) - s(\mu,t-\sigma)]^2 d\sigma \le C_\theta \mu^{\theta-1}|t-\tau|^\theta .$$

Hypothesis 4.3. There exists $M > 0$ such that

$$\begin{cases} |e_k(\xi)| \le M, & k \in \mathbb{N}, \quad \xi \in \mathcal{O}, \\ |\nabla e_k(\xi)| \le M\mu_k^{1/2}, & k \in \mathbb{N}, \quad \xi \in \mathcal{O}, \end{cases}$$

where $\mathcal{O}$ is a bounded open subset of $\mathbb{R}^d$.

Theorem 4.2 (Clément and Da Prato 1996). *Under Hypotheses 4.1 and 4.2, for every positive number $\alpha < \theta/2$, the trajectories of the convolution (12) are almost surely α-Hölder continuous.*

Theorem 4.3 (Clément and Da Prato 1996). *Under Hypotheses 4.1, 4.2 and 4.3, the trajectories of the convolution (12) are almost surely α-Hölder continuous in (t,θ) for any $\alpha \in (0,1/4)$.*

There are two next features of stochastic convolutions (10) and (11) which bring us some difficulties. Firstly, these stochastic convolutions are not semimartingales. In consequence, Doob's inequalities are not available directly.

Secondly, even the stochastic convolution (11) is not a Markov process unless $S(t), t \ge 0$, is a semigroup. Hence, we are not able to associate Kolmogorov equation to the above convolution.

However, some classical results in the case of Markov process can be generalized in the resolvent case, see Ph. Clément and G. Da Prato.[4]

If in (1) the kernel function $a \in W^{1,1}_{\text{loc}}(\mathbb{R}_+,\mathbb{R})$ and $a(0) = 1$, then taking $k(t) := a'(t)$ and $\Psi \equiv I$, the equation (1) can be written as

$$dX(t) = [AX(t) + \int_0^t k(t-\tau)AX(\tau)d\tau]dt + dW(t), \quad t \ge 0. \tag{13}$$

So, the mild solution corresponding to (13) has the form

$$X(t;x) := S(t)x + \int_0^t S(t-\tau)dW(\tau), \quad x \in H, \quad t > 0.$$

Let us define linear operators

$$P_t\varphi(x) := \mathbb{E}[\varphi(X(t;x))], \quad \varphi \in C_b(H). \tag{14}$$

Then the following result holds.

Theorem 4.4 (Clément and Da Prato, 2000). *Assume that Hypothesis 1 holds and let $P_t, t > 0$, be defined by (14). Then the following inclusion holds*

$$\text{Image}\, S(t) \subset \text{Image}\, Q_t^{1/2}, \quad t > 0,$$

where the operator Q_t is like in Theorem 2. Moreover, for any $t > 0$, $P_t\varphi(x)$ is differentiable in x.

Despite many problems and difficulties, it was possible to obtain several interesting results. Below, we recall the result providing the existence of strong solution to the equation (1).

Theorem 4.5 (Karczewska and Lizama, 2007/2009). *Let A be a generator of C_0-semigroup in H and let the function $a(t)$, $t \geq 0$ be completely positive. If Ψ and $A\Psi$ belong to $\mathcal{N}^2(0,T;L_2^0)$ and $\Psi(t)(U_0) \subset D(A)$, P-a.s., then the equation (1) has a strong solution.*

More precisely, the stochastic convolution (10) is the strong solution to (1).

Let us note that Theorem 4.5 generalizes results due to, *e.g.*, C. Tudor, G. Da Prato and J. Zabczyk. Moreover, it is worth to emphasize that the assumptions in Theorem 4.5, in the resolvent case, are the same like in the semigroup case.

For other results giving the existence of strong solutions we refer to.[19–21]

5. Remarks on resolvent approach and perspectives for stochastic Volterra equations

Now, let us summarize the difficulties presented and discussed above. Let us do this basing on the stochastic fractional Volterra equation of the form

$$X(t) = X_0 + \int_0^t \frac{(t-\tau)^{\alpha-1}}{\Gamma(\alpha)} A\,X(\tau)\,d\tau + W(t), \quad t \geq 0, \tag{15}$$

where $\alpha \in (0,1]$, Γ is the gamma function, A is the generator of a C_0-semigroup $T(t)$, $t \geq 0$, in H and W is an H-valued Wiener process.

It is worth noting that for $\alpha = 1$ the equation (15) becomes the integral form of a stochastic differential equation and semigroup tools can be used for the study of the stochastic convolution

$$W_T(t) := \int_0^t T(t-\tau)\, dW(\tau), \quad t \geq 0. \tag{16}$$

Particularly, we can use the factorization method already mentioned in the paper for obtaining time and spatial regularity of the convolution (16). For details we refer to the monograph G. Da Prato and J. Zabczyk.[10]

If $\alpha \in (0,1)$, the case is more complicated. Now, the following stochastic convolution

$$W_{S_\alpha}(t) := \int_0^t S_\alpha(t-\tau)\, dW(\tau), \quad t \geq 0, \tag{17}$$

corresponds to the equation (15), where $S_\alpha(t)$, $t \geq 0$, is the resolvent generated by the operator A and the kernel function $g_\alpha(t) := t^{\alpha-1}/\Gamma(\alpha)$, $t > 0$. Because the operators S_α do not create any semigroup we can not use to study the convolution (17) any tool basing on the semigroup property. We can overcome the difficulties using, e.g., results due to Ph. Clement and G. Da Prato, recalled in the paper, because the function g_α, for $\alpha \in (0,1)$ is completely positive.

We would like to emphasize that even in the resolvent approach to the Volterra equations (1) and (2) the role of semigroup generated by the operator A is crucial. It is used in proofs of the convergence (7) of the sequence of resolvents and in proofs of the existence of strong solutions to the equation (1). For more details we can refer to the papers.[19–21]

Let us finish the note with some general remarks concerning stochastic Volterra equations of convolution type. Such equations belong to an interdisciplinary area, where stochastic processes meet integral equations. In the last two decades stochastic integral equations have been one of the most dynamic branches of stochastic processes. Because of the resolvent approach to the linear Volterra equations, several new perspectives has been opened. Having results mentioned in this note and the references cited, some next interesting results seem to be possible to obtain. Particularly, several results can be extended for the equation (1) with other noises. Moreover, control of linear systems with memory is still incomplete area. Another direction of research is to consider equations of convolution type in more general spaces than Hilbert space.

Acknowledgements

The author greatly acknowledges financial support by Laboratorio de Analisis Estocástico, Proyecto Anillo ACT-13.

References

1. S. Bonaccorsi, L. Tubaro, *Stochastic Anal. Appl.* **21**, (2003) 61–78.
2. Ph. Clément, G. Da Prato, *Rend. Math. Acc. Lincei* **7**, (1996) 147–153.
3. Ph. Clément, G. Da Prato, *Dynamic Systems and Applications*, **6**, (1997) 441–460.
4. Ph. Clément, G. Da Prato, In *Stability Control Theory Methods Appl.* **10**, Gordon and Breach, Amsterdam (2000).
5. Ph. Clément, G. Da Prato, J. Prüss, *Rend. Inst. Mat. Univ. Trieste*, **29**, (1997) 207–220.
6. Ph. Clément, J.A. Nohel, *SIAM J. Math. Anal.* **10**, (1979) 365–388.
7. R.F. Curtain, P.L. Falb, *J. Differential Equations* **10**, (1971) 412–4130.
8. R.F. Curtain, A.J. Pritchard, *Infinite dimensional linear systems theory*, Springer-Verlag, Berlin (1978).
9. G. Da Prato, S. Kwapień, J. Zabczyk, *Stochastics*, **3**, (1987) 1–23.
10. G. Da Prato, J. Zabczyk, *Stochastic equations in infinite dimensions*, Cambridge University Press, Cambridge (1992).
11. D.A. Dawson, *J. Multivariate Anal.* **5**, (1975) 1–52.
12. W. Desch, J. Prüss, *J. Integral Equations Appl.* **5**, (1993) 29–45.
13. A. Friedman, M. Shinbrot, *Trans. Amer. Math. Soc.* **126**, (1967) 131–179.
14. Y. Fujita *J. Math. Phys.* **30**, (1989) 134–144.
15. W. Grecksch,C. Tudor, *Stochastic evolution equations*, Akademie Verlag, Berlin (1995).
16. G. Gripenberg, S.-O. Londen, O. Staffans, *Volterra integral and functional equations*, Cambridge University Press, Cambridge (1990).
17. T.V. Hromadka II, R.J. Whitley, *Stochastic Integral Equations in Rainfall-Runoff Modeling*, Springer, New York (1989)
18. A. Karczewska, *Int. J. Contemp. Math. Sci.* **21**, (2007) 1037–1052.
19. A. Karczewska, *Convolution type stochastic Volterra equations*, Lecture Notes in Nonlinear Analysis, Juliusz Schauder Center for Nonlinear Studies **10**, (2007) 1-101. http://www.uz.zgora.pl/∼akarczew/PRYW
20. A. Karczewska, C. Lizama, *J. Evol. Equ.* **7**, (2007) 373–386.
21. A. Karczewska, C. Lizama, *J. Math. Anal. Appl.* **349**, (2009) 301–310.
22. M. Li, Q. Zheng, *Semigroup Forum* **69**, (2004) 356–368.
23. R.L. Magin, *Fractional calculus in bioengineering*, Begel House Publishers, Connecticut (2006).
24. J. Prüss, *Evolutionary integral equations and applications*, Birkhäuser, Basel (1993).

ENTANGLEMENT PROTECTION AND GENERATION IN A TWO-ATOM SYSTEM

M. ORSZAG
Facultad de Física, Pontificia Universidad Católica de Chile,
Santiago, CHILE
Centro de Análisis Estocástico y Aplicaciones
Casill 306, Santiago 22, Chile
E-mail: morszag@fis.puc.cl

Entanglement is a fundamental concept in Quantum Mechanics. In this paper, we study various aspects of coherence and entanglement, illustrated by several examples. We relate the concepts of loss of coherence and disentanglement, via a model of two two-level atoms in different types of reservoir, including both cases of independent and common bath. We also relate decoherence and disentanglement, by focussing on the sudden death of the entanglement and the dependence of the death time with the "distance" of our initial condition, from the decoherence free subspace. In particular, we study the "sudden death of the entanglement", in a two-atom system with a common reservoir. Finally we study the creation of entanglement under various initial conditions.

Keywords: entanglement; sudden death; decoherence; reservoir.

1. Entanglement evolution

Decoherence and entanglement are closely related phenomena, mainly because decoherence is responsible for the fragility of the entanglement in systems interacting with reservoirs.[1] For this reason in the current decade, many papers have investigated extensively, the decoherence dynamics of entangled quantum systems under the influence of environmental noise by focusing mainly on the dynamical system of two parties.

Recently Yu and Eberly[2] investigated the dynamics of disentanglement of a bipartite qubit system due to spontaneous emission, where the two two level atoms (qubits) were coupled individually to two cavities (environments). They found that the quantum entanglement may vanish in a finite time, while local decoherence takes a infinite time. They called this phenomena "Entanglement Sudden Death" (ESD). In a previous work, Diósi[3] demonstrated, using Werner's criteria for separability that ESD can also occur in two-state quantum systems.

Since then, ESD has been examined in several model situations involving pairs of atomic, photonic and spin qubits,[4–6] continuous Gaussian states[7] and spin chains.[8] Also, ESD has been examined for different environments including random matrix environments,[9,10] thermal reservoir[11–13] and squeezed reservoir.[14] ESD is not unique to systems of independent atoms. It can also occur for atoms coupled to a common reservoir, in which case we also observe the effect of the revival of the entanglement that has already been destroyed.[15] The effect of global noise on entanglement decay may depend on whether the initial two-party state belongs to a decoherence free subspace (DFS) or not.

In a recent experiment by Almeida *et al*,[16] they used correlated horizontally and vertically polarized photons to show evidence of sudden death of entanglement, under the influence of independent reservoirs. They created the initial state using a down-conversion process.

As opposed to the ESD and against our intuition, it has been shown that under certain conditions, the process of spontaneous emission can entangle qubits that were initially unentangled,[17] and in some cases the creation of entanglement can occur some time after the system-reservoir interaction has been turned on. The authors in[18] call this phenomenon "delayed sudden birth of entanglement".

1.1. *Time evolution of a two-atom system*

In this section we study two-atom systems. However, in real systems, these are not isolated and experience interactions with the outside world. In order to build useful quantum information systems, it is necessary to understand and control such noise processes.

An useful mathematical description for the qubits (system of interest) is given by the density matrix. When the system is exposed to environmental noise, the density matrix will change in time. Such a time evolution is traditionally studied via a master equation. In this approach, the dynamics is studied in terms of the reduced density operator $\hat{\rho}_s$ of the atomic system interacting with the quantized electromagnetic field regarded as a reservoir. The reservoirs have many possible realizations. It can be modelled as a many mode vacuum , thermal or squeezed vacuum field. The major advantage of the master equation is that it allows us to consider the evolution of the atoms plus field system entirely in terms of average values of atomic operators.

We consider two situations. The first one consists of two two-level atoms initially entangled and coupled to uncorrelated reservoirs, and the second one of two two-level atoms initially entangled and interacting with a common reservoir. This coupling between the system and the reservoir originates the disentanglement.

We write now, a general master equation for the reduced density matrix in the interaction picture, assuming that the correlation time between the atoms and the reservoirs is much shorter than the characteristic time of the dynamical evolution of the atoms, so that the Markov approximation is valid, [a]

$$\begin{aligned}\frac{\partial\hat{\rho}}{\partial t} = \frac{\Gamma}{2}\sum_{i,j=1}^{2} &[(N+1)(2\sigma_i\hat{\rho}\sigma_j^{\dagger} - \sigma_i^{\dagger}\sigma_j\hat{\rho} - \hat{\rho}\sigma_i^{\dagger}\sigma_j) \\ &+ N(2\sigma_i^{\dagger}\hat{\rho}\sigma_j - \sigma_i\sigma_j^{\dagger}\hat{\rho} - \hat{\rho}\sigma_i\sigma_j^{\dagger}) \\ &- M(2\sigma_i^{\dagger}\hat{\rho}\sigma_j^{\dagger} - \sigma_i^{\dagger}\sigma_j^{\dagger}\hat{\rho} - \hat{\rho}\sigma_i^{\dagger}\sigma_j^{\dagger}) \\ &- M^*(2\sigma_i\hat{\rho}\sigma_j - \sigma_i\sigma_j\hat{\rho} - \hat{\rho}\sigma_i\sigma_j)], \end{aligned} \quad (1)$$

where Γ is the decay constant of the qubits, and $\sigma_i^{\dagger} = |1\rangle_i\langle 0|$ and $\sigma_i = |0\rangle_i\langle 1|$ are the raising and lowering operators of the ith atom, respectively. It should be pointed out that in Eq.(1), the $i = j$ terms describe the atoms interacting with independent local reservoirs, while the $i \neq j$ terms denote the couplings between the modes induced by the common bath, see Fig.1(a) and (b).Thus, for independent reservoirs, we take only the $i = j$ terms . On the other hand, in the common reservoir case, we take all the terms ($i = j$ and $i \neq j$)

The Eq.(1) has at least three possible realizations. These are vacuum, thermal, and squeezed reservoir.

For a **vacuum reservoir** we set $N \to 0$ and $M \to 0$ in Eq.(1).

For a **thermal reservoir** $N \to \bar{n}$ where $\bar{n}$ is the mean number of the thermal field (assumed to be the same for both qubits), and $M \to 0$. Additionally ,this reservoir can cause excitation of the qubits. Thus the first term on the right side of Eq.(1) corresponds to the depopulation of the atoms due to stimulated and spontaneous emission, and the second term

[a] For non-Markovian effects, see[19,20]

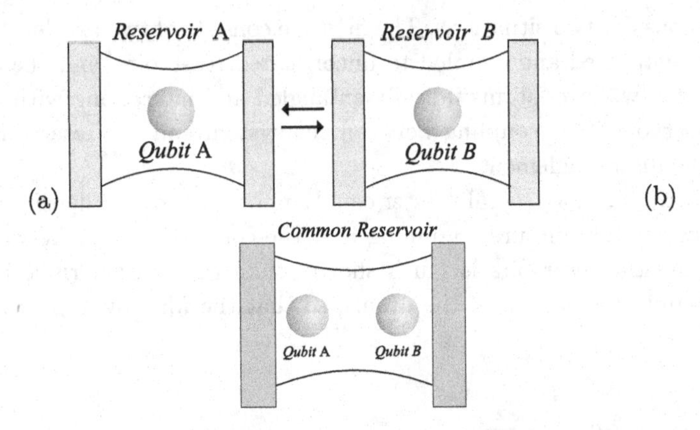

Fig. 1. Schematic of an assembly of two qubits A and B, located in (a) two independent and spatially separated, reservoirs. (b)a common reservoir.

describes the excitations caused by temperature.

For a **squeezed reservoir**, N represents the mean photon number of the reservoir and it is defined as $N = \sinh^2 r$, and M is a parameter related to the phase correlations of the squeezed reservoir defined as $M = -e^{i\theta} \sinh r \cosh r$. r is the squeezing parameter of reservoir and θ is the squeezing angle . The Heisenberg uncertainty relation imposes the constraint $|M|^2 \leq N(N+1)$, where the equality holds for a minimum-uncertainty squeezed state.

2. Entanglement with independent reservoirs

In the next sections we consider two two-level atoms A and B that represent a two-qubit system, each one interacting independently with their local environments, as is showed in Fig.1(a). There is no direct interaction between the atoms. The correlation between the atoms results only from an initial quantum entanglement between them.

2.1. *Vacuum Reservoir*

The vacuum environment can still have a noisy degrading effect through its quantum fluctuations. As a consequence, the atoms lose their excitation at a rate Γ. So, their stationary state is their ground state $|--\rangle$. In other words, the atoms experience a disentanglement process. This process can be completed in a finite-time (ESD) or in an infinite time. These different behaviors exhibited in the disentanglement times depend on the initial state. Various publications have considered mixed and pure states,[2,11,21,22] as well as quantum recoil effects.[23]

Given the solution of the master equation (1) for a vacuum reservoir, and considering as initial condition the matrix (X matrix or X state), in the standard basis,

$$\rho(0) = \begin{pmatrix} \rho_{11} & 0 & 0 & \rho_{14} \\ 0 & \rho_{22} & \rho_{23} & 0 \\ 0 & \rho_{32} & \rho_{33} & 0 \\ \rho_{41} & 0 & 0 & \rho_{44} \end{pmatrix}, \tag{2}$$

where $\rho_{11} + \rho_{22} + \rho_{33} + \rho_{44} = 1$,we are interested in calculating the entanglement. For a general mixed state ρ_{AB} of two qubits, we define $\widetilde{\rho}$ to be the spin-flipped state

$$\widetilde{\rho}_{AB} = (\sigma_y \otimes \sigma_y)\rho^*_{AB}(\sigma_y \otimes \sigma_y), \tag{3}$$

where ρ^* is the complex conjugate of ρ,and σ_y is the Pauli matrix.

The concurrence is defined as

$$C(\rho) = max\{0, \lambda_1 - \lambda_2 - \lambda_3 - \lambda_4\}, \tag{4}$$

where $\{\lambda_i\}$ are the square roots, in decreasing order of the eigenvalues of the non-hermitian matrix $\rho\widetilde{\rho}$. For separable qubits $C = 0$ and for maximally entangled ones $C = 1$. For the initial condition given above, the concurrence has a particularly simple form, given by[22] $C(t) = 2\max\{0, C_1(t), C_2(t)\}$, where

$$\begin{aligned} C_1(t) &= 2e^{-\Gamma t}[|\rho_{23}| - \sqrt{\rho_{11}(\rho_{44} + (\rho_{22} + \rho_{33})\omega^2 + \rho_{11}\omega^4)}], \\ C_2(t) &= 2e^{-\Gamma t}[|\rho_{14}| - \sqrt{\rho_{22}\rho_{33} + \rho_{11}(\rho_{22} + \rho_{33})\omega^2 + \rho_{11}^2\omega^4}], \end{aligned} \tag{5}$$

with $\omega = \sqrt{1 - e^{-\Gamma t}}$.

When $C_1(t)$ is maximal, then if $\rho_{11} > |\rho_{23}|^2$, the concurrence decays to zero in a finite time

$$t_{d-1} = \frac{1}{\Gamma}\ln\left(\frac{\rho_{11}(2\rho_{11} + \rho_{22} + \rho_{33} + \sqrt{(\rho_{22} + \rho_{33})^2 - 4(\rho_{11}\rho_{44} - |\rho_{23}|^2)})}{2(\rho_{11} - |\rho_{23}|^2)}\right). \tag{6}$$

However, if $\rho_{11} \leq |\rho_{23}|^2$, then the equation $C_1(t) = 0$ has no solution and the concurrence decays asymptotically to zero.

Similarly, when $C_2(t)$ is maximal, then if $\rho_{22}\rho_{33}+\rho_{11}(1-\rho_{44}) > |\rho_{14}|^2$, the concurrence vanishes in a finite time,

$$t_{d-2} = \frac{1}{\Gamma}\ln(\frac{\rho_{11}(2\rho_{11}+\rho_{22}+\rho_{33}+\sqrt{(\rho_{22}-\rho_{33})^2+4|\rho_{23}|^2}))}{2(\rho_{22}\rho_{33}+\rho_{11}(\rho_{11}+\rho_{22}+\rho_{33})-|\rho_{14}|^2)}), \quad (7)$$

and if $\rho_{22}\rho_{33}+\rho_{11}(1-\rho_{44}) \leq |\rho_{14}|^2$, then we have again the exponential behavior.

To summarize, the ranges of the respective initial conditions are

$$\begin{aligned} &\rho_{11} \leq |\rho_{23}|^2, \text{ The entanglement decays asimptotically;} \\ &\rho_{11} > |\rho_{23}|^2, \text{ The entanglement decays in a finite time,} \end{aligned} \quad (8)$$

for the case Eq.(6), and

$$\begin{aligned} &\rho_{22}\rho_{33}+\rho_{11}(1-\rho_{44}) \leq |\rho_{14}|^2, \text{ The entanglement decays asimptotically;} \\ &\rho_{22}\rho_{33}+\rho_{11}(1-\rho_{44}) > |\rho_{14}|^2, \text{ The entanglement decays in a finite time,} \end{aligned} \quad (9)$$

for Eq.(7).

Let us show some examples.

a) Consider the initial state of the form $\alpha|00\rangle+\beta|11\rangle$, with $|\alpha|^2+|\beta|^2=1$. For this initial state, the concurrence is,

$$C = \max\{0, 2|\beta|e^{-\Gamma t}(|\alpha|-|\beta|(e^{-\Gamma t}-1))\}, \quad (10)$$

and the disentanglement time occurs for $t_d = -\frac{1}{\Gamma}(1-\frac{\alpha}{\beta})$. Only the states with $|\beta| > |\alpha|$ i.e, $|\beta| > \frac{1}{\sqrt{2}}$, have a finite disentanglement time. Consider now the initial state $\gamma|01\rangle+\delta|10\rangle$, with $|\gamma|^2+|\delta|^2=1$. Its concurrence is $C=\max\{0, 2e^{-\Gamma t}\delta\gamma\}$. In this case the concurrence goes asymptotically to zero, for all values of δ.

b) The first and more traditional case of ESD mentioned in the literature, contains the double excitation and the ground state components, and also includes the state in which one of the atoms is excited. This initial condition is

$$\rho(0) = \frac{1}{3}\begin{pmatrix} a & 0 & 0 & 0 \\ 0 & 1 & 1 & 0 \\ 0 & 1 & 1 & 0 \\ 0 & 0 & 0 & 1-a \end{pmatrix}. \quad (11)$$

Yu and Eberly studied the entanglement sudden death in the case $a=1$. They showed that for $\frac{1}{3} \leq a \leq 1$, there is ESD. The concurrence is given by

$$C = \max\{0, \frac{2}{3}e^{-\Gamma t}(1 - \sqrt{a(3 - 2(1+a)e^{-\Gamma t} + ae^{-2\Gamma t})})\}, \quad (12)$$

and the time of disentanglement is ,

$$t_d = -\frac{1}{\Gamma}\ln(\frac{a+1-\sqrt{2-a+a^2}}{a}). \quad (13)$$

For the case $a = 1$ the results obtained in[2] are recovered. The time evolution of the concurrence for all values of a is shown in Fig.2(a). Also in Fig.2(b) we show the typical concurrence versus time, for initial condition leading to sudden death and asymptotic behavior.

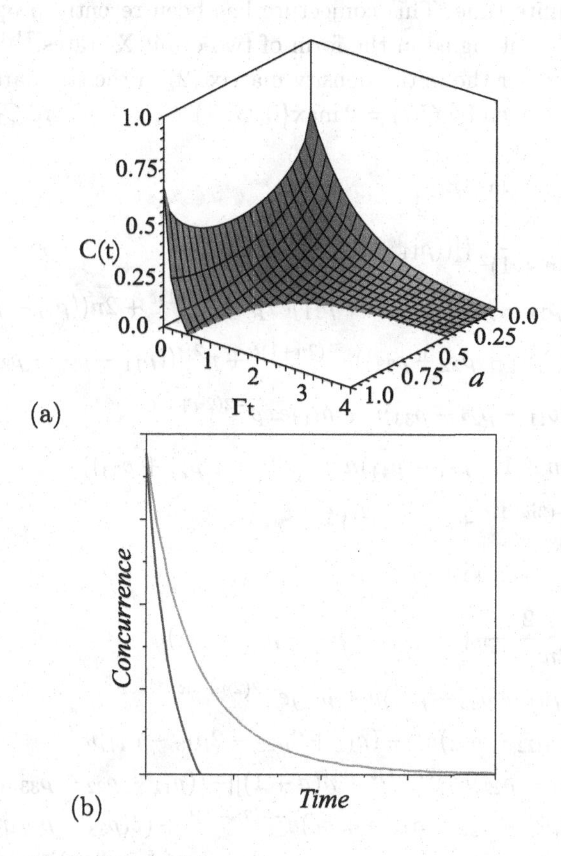

Fig. 2. (a) The time evolution of the concurrence in a vacuum reservoir when the atoms are initially in the entangled mixed state (Eq.11). (b) The figure shows the two typical behavior of sudden death and asymptotic decay in time, for qubits interacting with two independent vacuum reservoirs.

Clearly, to destroy the entanglement in a finite time, the spontaneous emission is not enough, and the sudden death of entanglement results from the decay of a mixed state with a doubly excited state component.

2.2. *Thermal Reservoir*

The interactions with a thermal reservoir lead typically to very rapid decoherence, thus one might expect the destruction of quantum entanglement. But, how quickly does it occur?. In a vacuum reservoir, the disentanglement time depends explicitly on the initial state of the atoms, specifically the ESD depends on the double excitation state $|++\rangle$. On the other hand, in the case of a thermal reservoir, the decay of the entanglement *always* occurs in a finite time. This conjecture has been recently proposed for all atoms initially entangled in the form of two-qubit X-states,[11,13]

Let us consider the initial density matrix (2) in the standard basis. The concurrence is given by $C(t) = 2\max\{0, C_1(t), C_2(t)\}$, where C_1 and C_2 are given by

$$
\begin{aligned}
C_1(t) &= 2|\rho_{23}|e^{-(2n+1)t} \\
&\quad - \frac{2}{(2n+1)^2}\{[((\rho_{11} - \rho_{22} - \rho_{33} + \rho_{44})n^2 \\
&\quad + (2\rho_{11} - \rho_{22} - \rho_{33})n + \rho_{11})exp^{-2(2n+1)t} + 2n((\rho_{11} - \rho_{44})n \\
&\quad + \frac{1}{2}(2\rho_{11} + \rho_{22} + \rho_{33}))e^{-(2n+1)t} + n^2][((\rho_{11} - \rho_{22} - \rho_{33} + \rho_{44})n^2 \\
&\quad + (2\rho_{11} - \rho_{22} - \rho_{33})n + \rho_{11})exp^{-2(2n+1)t} \\
&\quad - 2(n+1)((\rho_{11} - \rho_{44})n + \frac{1}{2}(2\rho_{11} + \rho_{22} + \rho_{33})) \\
&\quad \times e^{-(2n+1)t} + (n+1)^2]\}^{\frac{1}{2}}
\end{aligned} \tag{14}
$$

$$
\begin{aligned}
C_2(t) &= 2|\rho_{14}|e^{-(2n+1)t} \\
&\quad - \frac{2}{(2n+1)^2}\{[-((\rho_{11} - \rho_{22} - \rho_{33} + \rho_{44})n^2 \\
&\quad + (2\rho_{11} - \rho_{22} - \rho_{33})n + \rho_{11})e^{-2(2n+1)t} \\
&\quad + (2(\rho_{22} - \rho_{33})n^2 + (\rho_{11} + 2\rho_{22} - 2\rho_{33} - \rho_{44})n \\
&\quad + \rho_{11} + \rho_{22})e^{-(2n+1)t} + n(n+1)][-((\rho_{11} - \rho_{22} - \rho_{33} + \rho_{44})n^2 \\
&\quad + (2\rho_{11} - \rho_{22} - \rho_{33})n + a_0)e^{-2(2n+1)t} + (2(\rho_{33} - \rho_{22})n^2 \\
&\quad + (\rho_{11} - 2\rho_{22} + 2\rho_{33} - \rho_{44})n + \rho_{11} + \rho_{33})e^{-(2n+1)t} + n(n+1)]\}^{\frac{1}{2}},
\end{aligned} \tag{15}
$$

n being the average number of thermal photons.

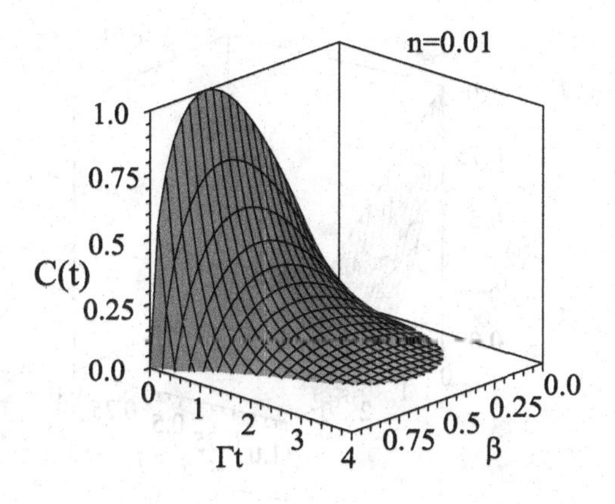

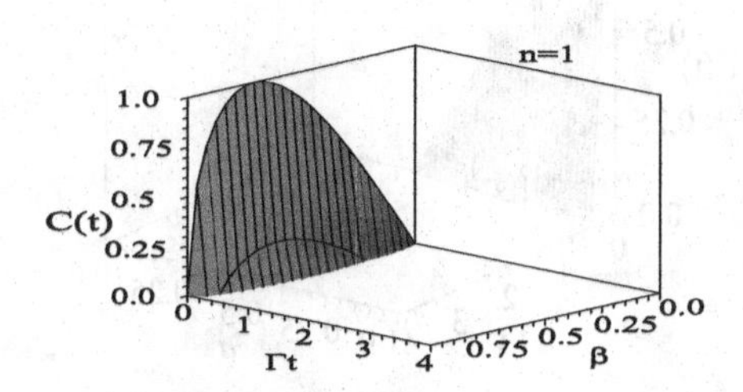

Fig. 3. The figures show the entanglement evolution for the initial state $\alpha|00\rangle \pm \beta|11\rangle$ with $n = 0.01$ (upper curve) and $n = 1$ (lower curve). In both cases, ESD occurs for all ranges of $0 < \beta < 1$.

In order to get the disentanglement time, we solved analytically the equations $C_1(t) = 0$, when $w(0) = 0$ and $C_2(t) = 0$ when $z(0) = 0$. In Fig.3, we plotted the time evolution of the concurrence for $n = 0.01$ and $n = 1$ when the atoms are initially in the state $\alpha|00\rangle \pm \beta|11\rangle$. The $\pm$ sign does not make any difference in the entanglement evolution.

Also we considered the atoms initially entangled in a Werner state, Fig.4. In this case, the initial condition is:

$$\rho(0) = \frac{1}{4}\begin{pmatrix} 1-a & 0 & 0 & 0 \\ 0 & 1+a & -2a & 0 \\ 0 & -2a & 1+a & 0 \\ 0 & 0 & 0 & 1-a \end{pmatrix} \tag{16}$$

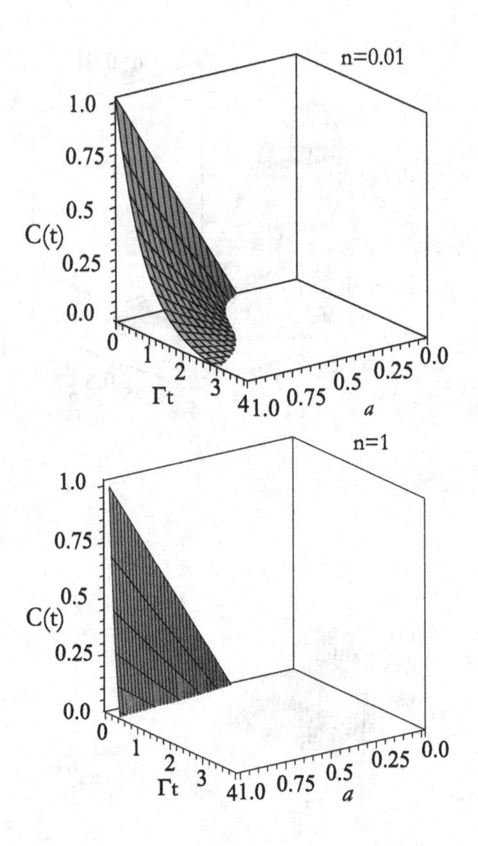

Fig. 4. The concurrence for Werner states (16) with $n = 0.01$ (upper figure) $n = 1$ (lower figure).

In the Fig.5 we show the disentanglement time for different values of n. The main observed effect is that the death time of the entanglement decreases with the mean thermal photon number.

As we can see, at finite (non-zero) temperatures, for various initial mixed states, the phenomena of entanglement sudden death is always present.

3. Entanglement in a common reservoir

In the next sections, we will explore the relation between the sudden death (and revival) of the entanglement between the two two-level atoms in a squeezed bath and the normal decoherence via the decoherence free subspace (DFS), which in this case is a two-dimensional plane,[14] We will also look at the special cases of the vacuum and thermal reservoirs. We should

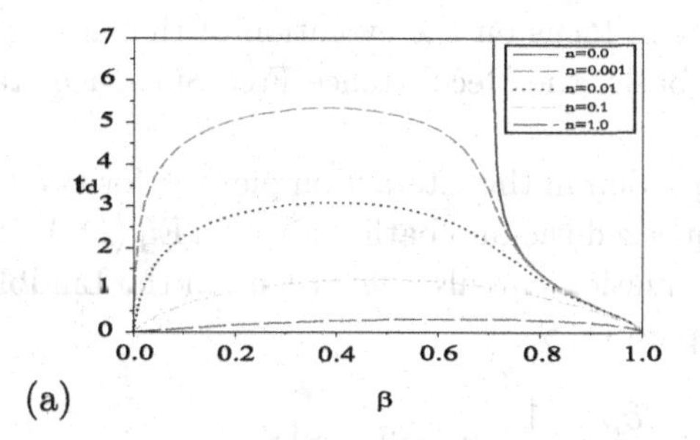

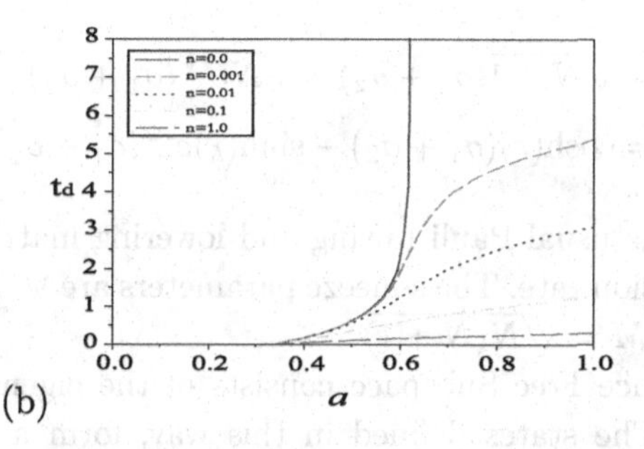

Fig. 5. Death time for the (a) $\alpha|00\rangle \pm \beta|11\rangle$ and (b) a Werner state. We plotted the disentanglement time for different values of the parameter n. In the case of $n = 0$, there exists a range in the initial conditions for which the ESD is not permitted, decaying asymptotically. For larger values of n, the disentanglement time decreases, i.e. the sudden death occurs faster.

point out, that since we have a common bath, it would imply that the atoms are rather close to each other, that is, at a distance which is smaller than the correlation length of the reservoir. This would mean that we cannot neglect some direct interaction between them, like a dipole or Ising-type coupling.

However, one can show that this type of coupling does not damage the decoherence free subspace, in this case a plane.[24]

The case we present next, is a good example, where we are able to study in detail, the effects of decoherence and disentanglement. Furthermore, we can relate the two phenomena, by studying the effect of the "distance" from the decoherence free subspace on the sudden death and revival times.

3.1. *The model*

We consider, two two-level atoms that interact with a *common* squeezed reservoir, and we will focus on the evolution of the entanglement between them, using as a basis, the Decoherence Free Subspace states, as defined in.[24,25]

The master equation in the interaction picture, for two two-level system in a broadband squeezed vacuum bath is given in Eq.(1). It is simple to show that this master equation can also be written in the Lindblad form with a single Lindblad operator S

$$\frac{\partial \rho}{\partial t} = \frac{1}{2}\Gamma(2S\rho S^\dagger - S^\dagger S\rho - \rho S^\dagger S), \tag{17}$$

with

$$\begin{aligned} S &= \sqrt{N+1}(\sigma_1 + \sigma_2) - \sqrt{N}e^{i\Psi}(\sigma_1^\dagger + \sigma_2^\dagger) \\ &= \cosh(r)(\sigma_1 + \sigma_2) - \sinh(r)e^{i\Psi}(\sigma_1^\dagger + \sigma_2^\dagger), \end{aligned} \tag{18}$$

where $\sigma^\dagger, \sigma$ are the usual Pauli raising and lowering matrices and Γ is the spontaneous emission rate. The squeeze parameters are Ψ, and $N = \sinh^2 r$. Here we consider $M = \sqrt{N(N+1)}$.

The Decoherence Free Subspace consists of the eigenstates of S with zero eigenvalue. The states defined in this way, form a two-dimensional plane in Hilbert Space. Two orthogonal vectors in this plane are:

$$|\phi_1\rangle = \frac{1}{\sqrt{N^2+M^2}}(N|++\rangle + Me^{-i\Psi}|--\rangle), \tag{19}$$

$$|\phi_2\rangle = \frac{1}{\sqrt{2}}(|-+\rangle - |+-\rangle). \tag{20}$$

We can also define the states $|\phi_3\rangle$ and $|\phi_4\rangle$ orthogonal to the $\{|\phi_1\rangle$, $|\phi_2\rangle\}$ plane:

$$|\phi_3\rangle = \frac{1}{\sqrt{2}}(|-+\rangle + |+-\rangle), \tag{21}$$

$$|\phi_4\rangle = \frac{1}{\sqrt{N^2+M^2}}(M|++\rangle - Ne^{-i\Psi}|--\rangle). \tag{22}$$

We solve analytically the master equation by using the $\{|\phi_1\rangle, |\phi_2\rangle, |\phi_3\rangle,$ $|\phi_4\rangle\}$ basis, however, we use the standard basis to calculate the concurrence. For simplicity we will consider $\Gamma = 1$ throughout this section.

3.2. *Solution for initial states in DFS.*

a) Consider $|\Phi_1(0)\rangle = |\phi_1\rangle$ as the initial state.
The solution of master equation is $\rho_1(t) = |\phi_1\rangle\langle\phi_1|$. This corresponds to an invariant state, and its concurrence is

$$C(\rho_1(t)) = \frac{2\sqrt{N(N+1)}}{2N+1},$$

which is a constant in time. The concurrence only depends of N.
For $N = 0$, $|\Phi_1(0)\rangle = |--\rangle$ we have a factorized state at all times, but as we increase N, we get a maximally entangled state in the large N limit.

b) If we now consider $|\Phi_2(0)\rangle = |\phi_2\rangle$ as the initial state.
The solution of the master equation is $\rho_2(t) = |\phi_2\rangle\langle\phi_2|$. This initial state is also an invariant state and its concurrence is independent of time,

$$C(\rho_2(t)) = 1.$$

3.3. *General Solution for vacuum reservoir. $N \longrightarrow 0$*

a) The third initial state considered is $|\Phi_3(0)\rangle = |\phi_3\rangle$.
Initially its concurrence is: $C(\rho_3(0)) = 1$. It corresponds to a maximally entangled state.
The solution of master equation for this initial condition and $N = 0$ is given by

$$\rho_3(t) = \begin{pmatrix} (e^{2t}-1)e^{-2t} & 0 & 0 & 0 \\ 0 & 0 & 0 & 0 \\ 0 & 0 & e^{-2t} & 0 \\ 0 & 0 & 0 & 0 \end{pmatrix}. \tag{25}$$

Since the matrix $\rho(t)\widetilde{\rho}(t)$ has only one nonzero eigenvalue, in this case we use the separability criterion.[26] According to this criterion, the necessary condition for separability is that the matrix ρ^{PT}, obtained by partial transposition of ρ, should have only non-negative eigenvalues. In this particular case, we observe a negative eigenvalue for all times, so the state stays entangled, (Fig.6).

b) Consider the initial state $|\Phi_4(0)\rangle = |\phi_4\rangle$.
When $N = 0$, $|\Phi_4(0)\rangle = |++\rangle$ and $C(\rho_4(0)) = 0$, since it is a factorized state.

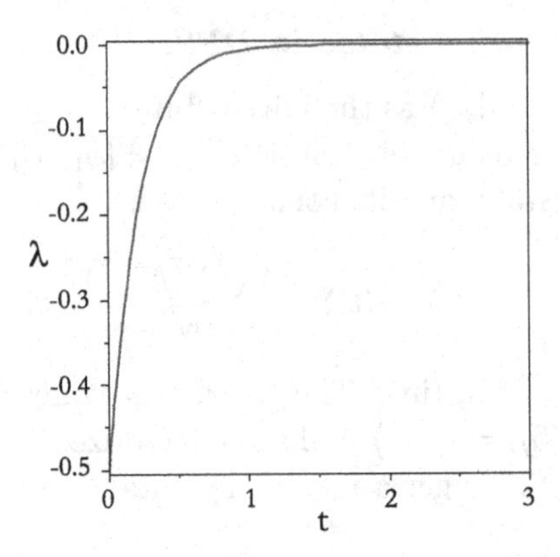

Fig. 6. Negative eigenvalue (λ) of ρ^{PT}, for the separability criterion, for $|\phi_3\rangle$ as the initial state. This eigenvalue is always negative , indicating entanglement at all times.

The solution of master equation for this initial condition is given by:

$$\rho_4(t) = \begin{pmatrix} (-1-2t+e^{2t})e^{-2t} & 0 & 0 & 0 \\ 0 & 0 & 0 & 0 \\ 0 & 0 & 2te^{-2t} & 0 \\ 0 & 0 & 0 & e^{-2t} \end{pmatrix}, \tag{26}$$

and its concurrence is $C(\rho_4(t)) = 0$.

In the following, we will consider superpositions with one component in the DFS and an orthogonal one (to the DFS), of the form:

$$|\Psi_a\rangle = \varepsilon|\phi_1\rangle + \sqrt{1-\varepsilon^2}|\phi_4\rangle, \tag{27}$$

$$|\Psi_b\rangle = \varepsilon|\phi_2\rangle + \sqrt{1-\varepsilon^2}|\phi_3\rangle, \tag{28}$$

where ε varies from zero to one. The idea is to increase ε and study the effect of having an increased component in the DFS on the death time of the entanglement.

c) We consider an initial superposition of $|\phi_1\rangle$ and $|\phi_4\rangle$, i.e. $|\Psi_a\rangle$ given by Eq.(27).
   For $N = 0$ we have:

$$|\Psi_a(0)\rangle = \varepsilon|--\rangle + \sqrt{1-\varepsilon^2}|++\rangle, \tag{29}$$

   and its initial concurrence is $C(\Psi_a(0)) = 2\varepsilon\sqrt{1-\varepsilon^2}$.

The solution of master equation for this initial condition is given by:

$$\rho_a(t) = \begin{pmatrix} \frac{(-2t-1+2t\varepsilon^2+\varepsilon^2+e^{2t})}{e^{2t}} & 0 & 0 & \frac{\varepsilon\sqrt{1-\varepsilon^2}}{e^t} \\ 0 & 0 & 0 & 0 \\ 0 & 0 & \frac{2t(1-\varepsilon^2)}{e^{2t}} & 0 \\ \frac{\varepsilon\sqrt{1-\varepsilon^2}}{e^t} & 0 & 0 & \frac{(1-\varepsilon^2)}{e^{2t}} \end{pmatrix}, \tag{30}$$

and the corresponding concurrence is given by:

$$C(\rho_a) = \max\{0, 2((\varepsilon\sqrt{1-\varepsilon^2})e^{-t} - te^{-2t}(1-\varepsilon^2))\} \tag{31}$$

which is shown in Fig.7 for various values of ε: For $\varepsilon > 0$, the initial

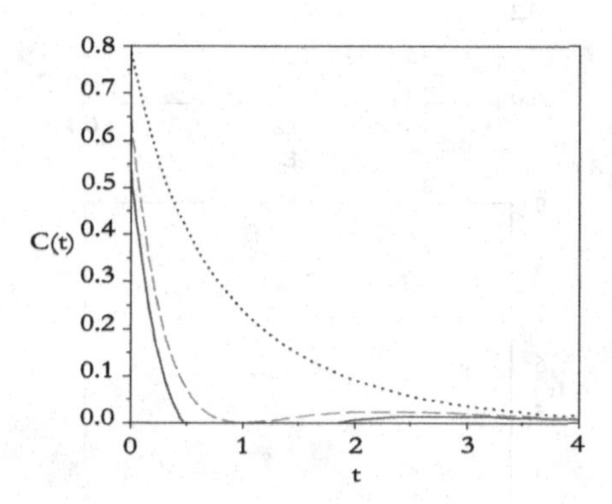

Fig. 7. Time evolution of the concurrence for initial $|\Psi_1(0)\rangle$ with : $\varepsilon = 0.28$ (solid line), $\varepsilon = 0.345$ (dashed line), $\varepsilon = 0.9$ (dotted line),

entanglement decreases in time, and the system becomes disentangled (sudden death) at a time satisfying the relation:

$$te^{-t} = \frac{\varepsilon}{\sqrt{1-\varepsilon^2}}. \tag{32}$$

- For $\varepsilon = 0$ and $\varepsilon = 1$, the concurrence is zero, therefore we have a non-entangled state.
- For $0 < \varepsilon < 0.34525$ the equation (32) has two solutions, namely, t_d when the system becomes separable, and $t_r \geq t_d$ when the entanglement revives. It should be noted that there is a critical ε for which $t_d = t_r$.

- For $0.34525 < \varepsilon < 1$ the above equation has no solution and the concurrence vanishes asymptotically in time.

Thus, when we are "not far" from $|\phi_4\rangle$ we observe a sudden death and revival, but when we get "near" $|\phi_1\rangle$ this phenomenon disappears. Fig.8 shows the behavior of the death and revival time as function of ε.

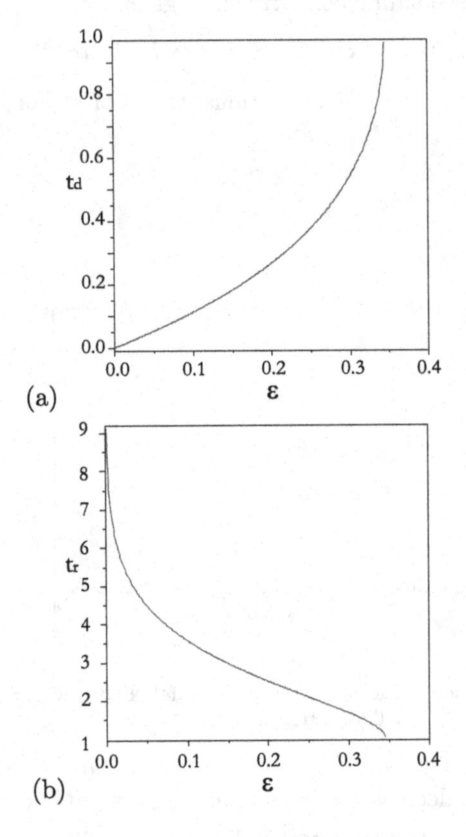

Fig. 8. **a)** The death time **b)** The revival time of the entanglement as a function of ε, with initial $|\Psi_a\rangle$.

d) Finally, we consider an initial superposition of $|\phi_2\rangle$ and $|\phi_3\rangle$: $|\Psi_b(0)\rangle = \varepsilon\ |\phi_2\rangle + \sqrt{1-\varepsilon^2}\ |\phi_3\rangle$, (28) which is independent of N. Here, like in the previous cases, as we increase ε, starting from $\varepsilon = 0$, we increase the initial projection onto the DFS. For $\varepsilon = 1$ the initial state is in the DFS plane.

For $N = 0$ we have

$$|\Psi_b(0)\rangle = \frac{1}{\sqrt{2}}[(\varepsilon + \sqrt{1-\varepsilon^2})|-+\rangle - (\varepsilon - \sqrt{1-\varepsilon^2})|+-\rangle], \tag{33}$$

and its initial concurrence is $C(\Psi_b(0)) = |2\varepsilon^2 - 1|$.
The solution of master equation for this initial condition is given by:

$$\rho_b(t) = \begin{pmatrix} \frac{(e^{2t} - \varepsilon^2 e^{2t} - 1 + \varepsilon^2)}{e^{2t}} & 0 & 0 & 0 \\ 0 & \varepsilon^2 & \frac{\varepsilon\sqrt{1-\varepsilon^2}}{e^t} & 0 \\ 0 & \frac{\varepsilon\sqrt{1-\varepsilon^2}}{e^t} & \frac{(1-\varepsilon^2)}{e^{2t}} & 0 \\ 0 & 0 & 0 & 0 \end{pmatrix}, \tag{34}$$

and the corresponding concurrence is:

$$C(\rho_b(t)) = \max\{0, e^{-2t}|\varepsilon^2 e^{2t} - 1 + \varepsilon^2|\}, \tag{35}$$

which is shown in Fig.9(a).
For $0 < \varepsilon < 0.707$ the initial entanglement decreases in time, and the system becomes disentangled at a time given by (See Fig.9(b)):

$$t = \frac{1}{2}\ln(\frac{1-\varepsilon^2}{\varepsilon^2}), \tag{36}$$

however, at the same time, the entanglement revives reaching asymptotically its stationary value. It means that the sudden death and revival happen simultaneously. The phenomena of one or periodical revivals have been obtained before, but always in the context of one single reservoir connecting both atoms, like in the present case.[15,27,28]
When we approach the decoherence free subspace this phenomenon disappears.

Next, we treat the cases with $N > 0$.

3.4. *General Solution for $N \neq 0$*

In general, for both $|\Psi_a\rangle$ and $|\Psi_b\rangle$ as initial states, the evolution of the concurrence $C(\rho(t))$ is calculated in the standard basis but written in terms of density matrix $\rho'(t)$ in the $\{\phi_i\}$ basis as $C(\rho'(t)) =$

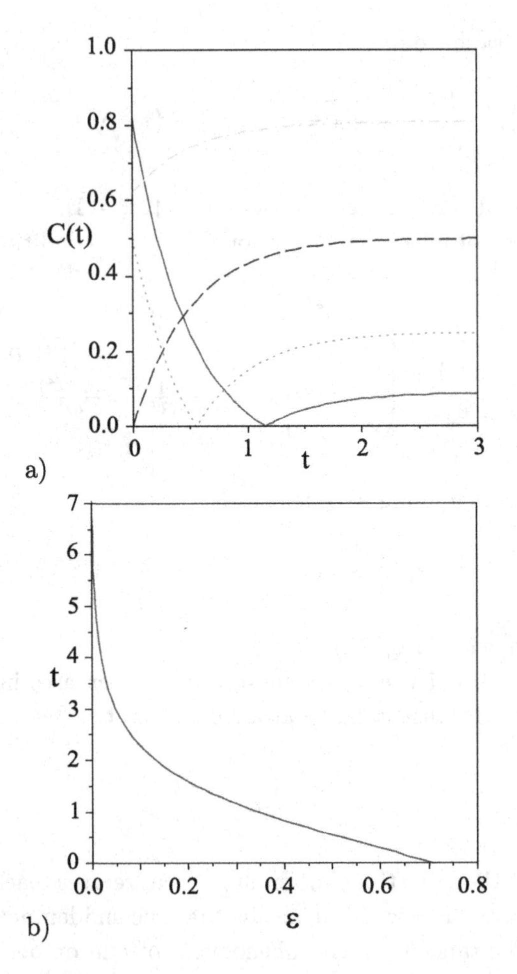

Fig. 9. a) Time evolution of the concurrence with initial $|\Psi_b\rangle$, for: $\varepsilon = 0.3$ (solid line), $\varepsilon = 0.5$ (dotted line), $\varepsilon = 0.707$ (dashed line), $\varepsilon = 0.9$ (dash dotted line). b) Death-revival time as given by Eq.(36), versus ε.

$max\{0, C_1(\rho'(t)), C_2(\rho'(t))\}$, where

$$C_1(\rho'(t)) = |\rho'_{33}(t) - \rho'_{22}(t)| - 2\sqrt{\frac{(N(\rho'_{11}(t) + \rho'_{44}(t)) + \rho'_{44}(t) + 2\rho'_{14}(t)\sqrt{N(N+1)})}{2N+1}} \times \sqrt{\frac{(N(\rho'_{11}(t) + \rho'_{44}(t)) + \rho'_{11}(t) - 2\rho'_{14}(t)\sqrt{N(N+1)})}{2N+1}}). \quad (37)$$

$$C_2(\rho'(t)) = \frac{2}{2N+1}|\sqrt{N(N+1)}(\rho'_{11}(t) - \rho'_{44}(t)) + \rho'_{14}(t)| - \sqrt{(\rho'_{22}(t) - 2\rho'_{23}(t) + \rho'_{33}(t))(\rho'_{22}(t) + 2\rho'_{23}(t) + \rho'_{33}(t))}. \quad (38)$$

where $\rho'_{ij}(t)$ are the density matrix elements in the $\{\phi_i\}$ basis.

a) Next, we consider again the case for initial $|\phi_3\rangle$. In Fig.10, we show the evolution of the concurrence for various values of N. We always observe sudden death in a finite time, then the concurrence remains zero for a period of time until the entanglement revives, and the concurrence reaches asymptotically its stationary value. Notice that this time period increases with N. In the fig.11 we show the death and revival times

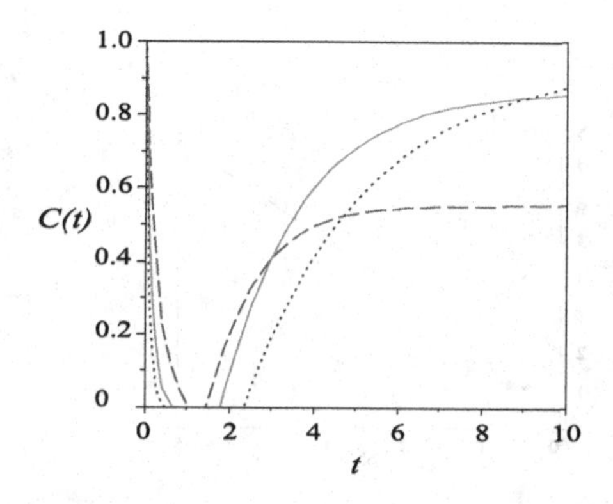

Fig. 10. Time evolution of concurrence for initial $|\phi_3\rangle$, with: N=0.1 (dashed line), N=0.5 (solid line), N=1 (dotted line).

versus N. They decrease and increase with N respectively.

b) We consider $|\phi_4\rangle$ as an initial state. The behavior of concurrence is similar as in $|\phi_3\rangle$ case. The initial entanglement quickly decays to zero, getting disentanglement for a finite time interval, then the entanglement revives and asymptotically it reaches its stationary value. However, unlike the case with initial state $|\phi_3\rangle$, the death time first increases reaching a maximum for $N = 0.421$, and subsequently it decreases, as shown in Fig 12. The revival time has the same behavior as in $|\phi_3\rangle$.

c) In the following case, we consider the superposition

$$|\Psi_a(0)\rangle = \varepsilon|\phi_1\rangle + \sqrt{1-\varepsilon^2}|\phi_4\rangle, \tag{39}$$

as the initial state. The solution of master equation for this initial condition depends on ε and N and also its concurrence. In this case,

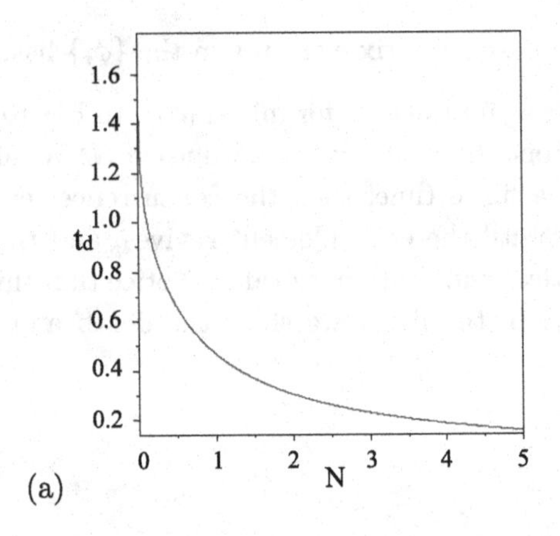

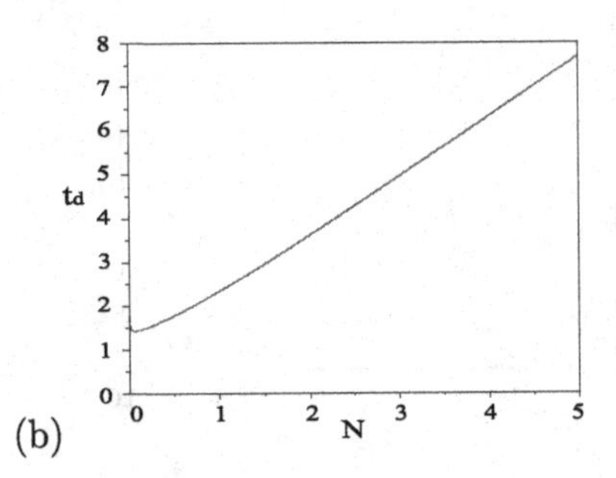

Fig. 11. a) Death time b) Revival time versus N for the initial state $| \phi_3\rangle$.

$\rho'_{23}(t) = \rho'_{22}(t) = 0$ thus, the concurrence is given by $C(\rho'_a(t)) = \max\{0, C_2(\rho'_a(t))\}$, where $C_2(\rho'_a(t))$ given in Eq.(38)
Initially, the concurrence is given by

$$C(\rho'_a(0)) = \frac{|2\varepsilon\sqrt{1-\varepsilon^2} + 4\sqrt{N(N+1)}(\varepsilon^2 - \frac{1}{2})|}{2N+1}, \tag{40}$$

we see that for certain pairs of N and ε, our initial state will be a non-separable one.
In the Fig. 13 we show the time evolution of the concurrence for $N = 0.1$ and several values of ε.
For $\varepsilon = 0$ and $\varepsilon = 1$ we retrieve $|\phi_4\rangle$ and $|\phi_1\rangle$ respectively.
For $0 < \varepsilon < 0.5$ the concurrence dies in a finite time, stays zero for a time interval and subsequently revives, going asymptotically to its stationary value.

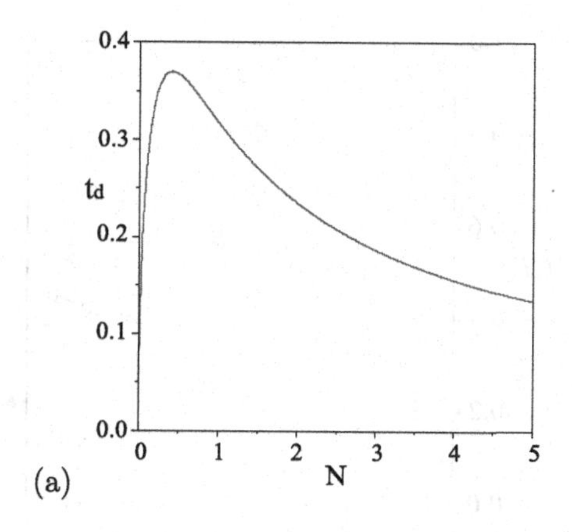

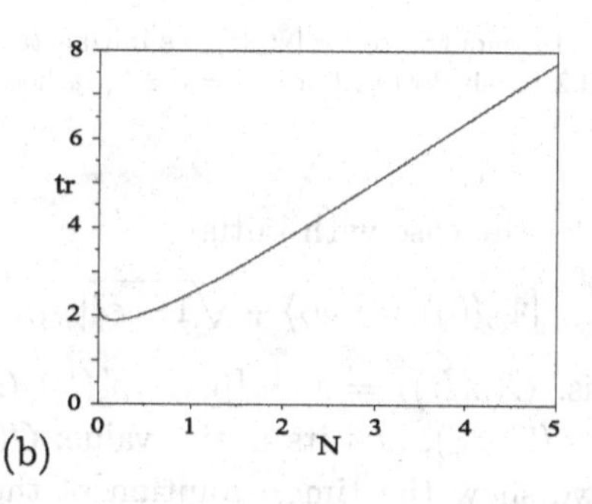

Fig. 12. (a) Death, and (b) Revival times versus N for the initial state $|\phi_4\rangle$.

For values larger than $\varepsilon = 0.5$, there is no more sudden death, since we are getting "close" to the DFS, and the concurrence goes asymptotically to its stationary value.

The Fig.14(a) shows the death times versus ε for $N = \{0, 0.1, 0.2\}$. In the case $N = 0$, we notice a steady increase of the death time up to some critical value of ε, where the death time becomes infinite. There is a curious effect, that for $N \neq 0$, as we increase ε, the death time first decreases up to the value $\varepsilon = \sqrt{\frac{N}{2N+1}}$, and subsequently it behaves "normally", by increasing with ε. In the Fig.14(b) we show the revival time as a function of ε for the same values of N. In all cases the revival time decreases with ε.

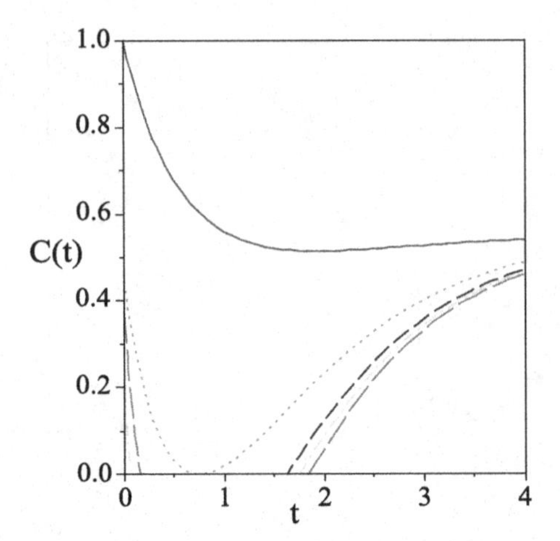

Fig. 13. Time evolution of Concurrence for $|\Psi_a(t)\rangle$ as initial state and $N = 0.1$: $\varepsilon = 0.1$ (long dashed line), $\varepsilon = 0.2$ (dash dotted line), $\varepsilon = 0.29$ (dashed line), $\varepsilon = 0.5$ (dotted line), $\epsilon = 0.9$ (solid line).

d) Finally, we consider the case with initial

$$|\Psi_b(0)\rangle = \varepsilon|\phi_2\rangle + \sqrt{1-\varepsilon^2}|\phi_3\rangle. \tag{41}$$

Its concurrence is: $C(\rho_b'(t)) = \max\{0, C_1(\rho_b'(t)), C_2(\rho_b'(t))\}$, with C_1 and C_2 defined in (37,38), and its initial value: $C(\rho_b'(0)) = |2\varepsilon^2 - 1|$. In the Fig. 15, we show the time evolution of the concurrence with $N = 0.1$ for several values of ε.
As we can see from the Fig. 15, this case is more complex, since there are more than one death and revival before reaching the critical value of ε. Such a situation has been described previously.[15,27] Like in the previous cases above a certain critical ε, when we get "close" to the DFS, these effects disappear and $C(t)$ goes asymptotically to its stationary value.

3.5. *Discussion*

The first observation is that if we start from an initial state that is in the DFS plane, the local and non-local coherences are not affected by the environment, thus it experiences no decoherence and the concurrence stays constant in time.

In the case of initial $|\phi_1\rangle$, the concurrence does increase with the squeeze parameter N, getting maximum entanglement for $N \to \infty$. So this reservoir is not acting as a thermal one, in the sense that introduces randomness. On

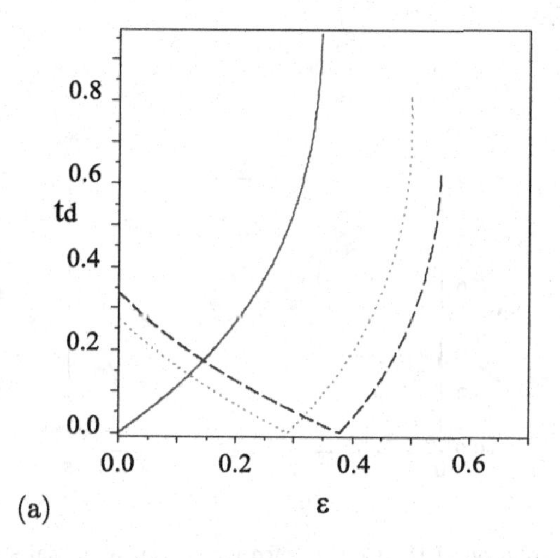

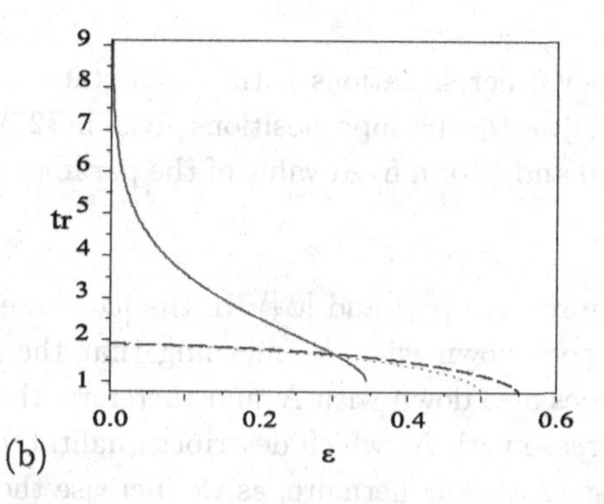

Fig. 14. a) Death time and b) Revival time, with initial state $|\Psi_a\rangle$ and: $N = 0$ (solid line), $N = 0.1$ (dotted line), $N = 0.2$ (dashed line).

the contrary, a common squeezed bath tends to enhance the entanglement, as we increase the parameter N.

This is clear if we observe that for $N \to \infty$:

$$|\phi_1\rangle \to \frac{1}{\sqrt{2}}(|++\rangle + |--\rangle), \tag{42}$$

which is a Bell state.

On the other hand, if we start with the initial state $|\phi_2\rangle$, this state is independent of N and it is also maximally entangled, so $C = 1$ for all times and all $N's$.

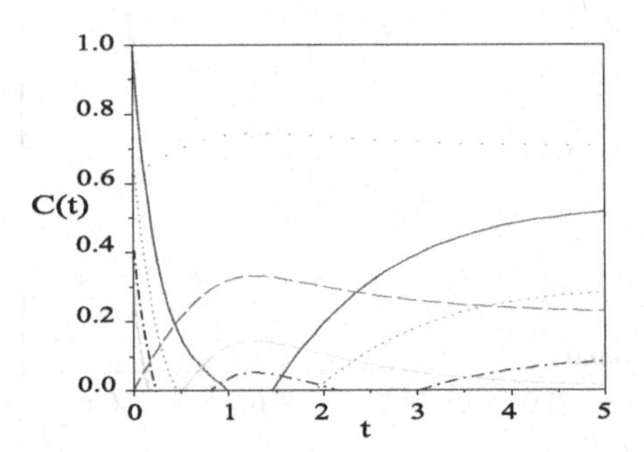

Fig. 15. Time evolution of the Concurrence for $|\Psi_b(t)\rangle$ as initial state and $N = 0.1$: $\varepsilon = 0$ (solid line), $\varepsilon = 0.4$ (dotted line), $\varepsilon = 0.54$ (dash dotted line), $\varepsilon = 0.6$ (long dashed line), $\varepsilon_c = 0.707$ (dashed line), $\varepsilon = 0.9$ (space dotted line).

Now, we consider other situations with initial states outside the DFS. We consider as initial states the superpositions given in (27) and (28), where we vary ε between 0 and 1 for a fixed value of the parameter N. We observe that:

- When $\varepsilon = 0$ we recover $|\phi_4\rangle$ and $|\phi_3\rangle$. In the $|\phi_4\rangle$ case, the population of the $|++\rangle$ goes down with N, meaning that the interaction with the reservoir goes also down with N and therefore, the death time will necessarily increase with N, which describes qualitatively the first part of the curve(fig 12-a). Furthermore, as we increase the average photon number N, other processes like the two photon absorption will be favored, and since there will be more photons and the $|--\rangle$ population tends to increase with N, this will enhance the system-bath interaction and therefore the death of the entanglement will occur faster, or the death time will decrease.
  In the $|\phi_3\rangle$ case, initially there is no $|++\rangle$ component, thus we expect a higher initial death time. However this case is different from the previous one in the sense that the state is independent of N, so there is no initial increase. However, as the state evolves in time, the $|++\rangle$ and $|--\rangle$ components will build up and the argument for the decrease of the death time with N follows the same logic as in the previous case. See Fig.11(a).

- For the interval $0 \leq \varepsilon < \varepsilon_c$ the initial entanglement decays to zero in a finite time, t_d. After a finite period of time during which the concurrence stay null, it revives a time t_r reaching asymptotically its steady-state value. This death and revival cycle occurs once for the initial state $|\Psi_a\rangle$, and twice for $|\Psi_b\rangle$.
  For $|\Psi_a\rangle$ as initial state, when ε is equal to the critical value ε_c, the entanglement dies and revives simultaneously and eventually goes to its steady-state value. For the initial state $|\Psi_b\rangle$, the critical value of ε is $\varepsilon_c = \frac{1}{\sqrt{2}}$, and, unlike to the $|\Psi_a\rangle$ case, it is independent of N.
- When we get *near* to the DFS ($\varepsilon_c < \varepsilon \leq 1$), the system shows no disentanglement and this phenomenon of sudden death and revival disappears.

The squeezed vacuum has only non-zero components for even number of photons, so the interaction between our system and the reservoir goes by pairs of photons. Now, for very small N, the average photon number is also small, so the predominant interaction with the reservoir will be the doubly excited state that would tend to decay via two photon spontaneous emission.

Let us consider again $|\Psi_a\rangle$, but now in terms of the standard basis

$$|\Psi_a\rangle = k_1|++\rangle + k_2|--\rangle,$$

with

$$k_1 = \frac{\varepsilon N + M\sqrt{1-\varepsilon^2}}{\sqrt{N^2+M^2}}, \quad k_2 = \frac{\varepsilon M - N\sqrt{1-\varepsilon^2}}{\sqrt{N^2+M^2}}, \tag{44}$$

We plotted k_1 versus ε for N between 0 to 2, Fig.16. Initially, k_1 increases with ε, thus favoring the coupling with the reservoir, or equivalently, producing a decrease in the death time. This is up to $\varepsilon = \sqrt{\frac{N}{2N+1}}$, where the curve shows a maxima. Beyond this point, k_1 starts to decrease and therefore our system is slowly decoupling from the bath and therefore the death time shows a steady increase.(Fig 14(a))

3.6. *A curious effect. Creation of delayed entanglement*

In some cases, if two systems do not interact directly, but share a common heath bath in thermal equilibrium, entanglement can be created, some time after the interaction is turned on. Furthermore, this entanglement may persist in time. So, contrary to intuition that spontaneous emission should have a destructive effect on the entanglement, it has been shown by several

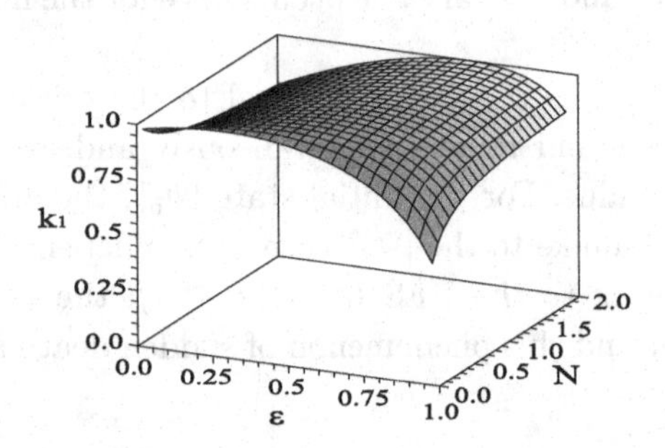

Fig. 16. k_1 versus ε for N between 0 to 2.

authors that under certain conditions, this irreversible process can even entangle initially unentangled qubits.[29–37]

Also, more recently, it was shown that a "sudden" feature in the creation of entanglement exists, in the sense that it takes some finite time after the system is connected to the bath, for the creation of entanglement to take place.[18]

In the following we considered the simplest composite quantum system of two qubits. This qubits are coupled to a common thermal reservoir at zero temperature. The qubits do not interact directly with each other but through the common environment, in that case there is a probability that a photon emitted by one atom will be absorbed by the other, and as a consequence of this photon exchange process can produce entanglement between atoms, which is larger than the decoherence caused by spontaneous emission. According to this assumption, is crucial to have one atom excited and the other in the ground state to create entanglement. The dynamics of the system is given by the Eq.(1) with N and M equal to zero.

For a given initial state $\rho(0) = \rho_{ij}(0)$ the solution of the master equation (1) in the $\{e_1 = |11\rangle, e_2 = |10\rangle, e_3 = |01\rangle, e_4 = |00\rangle\}$ basis, is given by the following matrix elements

$$
\begin{aligned}
\rho_{11}(t) &= \rho_{11}e^{-2\Gamma t}\\
\rho_{22}(t) &= \frac{1}{4}(\rho_{22}+\rho_{33}-\rho_{23}-\rho_{32}+2(\rho_{22}-\rho_{33})e^{-\Gamma t}\\
&\quad+(4\rho_{11}t+\rho_{22}+\rho_{33}+\rho_{23}+\rho_{32})e^{-2\Gamma t})\\
\rho_{33}(t) &= \frac{1}{4}(\rho_{22}+\rho_{33}-\rho_{23}-\rho_{32}+2(\rho_{33}-\rho_{22})e^{-\Gamma t}\\
&\quad+(4\rho_{11}t+\rho_{22}+\rho_{33}+\rho_{23}+\rho_{32})e^{-2\Gamma t})\\
\rho_{44}(t) &= \frac{1}{2}(\rho_{23}+\rho_{32}+\rho_{11}+\rho_{44}+1\\
&\quad-(4\rho_{11}t+2\rho_{11}+\rho_{22}+\rho_{33}+\rho_{23}+\rho_{32})e^{-2\Gamma t})\\
\rho_{12}(t) &= \frac{1}{2}((\rho_{12}-\rho_{13})e^{-\Gamma t}+(\rho_{12}+\rho_{13})e^{-2\Gamma t})\\
\rho_{13}(t) &= \frac{1}{2}(-(\rho_{12}-\rho_{13})e^{-\Gamma t}+(\rho_{12}+\rho_{13})e^{-2\Gamma t})\\
\rho_{14}(t) &= \rho_{14}e^{-\Gamma t}\\
\rho_{23}(t) &= \frac{1}{4}(\rho_{23}+\rho_{32}-\rho_{22}-\rho_{33}+2(\rho_{23}-\rho_{32})e^{-\Gamma t}\\
&\quad+(4\rho_{11}t+\rho_{22}+\rho_{33}+\rho_{23}+\rho_{32})e^{-2\Gamma t})\\
\rho_{24}(t) &= \frac{1}{2}(\rho_{24}-\rho_{34}-2(\rho_{12}+\rho_{13})e^{-2\Gamma t}+(2\rho_{12}+2\rho_{13}+\rho_{34}+\rho_{24})e^{\Gamma t})\\
\rho_{34}(t) &= \frac{1}{2}(\rho_{34}-\rho_{24}-2(\rho_{12}+\rho_{13})e^{-2\Gamma t}+(2\rho_{12}t+2\rho_{13}+\rho_{34}+\rho_{24})e^{-\Gamma t})
\end{aligned}
\tag{45}
$$

For simplicity, we write each matrix element $\rho_{ij}(0)$ as ρ_{ij} and take $\Gamma = 1$. The remaining matrix elements can be obtained from $\rho_{ij} = \rho^*_{ji}$.

In order to obtain a condition that determines when a state that was initially separable becomes non-separable after a certain period of time, we will study the stationary state. In the limit $t \to \infty$, the density matrix is given by

$$
\rho_{st} = \frac{1}{4}\begin{pmatrix} 0 & 0 & 0 & 0\\ 0 & \rho_{22}+\rho_{33}-(\rho_{23}+\rho_{32}) & \rho_{23}+\rho_{32}-(\rho_{22}+\rho_{33}) & 2(\rho_{24}-\rho_{34})\\ 0 & \rho_{23}+\rho_{32}-(\rho_{22}+\rho_{33}) & \rho_{22}+\rho_{33}-(\rho_{23}+\rho_{32}) & 2(\rho_{34}-\rho_{24})\\ 0 & 2(\rho_{42}-\rho_{43}) & 2(\rho_{43}-\rho_{42}) & 2(\rho_{23}+\rho_{32})+2(1+\rho_{11}+\rho_{44}) \end{pmatrix}
$$

Following the procedure to get the concurrence, we calculate the matrix $R_{st} = \rho_{st}(\sigma_y \otimes \sigma_y)\rho^*_{st}(\sigma_y \otimes \sigma_y)$. For real matrix elements, R_{st} has only one nonzero eigenvalue:

$$\lambda = \frac{1}{4}(\rho_{22} + \rho_{33} - 2Re(\rho_{23}))^2, \tag{46}$$

Thus, the condition to have a disentangled steady state is

$$\rho_{22} + \rho_{33} = 2\rho_{23}. \tag{47}$$

Now, we will consider some examples of different initial separable states.

a) Let us first have two atoms in excited state $|\Psi(0)\rangle = |11\rangle$. The concurrence is

$$C(t) = \max\{0, 2te^{-2t} - 2\sqrt{e^{-2t}(1-(1+2t)e^{-2t})}\} = 0. \tag{48}$$

Thus, this state is separable for all times. In the stationary state, both atoms go to the ground state. Thus its concurrence is $C(t) = 0$.

b) For the case when one atom is in an excited state $|1\rangle$ and the other one in the ground state $|0\rangle$, the concurrence is

$$C(t) = \max\{0, \frac{1}{2}|e^{-2t} - 1|\}, \tag{49}$$

this result being the same for $|10\rangle$ and $|01\rangle$. The initial value of the concurrence increases to its steady state value of $\frac{1}{2}$.

c) When both atoms are initially in the ground state, $|\Psi(0)\rangle = |00\rangle$, the system remains in its ground state for all times, and the concurrence is $C(t) = 0$.

As mentioned before, the only way to generate entanglement (for the vacuum reservoir) is when one atom is in the ground state and the other one in the excited state, since this combination makes it possible for photon exchanges between the atoms. In the cases when there is no photon exchange, the entanglement generation never occurs.

In order to study this effect we consider linear combinations of the $|0\rangle$ and $|1\rangle$ states, for one of the atoms, and the ground or the excited state for the other one, respectively

d) Consider first

$$|\Psi(0)\rangle = |0\rangle \otimes (\alpha|1\rangle + \beta|0\rangle) = \alpha|01\rangle + \beta|00\rangle, \tag{50}$$

with $\alpha^2 + \beta^2 = 1$. The Fig.17 (a) shows that the entanglement reaches a higher value when α is greater.

e) The second possible initial condition is

$$|\Psi(0)\rangle = (\alpha|0\rangle + \beta|1\rangle) \otimes |1\rangle = \alpha|01\rangle + \beta|11\rangle, \tag{51}$$

with $\alpha^2 + \beta^2 = 1$. The Fig.17 (b) shows the time evolution of the concurrence for different values of α. As in the previous case, when α decreases, the maximum value of the entanglement decreases, but unlike the previous case, the entanglement creation has a time delay.

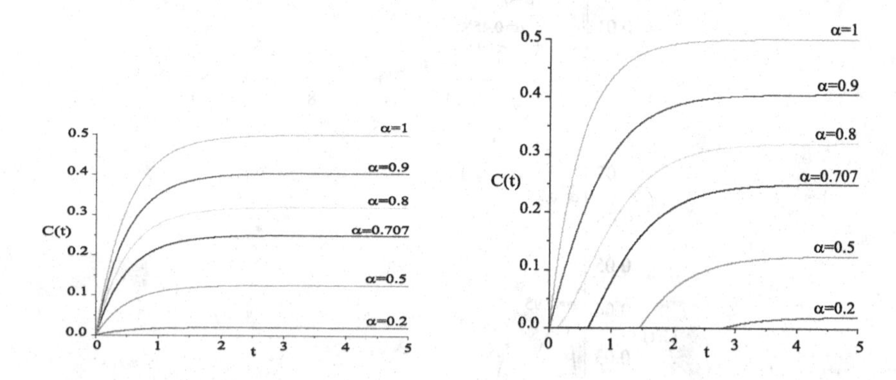

Fig. 17. (a) Time evolution of concurrence for the initial $|0\rangle\otimes(\alpha|1\rangle+\beta|0\rangle)$ with different values of α. (b) Time evolution of concurrence for the initial $(\alpha|0\rangle + \beta|1\rangle) \otimes |1\rangle$ with different values of α. The reservoir is at T=0

f) Consider a more general initial condition

$$(\alpha_1|0\rangle + \beta_1|1\rangle) \otimes (\alpha_2|0\rangle + \beta_2|1\rangle). \tag{52}$$

According to the Eq.(46), the condition to get a completely disentangled stationary state is

$$\frac{\alpha_1}{\beta_1} = \frac{\alpha_2}{\beta_2}, \tag{53}$$

implying that when both qubits are in the same state, the created entanglement is eventually destroyed, at large times. We show this effect in the Fig.(18), where we plotted the evolution of the concurrence for the various initial conditions in Eq.(52), when $\alpha_1 = \alpha_2 \equiv \alpha$.
Finally, we analyze the effects of having a reservoir with a finite (non zero) temperature. In the cases $|\Psi(0)\rangle = |10\rangle$ and $|\Psi(0)\rangle = |01\rangle$,thermal effects tend to destroy the entanglement (see Fig. 19).

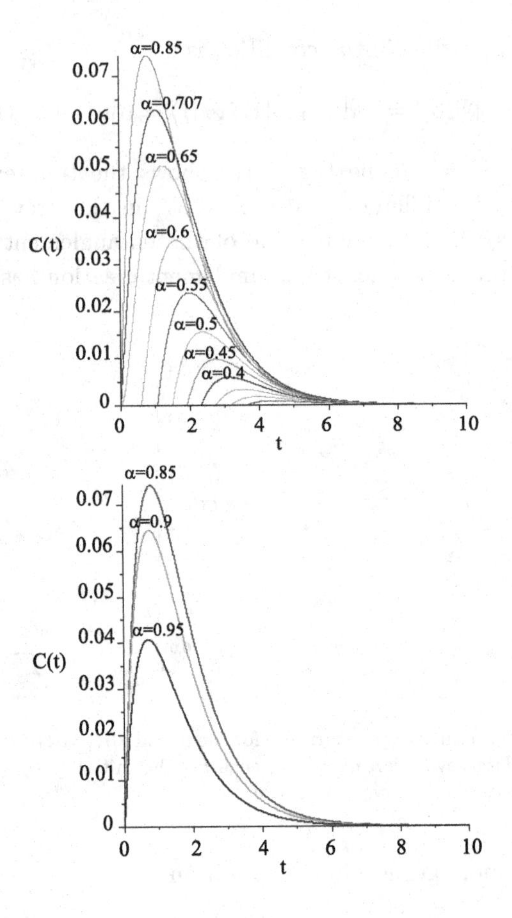

Fig. 18. Concurrence versus time for different initial conditions $\alpha_1 = \alpha_2 \equiv \alpha$. The reservoir is at T=0.

4. Conclusion

Entanglement is in the heart of quantum Mechanics. Also, besides its conceptual relevance, it is also crucial in many applications in Quantum communications and quantum computing. On the other hand, it is a subtle and delicate effect that can be easily altered or destroyed. In the various cases studying the dynamics of entanglement, we encountered different situations, described in the Fig.(20), where we show "trajectories" between the PPT (Partial Positive Transpose) and the NPT (Negative Partial Transpose) areas. The first case corresponds to a system that has both the initial and final states entangled. In the second case, the initial entanglement goes

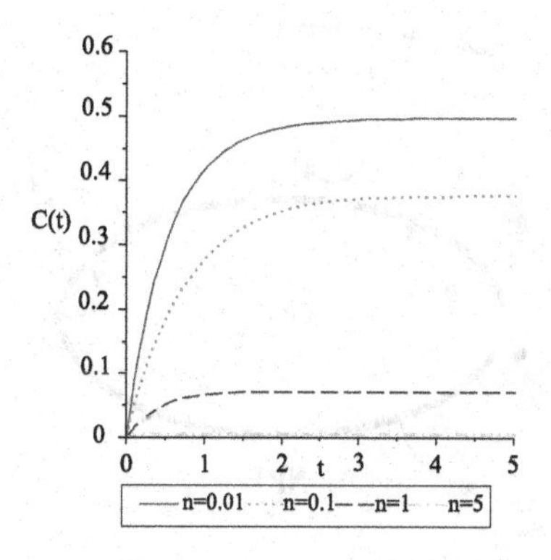

Fig. 19. Concurrence versus time for the initial conditions $|\Psi(0)\rangle = |10\rangle$ and $|\Psi(0)\rangle = |01\rangle$. The parameter of each curve is the reservoir temperature (average thermal photon number).

asymptotically to zero. In the third and fourth cases, we observe sudden death and death with revival respectively. Case number 5 corresponds to a periodic Hamiltonian system (for example, the Jaynes-Cummings Model), where death and revival appear periodically. Death and revival are intriguing effects that basically depend on the initial conditions as well as the nature of the reservoir. Since, for many applications, the sudden death is an undesired effect, one can try to protect these states, using for example, the distillation procedures.

Finally, in the cases 6 and 7, we have generation of entanglement without and with time delay respectively, starting from a separable state. These are very interesting effects, where, in spite of the fact that there is no coupling between the atoms, the common reservoir acts as an effective coupling, with particularly good results, in the case of a vacuum reservoir, when the initial atomic state combines an excited atom with the other one in the ground state, thus allowing, via a common reservoir, a photon exchange. When the reservoir is at a finite temperature, this effect is reduced.

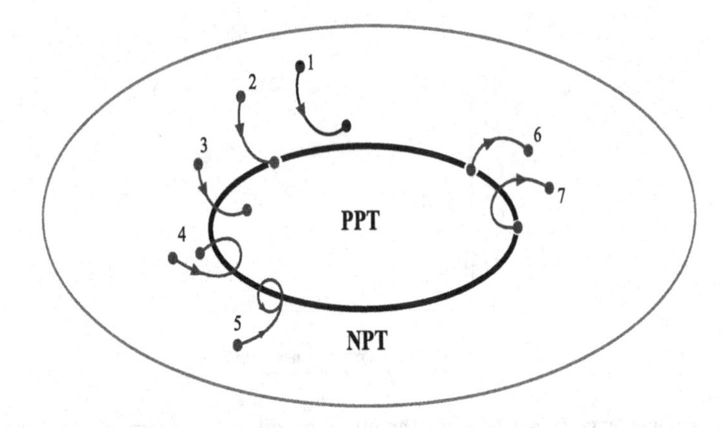

Fig. 20. "Trajectories" between the "positive partial transpose" (separable) and the "negative partial transpose" (entangled) areas: 1: Initial and final states are both entangled, 2: Initial entanglement going asymptotically to zero, 3: Initial entanglement with sudden death, 4: Initial entanglement with sudden death and revival, 5: Periodic death and revival, 6: Entanglement generation starting from a separable state, 7: Time delayed generation of entanglement starting from a separable state.

Acknowledgments

The author was supported by a Fondecyt grant (1100039).

References

1. W.H.Zurek, *Decoherence, einselection, and the quantum origins of the classical*, Rev.Mod.Phys. **75**, 715 (2003).
2. Ting Yu and J.H.Eberly, *Finite-Time Disentanglement Via Spontaneous Emission*, Phys.Rev.Lett. **93**,140404 (2004).
3. L.Diósi, *Progressive Decoherence and Total Environmental Disentanglement*, Lect. Notes Phys. 622, 157 (2003).
4. P.Marek, J.Lee, M.S.Kim, *Vacuum as a less hostile environment to entanglement*, Phys.Rev.A, **77**, 032302 (2008).
5. Y.X. Gong, Yong-Sheng Zhang, Yu-Li Dong, Xiao-Ling Niu, Yun-Feng Huang, Guan-Can Guo, *Dependence of the decoherence of polarization states in phase-damping channels on the frequency spectrum envelope of photons*, Phys.Rev.A. **78**, 042103 (2008).
6. D.Tolkunov, V.Privman, P.K.Aravind, *Decoherence of a measure of entanglement*, Phys.Rev.A. **71**, 060308(R) (2005).
7. Juan Pablo Paz and Augusto J. Roncaglia, *Dynamical phases for the evolution of the entanglement between two oscillators coupled to the same environment*, Phys.Rev.A. **79**, 032102 (2009).
8. Cecilia Cormick and Juan Pablo Paz, *Decoherence of Bell states by local interactions with a dynamic spin environment*, Phys.Rev.A. **78**, 012357 (2008).

9. T.Gorin, C.Pineda and T.H.Seligman, *Decoherence of an n-Qubit Quantum Memory*, Phys.Rev.Lett. **99**, 240405 (2007).
10. C.Pineda, T. Gorin, and T. H. Seligman, *Decoherence of two-qubit systems: a random matrix description*, New J. Phys. **9**, 106 (2007).
11. M.Ikram, Fu-li Li and M.Zubairy, *Disentanglement in a two-qubit system subjected to dissipation environments*, Phys.Rev.A. **75**, 062336 (2007).
12. Kuan-Liang Liu and Hsi-Sheng Goan *et.al*, *Non-Markovian entanglement dynamics of quantum continuous variable systems in thermal environments*, Phys.Rev.A. **76**, 022312 (2007).
13. Asma Al-Qasimi and Daniel F. V. James, *Sudden death of entanglement at finite temperature*, Phys.Rev.A. **77**, 012117 (2008).
14. Maritza Hernandez and Miguel Orszag, *Decoherence and disentanglement for two qubits in a common squeezed reservoir*, Phys.Rev.A. **78**, 042114 (2008).
15. Z.Ficek and R.Tanás *Dark periods and revivals of entanglement in a two qubit system*, Phys.Rev.A. **74**,024304 (2006).
16. M.P.Almeida, F.de Melo, M.Hor-Meyll, A.Salles, S.P.Walborn, P.H.Souto-Ribeiro, L.Davidovich, *Environment induced death of entanglement*, Science, **316**, 579 (2007); A.Salles,F.de Melo, M.P.Almeida, M.Hor-Meyll, S.P.Walborn, P.H.Souto-Ribeiro, L.Davidovich, *Experimental investigation of the dynamics of entanglement:sudden death, complementarity and continuous monitoring of the environment*, Phys.Rev.A. **78**, 022322 (2008).
17. D.Braun, *Creation of Entanglement by Interaction with a Common Heat Bath*, Phys.Rev.Lett. **89**, 277901 (2002); Z.Ficek and R.Tanas, *Entanglement induced by spontaneous emission in spatially extended two-atom systems* J.Mod. Opt. **50**, 2765 (2003); F.Benatti, R.Floreanini and M.Piani, *Environment Induced Entanglement in Markovian Dissipative Dynamics*, Phys.Rev.Lett. **91**, 070402 (2003).
18. Z.Ficek and R.Tanás, *Delayed sudden birth of entanglement*, Phys.Rev.A. **77**, 054301 (2008).
19. B.Bellomo, R. Lo Franco, G.Compagno, *Non-Markovian Effects on the Dynemics of Entanglement*, Phys.Rev.Lett, **99**, 160502 (2007)
20. S.Maniscalco, S.Olivares, M.G.A.Paris, *Entanglement Oscillations in non-Markovian quantum channels*, Phys.Rev.A, **75**, 062119 (2007); See also the related work: L.Mazzola, S.Maniscalco, J.Piilo, K-A. Suominen, B.M.Garraway, *Sudden death and sudden birth of entanglement in common structured reservoirs*, Phys.Rev.A, **79**, 042302 (2009); B.Bellomo, R.Lo Franco, S.Maniscalco, G.Compagno, *Entanglement trapping in structured environments*, Phys.Rev.A, **78**, 060302 (2008), and for multipartite systems: C.E.Lopez, G.Romero, F.Lastra, E.Solano, J.C.Retamal, *Sudden birth versus sudden death of Entanglement in Multipartite systems*, Phys.Rev.Lett, 101, 080503 (2008)
21. M.Franca-Santos, P.Milman, L.Davidovich, N.Zagury, *Direct Measurement of finite-time disentanglement induced by a reservoir*, Phys.Rev.A, **73**, 040305 (2006)
22. T.Yu, J.H.Eberly, *Evolution from entanglement to decoherenceof standard bipartite mixed states*, arxiv:quant-ph/0503089(2006)

23. F.Lastra, S.Wallentowitz, M.Orszag, M.Hernandez, *Quantum recoil effects in finite-time disentanglement of two distinguishable atoms*, J.Phys.B, At.Mol.Opt.Phys, **42**, 065504(2009)
24. D.Mundarain and M. Orszag, *Decoherence-free subspace and entanglement by interaction with a common squeezed bath*, Phys.Rev.A **75**, 040303(R) (2007).
25. D.Mundarain and M. Orszag, J.Stephany, *Total quantum Zeno effect and intelligent states for a two-level system in a squeezed bath*, Phys.Rev.A **74**, 052107 (2006).
26. A.Peres, *"separability criterion for density matrices"*, Phys.Rev.Lett,**77**, 1413(1996); M.Horodecki, P.Horodecki, and R.Horodecki, *"separability of mixed states; necessary and sufficient conditions"*, Phys.Lett, A, **223**, 1-8 (1996)
27. T.Yu, J.H.Eberly, *Negative entanglement measure and what it implies*, e-print arXiv:quant-ph/0703083.
28. R.Tanas, Z.Ficek, *Stationary two-atom entanglement induced by non-classical two-photon correlations*, J.Opt.B: Quantum.Semiclass, **6**, S610 (2004).
29. A.Beige et al, *Entangling atoms and ions in dissipative environments*, J.Mod.Opt, **47**, 2583(2000).
30. D.Braun, *Creation of entanglement by interaction with a common heath bath*, Phys.Rev.Lett, **89**, 277901 (2002).
31. A.M.Basharov, *Entanglement of atomic states upon collective radiative decay*, JETP Lett, **75**, 123 (2002).
32. L.Jakobczyk, *Entangling two qubits by dissipation*, J.Phys.A, **35**, 6383 (2002).
33. Z,Ficek, R.Tanas, *Entanglement induced by spontaneous emission in spatially extended two atom systems*, J.Mod.Opt, **50**, 2765 (2003).
34. F,Benatti, R.Floreanini, M.Piani, *Environment induced Entanglement in Markovian dissipative dynamics*, Phys.Rev. Lett, **91**, 070402 (2003).
35. M.Paternostro, S.M.Tame, G.M.Palma, M.S.Kim, *Entanglement generation and protection by detuning modulation*, Phys.Rev.A, **74**, 052317 (2006).
36. S.Natali, Z.Ficek, *Temporal and diffraction effects in entanglement creation in an optical cavity*, Phys.Rev.A, **75**, 042307 (2007).
37. L.Derkacz, L.Jakobczyk, *Vacuum induced stationary entanglement in radiatively coupled three-level atoms*, e-print arXiv: 0710.5048.

INTRODUCTION TO QUANTUM FISHER INFORMATION

D. PETZ
Alfréd Rényi Institute of Mathematics
H-1364 Budapest, POB 127, Hungary
E-mail: petz@math.bme.hu

C. GHINEA
Department of Mathematics and its Applications
Central European University, 1051 Budapest, Nádor utca 9, Hungary
E-mail: ghinea_catalin@ceu-budapest.edu

The subject of this paper is a mathematical transition from the Fisher information of classical statistics to the matrix formalism of quantum theory. If the monotonicity is the main requirement, then there are several quantum versions parametrized by a function. In physical applications the minimal is the most popular. There is a one-to-one correspondence between Fisher informations (called also monotone metrics) and abstract covariances. The skew information and the χ^2-divergence are treated here as particular cases.

Keywords: Quantum state estimation, Fisher information, Cramér-Rao inequality, monotonicity, covariance, operator monotone function, skew information.

Introduction

Parameter estimation of probability distributions is one of the most basic tasks in information theory, and has been generalized to quantum regime[20,22] since the description of quantum measurement is essentially probabilistic. First let us have a look at the classical Fisher information.

Let $(X, \mathcal{B}, \mu)$ be a probability space. If $\theta = (\theta^1, \dots, \theta^n)$ is a parameter vector in a neighborhood of $\theta_0 \in \mathbb{R}^n$, then we should have a smooth family μ_θ of probability measures with probability density f_θ:

$$\mu_\theta(H) = \int_H f_\theta(x)\, d\mu(x) \qquad (H \in \mathcal{B}).$$

The Fisher information matrix at θ_0 is

$$J(\mu_\theta;\theta_0)_{ij} := \int_X f_{\theta_0}(x)\frac{\partial}{\partial\theta^i}\log f_\theta(x)\Big|_{\theta=\theta_0}\frac{\partial}{\partial\theta^j}\log f_\theta(x)\Big|_{\theta=\theta_0} d\mu(x) \qquad (1)$$

$$= \int_X \frac{1}{f_{\theta_0}(x)}\partial_i f_{\theta_0}(x)\Big|_{\theta=\theta_0}\partial_j f_{\theta_0}(x)\Big|_{\theta=\theta_0} d\mu(x)$$

$$= -\int_X f_{\theta_0}(x)\partial_{ij}\log f_\theta(x)\Big|_{\theta=\theta_0} d\mu(x) \qquad (1 \le i,j \le n).$$

Note that $\log f_\theta(x)$ is usually called **log likelihood** and its derivative is the **score function**.

The Fisher information matrix is positive semidefinite. For example. if the parameter $\theta = (\theta^1, \theta^2)$ is two dimensional, then the Fisher information is a 2×2 matrix. From the Schwarz inequality

$$J(\mu_\theta;\theta_0)_{12}^2 \le \int_X \left[\frac{1}{\sqrt{f_{\theta_0}(x)}}\partial_1 f_{\theta_0}(x)\right]^2 d\mu(x) \int_X \left[\frac{1}{\sqrt{f_{\theta_0}(x)}}\partial_2 f_{\theta_0}(x)\right]^2 d\mu(x)$$

$$= J(\mu_\theta;\theta_0)_{11} J(\mu_\theta;\theta_0)_{22}$$

Therefore the matrix $J(\mu_\theta;\theta_0)$ is positive semidefinite.

Assume for the sake of simplicity, that θ is a single parameter. The random variable $\hat\theta$ is an **unbiased estimator** for the parameter θ if

$$\mathbb{E}_\theta(\hat\theta) := \int \hat\theta(x) f_\theta(x)\, d\mu(x) = \theta$$

for all θ. This means that the expectation value of the estimator is the parameter. The **Cramér-Rao inequality**

$$\mathrm{Var}(\hat\theta) := \mathbb{E}_\theta((\hat\theta - \theta)^2) \ge \frac{1}{J(\mu_\theta;\theta)}$$

gives a lower bound for the variance of an unbiased estimator. (For more parameters we have an inequality between positive matrices.)

In the quantum formalism a probability measure is replaced by a positive matrix of trace 1. (Its eigenvalues form a probability measure, but to determine the so-called density matrix a basis of the eigenvectors is also deterministic.) If a parametrized family of density matrices D_θ is given, then there is a possibility for the quantum Fisher information. This quantity is not unique, the possibilities are determined by linear mappings. The analysis of the linear mappings is the main issue of the paper. In physics $\theta \in \mathbb{R}$ mostly, but if it is an n-tuple, then Riemannian geometries appear. A coarse-graining gives a monotonicity of the Fisher informations and this is the second main subject of the present overview.

Fisher information has a big literature both in the classical and in the quantum case. The reference of the papers is not at all complete here. The aim is to have an introduction.

1. A general quantum setting

The Cramér-Rao inequality belongs to the basics of estimation theory in mathematical statistics. Its quantum analog appeared in the 1970's, see the book[20] of Helstrom and the book[22] of Holevo. Although both the classical Cramér-Rao inequality and its quantum analog are mathematically as trivial as the Schwarz inequality, the subject takes a lot of attention because it is located on the boundary of statistics, information and quantum theory. As a starting point we give a very general form of the quantum Cramér-Rao inequality in the simple setting of finite dimensional quantum mechanics. The paper[43] is followed here.

For $\theta \in (-\varepsilon, \varepsilon) \subset \mathbb{R}$ a statistical operator $\rho(\theta)$ is given and the aim is to estimate the value of the parameter θ close to 0. Formally $\rho(\theta)$ is an $n \times n$ positive semidefinite matrix of trace 1 which describes a mixed state of a quantum mechanical system and we assume that $\rho(\theta)$ is smooth (in θ). Assume that an estimation is performed by the measurement of a self-adjoint matrix A playing the role of an observable. A is called **locally unbiased estimator** if

$$\frac{\partial}{\partial \theta} \operatorname{Tr} \rho(\theta) A \bigg|_{\theta=0} = 1 . \tag{2}$$

This condition holds if A is an unbiased estimator for θ, that is

$$\operatorname{Tr} \rho(\theta) A = \theta \qquad (\theta \in (-\varepsilon, \varepsilon)). \tag{3}$$

To require this equality for all values of the parameter is a serious restriction on the observable A and we prefer to use the weaker condition (2).

Let $[K, L]_\rho$ be an inner product (or quadratic cost function) on the linear space of self-adjoint matrices. This inner product depends on a density matrix and its meaning is not described now. When $\rho(\theta)$ is smooth in θ, as already was assumed above, then

$$\frac{\partial}{\partial \theta} \operatorname{Tr} \rho(\theta) B \bigg|_{\theta=0} = [B, L]_{\rho(0)} \tag{4}$$

with some $L = L^*$. From (2) and (4), we have $[A, L]_{\rho(0)} = 1$ and the Schwarz inequality yields

$$[A, A]_{\rho(0)} \geq \frac{1}{[L, L]_{\rho(0)}} . \tag{5}$$

This is the celebrated **inequality of Cramér-Rao type** for the locally unbiased estimator.

The right-hand-side of (5) is independent of the estimator and provides a lower bound for the quadratic cost. The denominator $[L, L]_{\rho(0)}$ appears to be in the role of Fisher information here. We call it **quantum Fisher information** with respect to the cost function $[\cdot, \cdot]_{\rho(0)}$. This quantity depends on the tangent of the curve $\rho(\theta)$ and on the inner product $[\cdot, \cdot]_{\rho(0)}$. There are several possibilities to choose a reasonable inner product $[\cdot, \cdot]_{\rho(0)}$. Note that $[A, A]_{\rho(0)}$ should have the interpretation of "variance" (if $\mathrm{Tr}\, \rho_0 A = 0$.)

Another approach is due to Braunstein and Caves[4] in physics, but Nagaoka considered a similar approach.[34]

1.1. *From classical Fisher information via measurement*

The observable A has a spectral decomposition

$$A = \sum_{i=1}^{k} \lambda_i E_i \, .$$

(Actually the property $E_i^2 = E_i$ is not so important, only $E_i \geq 0$ and $\sum_i E_i = I$. Hence $\{E_i\}$ can be a so-called POVM as well.) On the set $X = \{1, 2, \dots, k\}$ we have probability distributions

$$\mu_\theta(\{i\}) = \mathrm{Tr}\, \rho(\theta) E_i.$$

Indeed,

$$\sum_{i=1}^{k} \mu_\theta(\{i\}) = \mathrm{Tr}\, \rho(\theta) \sum_{i=1}^{k} E_i = \mathrm{Tr}\, \rho(\theta) = 1.$$

Since

$$\mu_\theta(\{i\}) = \frac{\mathrm{Tr}\, \rho(\theta) E_i}{\mathrm{Tr}\, D E_i} \mathrm{Tr}\, D E_i$$

we can take

$$\mu(\{i\}) = \mathrm{Tr}\, D E_i$$

where D is a statistical operator. Then

$$f_\theta(\{i\}) = \frac{\mathrm{Tr}\, \rho(\theta) E_i}{\mathrm{Tr}\, D E_i} \tag{6}$$

and we have the classical Fisher information defined in (1):

$$\sum_i \frac{\mathrm{Tr}\, \rho(\theta) E_i}{\mathrm{Tr}\, D E_i} \left[\frac{\mathrm{Tr}\, \rho(\theta)' E_i}{\mathrm{Tr}\, D E_i} : \frac{\mathrm{Tr}\, \rho(\theta) E_i}{\mathrm{Tr}\, D E_i} \right]^2 \mathrm{Tr}\, D E_i = \sum_i \frac{[\mathrm{Tr}\, \rho(\theta)' E_i]^2}{\mathrm{Tr}\, \rho(\theta) E_i}$$

(This does not depend on D.) In the paper[4] the notation

$$F(\rho(\theta);\{E(\xi)\}) = \int \frac{[\mathrm{Tr}\,\rho(\theta)'E(\xi)]^2}{\mathrm{Tr}\,\rho(\theta)E(\xi)}\,d\xi$$

is used, this is an integral form, and for Braunstein and Caves the quantum Fisher information is the supremum of these classical Fisher informations.[4]

Theorem 1.1 *Assume that D is a positive definite density matrix, $B = B^*$ and $\mathrm{Tr}\,B = 0$. If $\rho(\theta) = D + \theta B + o(\theta^2)$, then the supremum of*

$$F(\rho(0);\{E_i\}) = \sum_i \frac{[\mathrm{Tr}\,BE_i]^2}{\mathrm{Tr}\,DE_i} \tag{7}$$

over the measurements $A = \sum_{i=1}^k \lambda_i E_i$ is

$$\mathrm{Tr}\,B\mathbb{J}_D^{-1}(B), \qquad \textit{where} \qquad \mathbb{J}_D C = (DC + CD)/2. \tag{8}$$

Proof: The linear mapping $\mathbb{J}_D$ is invertible, so we can replace B in (7) by $\mathbb{J}_D(C)$. We have to show

$$\sum_i \frac{[\mathrm{Tr}\,J_D(C)E_i]^2}{\mathrm{Tr}\,DE_i}$$
$$= \frac{1}{4}\sum_i \frac{(\mathrm{Tr}\,CDE_i)^2 + (\mathrm{Tr}\,DCE_i)^2 + 2(\mathrm{Tr}\,CDE_i)(\mathrm{Tr}\,DCE_i)}{\mathrm{Tr}\,DE_i} \le \mathrm{Tr}\,DC^2.$$

This follows from

$$\begin{aligned}(\mathrm{Tr}\,CDE_i)^2 &= \left(\mathrm{Tr}\,(E_i^{1/2}CD^{1/2})(D^{1/2}E_i^{1/2})\right)^2 \\ &\le \mathrm{Tr}\,E_iCDC\ \mathrm{Tr}\,D^{1/2}E_iD^{1/2} = \mathrm{Tr}\,E_iCDC\ \mathrm{Tr}\,DE_i.\end{aligned}$$

and

$$\begin{aligned}(\mathrm{Tr}\,DCE_i)^2 &= \left(\mathrm{Tr}\,(E_i^{1/2}D^{1/2})(D^{1/2}CE_i^{1/2})\right)^2 \\ &\le \mathrm{Tr}\,D^{1/2}E_iD^{1/2}\ \mathrm{Tr}\,D^{1/2}CE_iCD^{1/2} = \mathrm{Tr}\,E_iCDC\ \mathrm{Tr}\,DE_i.\end{aligned}$$

So $F(\rho(0);\{E_i\}) \le \mathrm{Tr}\,DC^2$ holds for any measurement $\{E_i\}$.

Next we want to analyze the condition for equality. Let $\mathbb{J}_D^{-1}B = C = \sum_k \lambda_k P_k$ be the spectral decomposition. In the Scwarz inequalities the condition of equality is

$$D^{1/2}E_i^{1/2} = c_i D^{1/2}CE_i^{1/2}$$

which is

$$E_i^{1/2} = c_i CE_i^{1/2}.$$

So $E_i^{1/2} \le P_{j(i)}$ for a spectral projection $P_{j(i)}$. This implies that all projections P_i are the sums of certain E_i's. (The simplest measurement for equality corresponds to the observable C.) □

Note that $\mathbb{J}_D^{-1}$ is in Example 1. It is an exercise to show that for

$$D = \begin{bmatrix} r & 0 \\ 0 & 1-r \end{bmatrix}, \qquad B = \begin{bmatrix} a & b \\ \bar{b} & -a \end{bmatrix}$$

the optimal observable is

$$C = \begin{bmatrix} \dfrac{a}{r} & 2b \\ 2\bar{b} & -\dfrac{a}{1-r} \end{bmatrix}.$$

The quantum Fisher information (8) is a particular case of the general approach of the previous session, $\mathbb{J}_D$ is in Example 1 below, this is the minimal quantum Fisher information which is also called **SLD Fisher information**. The inequality between (7) and (8) is a particular case of the monotonicity, see[40,42] and Theorem 1.2 below.

If $D = \mathrm{Diag}\,(\lambda_1, \dots, \lambda_n)$, then

$$F_{min}(D; B) := \mathrm{Tr}\, B\mathbb{J}_D^{-1}(B) = \sum_{ij} \frac{2}{\lambda_i + \lambda_j} |B_{ij}|^2.$$

In particularly,

$$F_{min}(D; \mathrm{i}[D, X]) = \sum_{ij} \frac{2(\lambda_i - \lambda_j)^2}{\lambda_i + \lambda_j} |X_{ij}|^2$$

and for commuting D and B we have

$$F_{min}(D; B) = \mathrm{Tr}\, D^{-1} B^2.$$

The minimal quantum Fisher information corresponds to the inner product

$$[A, B]_\rho = \tfrac{1}{2} \mathrm{Tr}\, \rho(AB + BA) = \mathrm{Tr}\, A\mathbb{J}_\rho(B).$$

Assume now that $\theta = (\theta^1, \theta^2)$. The formula (6) is still true. If

$$\partial_i \rho(\theta) = B_i,$$

then the classical Fisher information matrix $F(\rho(0); \{E_k\})_{ij}$ has the entries

$$F(\rho(0); \{E_k\})_{ij} = \sum_k \frac{\mathrm{Tr}\, B_i E_k \mathrm{Tr}\, B_j E_k}{\mathrm{Tr}\, \rho(0) E_k} \tag{9}$$

and the quantum Fisher information matrix is

$$\begin{bmatrix} \operatorname{Tr} B_1 \mathbb{J}_D^{-1}(B_1) & \operatorname{Tr} B_1 \mathbb{J}_D^{-1}(B_2) \\ \operatorname{Tr} B_2 \mathbb{J}_D^{-1}(B_1) & \operatorname{Tr} B_2 \mathbb{J}_D^{-1}(B_2) \end{bmatrix}. \tag{10}$$

Is there any inequality between the two matrices?

Let $\beta(A) = \sum_k E_k A E_k$. This is a completely positive trace preserving mapping. In the terminology of Theorem 2.3 the matrix (10) is J_1 and

$$J_2 = F(\rho(0); \{E_k\}).$$

The theorem states the inequality $J_2 \leq J_1$.

1.2. *The linear mapping $\mathbb{J}_D$*

Let $D \in \mathbf{M}_n$ be a positive invertible matrix. The linear mapping $\mathbb{J}_D^f : \mathbf{M}_n \to \mathbf{M}_n$ is defined by the formula

$$\mathbb{J}_D^f = f(\mathbb{L}_D \mathbb{R}_D^{-1})\mathbb{R}_D,$$

where $f : \mathbb{R}^+ \to \mathbb{R}^+$,

$$\mathbb{L}_D(X) = DX \qquad \text{and} \qquad \mathbb{R}_D(X) = XD.$$

(The operator $\mathbb{L}_D \mathbb{R}_D^{-1}$ appeared in the modular theory of von Neumann algebras.)

Lemma 1.1 *Assume that $f : \mathbb{R}^+ \to \mathbb{R}^+$ is continuous and $D = Diag(\lambda_1, \lambda_2, \ldots, \lambda_n)$. Then*

$$(\mathbb{J}_D^f B)_{ij} = \lambda_j f\left(\frac{\lambda_i}{\lambda_j}\right) B_{ij}$$

Moreover, if $f_1 \leq f_2$, then $0 \leq \mathbb{J}_D^{f_1} \leq \mathbb{J}_D^{f_2}$.

Proof: Let $f(x) = x^k$. Then

$$\mathbb{J}_D^f B = D^k B D^{1-k}$$

and

$$(\mathbb{J}_D^f B)_{ij} = \lambda_i^k \lambda_j^{1-k} B_{ij} = \lambda_j f\left(\frac{\lambda_i}{\lambda_j}\right) B_{ij}.$$

This is true for polynomials and for any continuous f by approximation. □

It follows from the lemma that

$$\langle A, \mathbb{J}_D^f B \rangle = \langle B^*, \mathbb{J}_D^f A^* \rangle \tag{11}$$

if and only if

$$\lambda_j f\left(\frac{\lambda_i}{\lambda_j}\right) = \lambda_i f\left(\frac{\lambda_j}{\lambda_i}\right),$$

which means $xf(x^{-1}) = f(x)$. Condition (11) is equivalent to the property that $\langle X, \mathbb{J}_D^f Y\rangle \in \mathbb{R}$ when X and Y are self-adjoint.

The functions $f : \mathbb{R}^+ \to \mathbb{R}^+$ used here are the **standard operator monotone functions** defined as

(i) if for positive matrices $A \le B$, then $f(A) \le f(B)$,
(ii) $xf(x^{-1}) = f(x)$ and $f(1) = 1$.

These functions are between the arithmetic and harmonic means:[27,44]

$$\frac{2x}{x+1} \le f(x) \le \frac{1+x}{2}.$$

Given f,

$$m_f(x, y) = yf\left(\frac{x}{y}\right)$$

is the corresponding mean and we have

$$(\mathbb{J}_D^f B)_{ij} = m_f(\lambda_i, \lambda_j) B_{ij}. \tag{12}$$

Hence

$$\mathbb{J}_D^f B = X \circ B$$

is a Hadamard product with $X_{ij} = m_f(\lambda_i, \lambda_j)$. Therefore the linear mapping $\mathbb{J}_D^f$ is positivity preserving if and only if the above X is positive.

The inverse of $\mathbb{J}_D^f$ is the mapping

$$\frac{1}{f}(\mathbb{L}_D \mathbb{R}_D^{-1})\mathbb{R}_D^{-1}$$

which acts as $B \mapsto Y \circ B$ with $Y_{ij} = 1/m_f(\lambda_i, \lambda_j)$. So $(\mathbb{J}_D^f)^{-1}$ is positivity preserving if and only if Y is positive.

A necessary condition for the positivity of $\mathbb{J}_D^f$ is $f(x) \le \sqrt{x}$, while the necessary condition for the positivity of $(\mathbb{J}_D^f)^{-1}$ is $f(x) \ge \sqrt{x}$. So only $f(x) = \sqrt{x}$ is the function which can make both mappings positivity preserving.

Example 1 If $f(x) = (x+1)/2$ (arithmetic mean), then

$$\mathbb{J}_D B = \frac{1}{2}(DB + BD) \quad \text{and} \quad \mathbb{J}_D^{-1} B = \int_0^\infty \exp(-tD/2) B \exp(-tD/2)\, dt.$$

This is from the solution of the equation $DB + BD = 2B$. □

Example 2 If $f(x) = 2x/(x+1)$ (harmonic mean), then

$$\mathbb{J}_D B = \int_0^\infty \exp(-tD^{-1}/2) B \exp(-tD^{-1}/2)\, dt$$

and

$$\mathbb{J}_D^{-1} B = \frac{1}{2}(D^{-1}B + BD^{-1}).$$

This function f is the minimal and it generates the maximal Fisher information which is also called **right information matrix.** □

Example 3 For the logarithmic mean

$$f(x) = \frac{x-1}{\log x} \tag{13}$$

we have

$$\mathbb{J}_D(B) = \int_0^1 D^t B D^{1-t}\, dt \quad \text{and} \quad \mathbb{J}_D^{-1}(B) = \int_0^\infty (D+t)^{-1} B (D+t)^{-1}\, dt$$

This function induces an importan Fisher information. □

Example 4 For the geometric mean $f(x) = \sqrt{x}$ and

$$\mathbb{J}_D(B) = D^{1/2} B D^{1/2} \quad \text{and} \quad \mathbb{J}_D^{-1}(B) = D^{-1/2} B D^{-1/2}.$$

□

$\mathbb{J}_D^f$ is the largest if f is the largest which is described in Example 1 and the smallest is in Example 2.

Theorem 1.2 *Let $\beta : \mathbf{M}_n \to \mathbf{M}_m$ be a completely positive trace preserving mapping and $f : \mathbb{R}^+ \to \mathbb{R}^+$ be a matrix monotone function. Then*

$$\beta^* (\mathbb{J}_{\beta(D)}^f)^{-1} \beta \le (\mathbb{J}_D^f)^{-1} \tag{14}$$

and

$$\beta \mathbb{J}_D^f \beta^* \le \mathbb{J}_{\beta(D)}^f. \tag{15}$$

Actually (14) and (15) are equivalent and they are equivalent to the matrix monotonicity of f.[43]

In the rest f is always assumed to be a standard matrix monotone function. Then $\operatorname{Tr} \mathbb{J}_D B = \operatorname{Tr} DB$.

Example 5 Here we want to study $\mathbb{J}_D^f$, when D can have 0 eigenvalues. Formula (12) makes sense. For example, if $D = \text{Diag}\,(0, \lambda, \lambda, \mu)$ $(\lambda, \mu > 0, \lambda \neq \mu)$, then

$$\mathbb{J}_D^f B = \begin{bmatrix} 0 & m(0,\lambda)B_{12} & m(0,\mu)B_{13} & m(0,\mu)B_{14} \\ m(0,\lambda)B_{21} & \lambda B_{22} & m(\lambda,\mu)B_{23} & m(\lambda,\mu)B_{24} \\ m(0,\mu)B_{31} & m(\lambda,\mu)B_{32} & \mu B_{34} & \mu B_{34} \\ m(0,\mu)B_{41} & m(\lambda,\mu)B_{42} & \mu B_{43} & \mu B_{43} \end{bmatrix}.$$

If $f(0) > 0$, then this matrix has only one 0 entry. If $f(0) = 0$, then

$$\mathbb{J}_D^f B = \begin{bmatrix} 0 & 0 & 0 & 0 \\ 0 & \lambda B_{22} & m(\lambda,\mu)B_{23} & m(\lambda,\mu)B_{24} \\ 0 & m(\lambda,\mu)B_{32} & \mu B_{34} & \mu B_{34} \\ 0 & m(\lambda,\mu)B_{42} & \mu B_{43} & \mu B_{43} \end{bmatrix}.$$

and the kernel of $\mathbb{J}_D$ is larger. We have

$$\langle B, \mathbb{J}_D^f B\rangle = \sum_{ij} m_f(\lambda_i, \lambda_j)|B_{ij}|^2$$

and some terms can be 0 if D is not invertible.

The inverse of $\mathbb{J}_D^f$ exists in the generalized sense

$$[(\mathbb{J}_D^f)^{-1}B]_{ij} = \begin{cases} \dfrac{1}{m_f(\lambda_i,\lambda_j)} B_{ij} & \text{if} \quad m_f(\lambda_i,\lambda_j) \neq 0, \\ 0 & \text{if} \quad m_f(\lambda_i,\lambda_j) = 0. \end{cases}$$

(This is the Moore-Penrose generalized inverse.) □

It would be interesting to compare the functions which non-zero at 0 with the others.

2. Fisher information and covariance

Assume that f is a standard matrix monotone function. The operators $\mathbb{J}_D^f$ are used to define Fisher information and the covariance. (The latter can be called also quadratic cost.) The operator $\mathbb{J}_D^f$ depends on the function f, but f will be not written sometimes.

Let $A = A^*, B = B^* \in \mathbf{M}_n$ be observables and $D \in \mathbf{M}_n$ be a density matrix. The covariance of A and B is

$$\text{Cov}_D^f(A,B) := \langle A, \mathbb{J}_D^f(B)\rangle - (\text{Tr}\,DA)(\text{Tr}\,DB). \tag{16}$$

Since

$$\text{Cov}_D^f(A,A) = \langle (A - I\text{Tr}\,DA), \mathbb{J}_D(A - I\text{Tr}\,DA\rangle$$

and $\mathbb{J}_D \geq 0$, we have for the variance $\text{Var}_D^f(A) := \text{Cov}_D^f(A, A) \geq 0$.

The monotonicity (15) gives

$$\text{Var}_D^f(\beta^* A) \leq \text{Var}_{\beta D}^f(A).$$

for a completely positive trace preserving mapping β.

The usual **symmetrized covariance** corresponds to the function $f(t) = (t+1)/2$:

$$\text{Cov}_D(A, B) := \frac{1}{2}\text{Tr}\,(D(A^*B + BA^*)) - (\text{Tr}\,DA^*)(\text{Tr}\,DB).$$

Let $A_1, A_2, \ldots, A_k$ be self-adjoint matrices and let D be a statistical operator. The covariance is a $k \times k$ matrix $C(D)$ defined as

$$C(D)_{ij} = \text{Cov}_D^f(A_i, A_j). \tag{17}$$

$C(D)$ is a positive semidefinite matrix and positive definite if the observables $A_1, A_2, \ldots, A_k$ are linearly independent. It should be remarked that this matrix is only a formal analogue of the classical covariance matrix and it is not related to a single quantum measurement.[33]

The variance is defined by $\mathbb{J}_D$ and the Fisher information is formulated by the inverse of this mapping:

$$\gamma_D(A, B) = \text{Tr}\,A\mathbb{J}_D^{-1}(B^*). \tag{18}$$

Here A and B are self-adjoint. If A and B are considered as tangent vectors at the footpoint D, then $\text{Tr}\,A = \text{Tr}\,B = 0$. In this approach $\gamma_D(A, B)$ is a an inner product in a Riemannian geometry.[2,21] It seems that this approach is not popular in quantum theory. It happens also that the condition $\text{Tr}\,D = 1$ is neglected and only $D > 0$. Then formula (18) can be extended.[26]

If $DA = AD$ for a self-adjoint matrix A, then

$$\gamma_D(A, A) = \text{Tr}\,D^{-1}A^2$$

does not depend on the function f. (The dependence is characteristic on the orthogonal complement, this will come later.)

Theorem 2.1 *Assume that $(A, B) \mapsto \gamma_D(A, B)$ is an inner product for $A, B \in \mathbf{M}_n$, for positive definite density matrix $D \in \mathbf{M}_n$ and for every n. Suppose the following properties:*

(i) For commuting D and $A = A^*$ we have $\gamma_D(A, A) = \text{Tr}\,D^{-1}A^2$.

(ii) If $\beta : \mathbf{M}_n \to \mathbf{M}_m$ is a completely positive trace preserving mapping, then

$$\gamma_{\beta(D)}(\beta(A), \beta(A)) \leq \gamma_D(A, A). \tag{19}$$

(iii) If $A = A^*$ and $B = B^*$, then $\gamma_D(A, B)$ is a real number.
(iv) $D \mapsto \gamma_D(A, B)$ is continuous.

Then

$$\gamma_D(A, B) = \langle A, (\mathbb{J}_D^f)^{-1} B \rangle \tag{20}$$

for a standard matrix monoton function f.

Example 6 In quantum statistical mechanics, perturbation of a density matrix appears. Suppose that $D = e^H$ and $A = A^*$ is the perturbation

$$D_t = \frac{e^{H+tA}}{\operatorname{Tr} e^{H+tA}} \qquad (t \in \mathbb{R}).$$

The quantum analog of formula (1) would be

$$-\operatorname{Tr} D_0 \frac{\partial^2}{\partial t^2} \log D_t \Big|_{t=0}.$$

A simple computation gives

$$\int_0^1 \operatorname{Tr} e^{sH} A e^{(1-s)H} A \, ds - (\operatorname{Tr} DA)^2$$

This is a kind of variance. □

Let $\mathcal{M} := \{D_\theta : \theta \in G\}$ be a smooth m-dimensional **manifold** of $n \times n$ density matrices. Formally $G \subset \mathbb{R}^m$ is an open set including 0. If $\theta \in G$, then $\theta = (\theta_1, \theta_2, \dots, \theta_m)$. The **Riemannian structure** on $\mathcal{M}$ is given by the inner product (18) of the tangent vectors A and B at the foot point $D \in \mathcal{M}$, where $\mathbb{J}_D : \mathbf{M}_n \to \mathbf{M}_n$ is a positive mapping when $\mathbf{M}_n$ is regarded as a Hilbert space with the Hilbert-Schmidt inner product. (This means $\operatorname{Tr} A\mathbb{J}_D(A)^* \geq 0$.)

Assume that a collection $A = (A_1, \dots, A_m)$ of self-adjoint matrices is used to estimate the true value of θ. The expectation value of A_i with respect to the density matrix D is $\operatorname{Tr} DA_i$. A is an **unbiased estimator** if

$$\operatorname{Tr} D_\theta A_i = \theta_i \qquad (1 \leq i \leq n). \tag{21}$$

(In many cases unbiased estimator $A = (A_1, \dots, A_m)$ does not exist, therefore a weaker condition is more useful.)

The **Fisher information** matrix of the estimator A is a positive definite matrix

$$J(D)_{ij} = \operatorname{Tr} L_i \mathbb{J}_D(L_j), \quad \text{where} \quad L_i = \mathbb{J}_D^{-1}(\partial_i D_\theta).$$

Both $C(D)$ and $J(D)$ depend on the actual state D.

The next theorem is the the **Cramér-Rao inequality** for matrices. The point is that the right-hand-side does not depend on the estimators.

Let $A = (A_1, \dots, A_m)$ be an unbiased estimator of θ. Then for the above defined matrices the inequality

$$C(D_\theta) \geq J(D_\theta)^{-1}$$

holds.

Proof: In the proof the block-matrix method is used and we restrict ourselves for $m = 2$ for the sake of simplicity and assume that $\theta = 0$. Instead of D_0 we write D.

The matrices A_1, A_2, L_1, L_2 are considered as vectors and from the inner product $\langle A, B\rangle = \operatorname{Tr} A\mathbb{J}_D(B)^*$ we have the positive matrix

$$X := \begin{bmatrix} \operatorname{Tr} A_1\mathbb{J}_D(A_1) & \operatorname{Tr} A_1\mathbb{J}_D(A_2) & \operatorname{Tr} A_1\mathbb{J}_D(L_1) & \operatorname{Tr} A_1\mathbb{J}_D(L_2) \\ \operatorname{Tr} A_2\mathbb{J}_D(A_1) & \operatorname{Tr} A_2\mathbb{J}_D(A_2) & \operatorname{Tr} A_2\mathbb{J}_D(L_1) & \operatorname{Tr} A_2\mathbb{J}_D(L_2) \\ \operatorname{Tr} L_1\mathbb{J}_D(A_1) & \operatorname{Tr} L_1\mathbb{J}_D(A_2) & \operatorname{Tr} L_1\mathbb{J}_D(L_1) & \operatorname{Tr} L_1\mathbb{J}_D(L_2) \\ \operatorname{Tr} L_2\mathbb{J}_D(A_1) & \operatorname{Tr} L_2\mathbb{J}_D(A_2) & \operatorname{Tr} L_2\mathbb{J}_D(L_1) & \operatorname{Tr} L_2\mathbb{J}_D(L_2) \end{bmatrix}.$$

From the condition (21), we have

$$\operatorname{Tr} A_i\mathbb{J}_D(L_i) = \frac{\partial}{\partial\theta_i}\operatorname{Tr} D_\theta A_i = 1$$

for $i = 1, 2$ and

$$\operatorname{Tr} A_i\mathbb{J}_D(L_j) = \frac{\partial}{\partial\theta_j}\operatorname{Tr} D_\theta A_i = 0$$

if $i \neq j$. Hence the matrix X has the form

$$\begin{bmatrix} C(D) & I_2 \\ I_2 & J(D) \end{bmatrix}, \tag{22}$$

where

$$C(D) = \begin{bmatrix} \operatorname{Tr} A_1\mathbb{J}_D(A_1) & \operatorname{Tr} A_1\mathbb{J}_D(A_2) \\ \operatorname{Tr} A_2\mathbb{J}_D(A_1) & \operatorname{Tr} A_2\mathbb{J}_D(A_2) \end{bmatrix}$$

and

$$J(D) = \begin{bmatrix} \operatorname{Tr} L_1\mathbb{J}_D(L_1) & \operatorname{Tr} L_1\mathbb{J}_D(L_2) \\ \operatorname{Tr} L_2\mathbb{J}_D(L_1) & \operatorname{Tr} L_2\mathbb{J}_D(L_2) \end{bmatrix}.$$

The positivity of (22) implies the statement of the theorem. □

We have have the orthogonal decomposition

$$\{B = B^* : [D, B] = 0\} \oplus \{\mathrm{i}[D, A] : A = A^*\} \tag{23}$$

of the self-adjoint matrices and we denote the two subspaces by $\mathcal{M}_D$ and $\mathcal{M}_D^c$, respectively.

Example 7 The Fisher information and the covariance are easily handled if D is diagonal, $D = \mathrm{Diag}\,(\lambda_1, \dots, \lambda_n)$ or formulated by the matrix units $E(ij)$

$$D = \sum_i \lambda_i E(ii).$$

The general formulas in case of diagonal D are

$$\gamma_D(A,A) = \sum_{ij} \frac{1}{\lambda_j f(\lambda_i/\lambda_j)} |A_{ij}|^2, \quad \mathrm{Cov}_D(A,A) = \sum_{ij} \lambda_j f(\lambda_i/\lambda_j)|A_{ij}|^2.$$

Moreover,

$$\gamma_D^f(\mathrm{i}[D,X], \mathrm{i}[D,X]) = \sum_{ij} \frac{(\lambda_i - \lambda_j)^2}{\lambda_j f(\lambda_i/\lambda_j)} |X_{ij}|^2. \tag{24}$$

Hence for diagonal D all Fisher informations have simple explicit formula.

The description of the commutators is more convenient if the eigenvalues are different. Let

$$S_1(ij) := E(ij) + E(ji), \qquad S_2(ij) := -\mathrm{i}E(ij) + \mathrm{i}E(ji)$$

for $i < j$. (They are the generalization of the Pauli matrices σ_1 and σ_2.) We have

$$\mathrm{i}[D, S_1(ij)] = (\lambda_i - \lambda_j) S_2(ij), \qquad \mathrm{i}[D, S_2(ij)] = (\lambda_j - \lambda_i) S_1(ij).$$

In Example 1 we have $f(x) = (1+x)/2$. This gives the minimal Fisher information described in Theorem 1.1:

$$\gamma_D(A,B) = \int_0^\infty \mathrm{Tr}\, A \exp(-tD/2) B \exp(-tD/2)\, dt.$$

The corresponding covariance is the symmetrized $\mathrm{Cov}_D(A,B)$. This is maximal among the variances.

From Example 2 we have the maximal Fisher information

$$\gamma_D(A,B) = \frac{1}{2} \mathrm{Tr}\, D^{-1}(AB + BA)$$

The corresponding covariance is a bit similar to the minimal Fisher information:

$$\mathrm{Cov}_D(A,B) = \int_0^\infty \mathrm{Tr}\, A \exp(-tD^{-1}/2) B \exp(-tD^{-1}/2)\, dt - \mathrm{Tr}\, DA\, \mathrm{Tr}\, DB.$$

Example 3 leads to the Boguliubov-Kubo-Mori inner product as Fisher information:[41,42]

$$\gamma_D(A,B) = \int_0^\infty \mathrm{Tr}\, A(D+t)^{-1} B (D+t)^{-1}\, dt$$

It is also called BKM Fisher information, the characterization is in the paper[14] and it is also proven that this gives a large deviation bound of consistent superefficient estimators.[17] □

Let $\mathcal{M} := \{\rho(\theta) : \theta \in G\}$ be a smooth k-dimensional manifold of invertible density matrices. The *quantum score operators* (or logarithmic derivatives) are defined as

$$L_i^f(\theta) := (\mathbb{J}^f_{\rho(\theta)})^{-1}(\partial_{\theta_i}\rho(\theta)) \qquad (1 \le i \le m) \tag{25}$$

and

$$J(\theta)_{ij} := \operatorname{Tr} L_i^f(\theta)\mathbb{J}^f_{\rho(\theta)}(L_j(\theta)) = \operatorname{Tr}(\mathbb{J}^f_{\rho(\theta)})^{-1}(\partial_{\theta_i}\rho(\theta))(\partial_{\theta_j}\rho(\theta)) \qquad (1 \le i, j \le k) \tag{26}$$

is the **quantum Fisher information matrix** (depending on the function f). The function $f(x) = (x+1)/2$ yields the symmetric logarithmic derivative (SLD) Fisher information.

Theorem 2.3 *Let $\beta : \mathbf{M}_n \to \mathbf{M}_m$ be a completely positive trace preserving mapping and let $\mathcal{M} := \{\rho(\theta) \in \mathbf{M}_n : \theta \in G\}$ be a smooth k-dimensional manifold of invertible density matrices. For the Fisher information matrix $J_1(\theta)$ of $\mathcal{M}$ and for Fisher information matrix $J_2(\theta)$ of $\beta(\mathcal{M}) := \{\beta(\rho(\theta)) : \theta \in G\}$ we have the monotonicity relation*

$$J_2(\theta) \le J_1(\theta).$$

Proof: We set $B_i(\theta) := \partial_{\theta_i}\rho(\theta)$. Then $\mathbb{J}^{-1}_{\beta(\rho(\theta))}\beta(B_i(\theta))$ is the score operator of $\beta(\mathcal{M})$ and we have

$$\begin{aligned}
\sum_{ij} J_2(\theta)_{ij} a_i \overline{a_j} &= \operatorname{Tr} \mathbb{J}^{-1}_{\beta(\rho(\theta))}\beta\Big(\sum_i a_i B_i(\theta)\Big)\beta\Big(\sum_j \overline{a_j} B_j(\theta)\Big) \\
&= \left\langle \sum_i a_i B_i, (\beta^* \mathbb{J}^{-1}_{\beta)(\rho(\theta))}\beta \sum_j a_j B_j(\theta) \right\rangle \\
&\le \left\langle \sum_i a_i B_i, \mathbb{J}^{-1}_{\rho(\theta)} \sum_j a_j B_j(\theta) \right\rangle \\
&= \operatorname{Tr} \mathbb{J}^{-1}_{\rho(\theta))}\Big(\sum_i a_i B_i(\theta)\Big)\Big(\sum_j \overline{a_j} B_j(\theta)\Big) \\
&= \sum_{ij} J_1(\theta)_{ij} a_i \overline{a_j},
\end{aligned}$$

where (14) was used. □

The monotonicity of the Fisher information matrix in some particular cases appeared already in the literature:[38] treated the case of the Kubo-Mori inner product and[4] considered the symmetric logarithmic derivative and measurement in the role of coarse graining.

Example 8 The function

$$f_\beta(t) = \beta(1-\beta)\frac{(x-1)^2}{(x^\beta - 1)(x^{1-\beta} - 1)} \tag{27}$$

is operator monotone if $0 < \beta < 2$. Formally $f(1)$ is not defined, but as a limit it is 1. The property $xf(x^{-1}) = f(x)$ also holds. Therefore this function determines a Fisher information.[39] If $\beta = 1/2$, then the variance has a simple formula:

$$\mathrm{Var}_D A = \frac{1}{2}\mathrm{Tr}\, D^{1/2}(D^{1/2}A + AD^{1/2})A - (\mathrm{Tr}\, DA)^2.$$

□

Example 9 The functions $x^{-\alpha}$ and $x^{\alpha-1}$ are matrix monotone decreasing and so is their sum. Therefore

$$f_\alpha(x) = \frac{2}{x^{-\alpha} + x^{\alpha-1}}$$

is a standard operator monotone function.

$$\gamma_\sigma^f(\rho,\rho) = 1 + \mathrm{Tr}\ \big((\rho-\sigma)\sigma^{-\alpha}(\rho-\sigma)\sigma^{\alpha-1}\big)$$

may remind us to the abstract Fisher information, however now ρ and σ are positive definite density matrices. In the paper[46]

$$\chi_\alpha^2(\rho,\sigma) = \mathrm{Tr}\ \big((\rho-\sigma)\sigma^{-\alpha}(\rho-\sigma)\sigma^{\alpha-1}\big)$$

is called **quantum χ^2-divergence**. (If ρ and σ commute, then the formula is independent of α.) Up to the constant 1, this is an interesting and important particular case of the monotone metric. The general theory (19) implies the monotonicity of the χ^2-divergence. □

3. Extended monotone metrics

As an extension of the papers[5,40] Kuamagai made the following generalization.[26] Now H_n^+ denotes the strictly positive matrices in $\mathbf{M}_n$. Formally $K_\rho(A,B) \in \mathbb{C}$ is defined for all $\rho \in H_n^+$, $A, B \in \mathbf{M}_n$ and $n \in \mathbb{N}$ and it is assumed that

(i) $(A,B) \mapsto K_\rho(A,B)$ is an inner product on $\mathbf{M}_n$ for every $\rho \in H_n^+$ and $n \in \mathbb{N}$.
(ii) $\rho \mapsto K_\rho(A,B)$ is continuous.
(iii) For a trace-preserving completely positive mapping β

$$K_{\beta(\rho)}(\beta(A),\beta(A)) \le K_\rho(A,A)$$

holds.

In the paper[26] such $K_\rho(A,B)$ is called **extended monotone metric** and the description is

$$K_\rho(A,B) = b(\mathrm{Tr}\,\rho)\mathrm{Tr}\,A^*\mathrm{Tr}\,B + c\langle A, (\mathbb{J}_\rho^f)^{-1}(B)\rangle,$$

where $f : \mathbb{R}^+ \to \mathbb{R}^+$ is matrix monotone, $f(1) = 1$, $b : \mathbb{R}^+ \to \mathbb{R}^+$ and $c > 0$. Note that

$$(A,B) \mapsto b(\mathrm{Tr}\,\rho)\mathrm{Tr}\,A^*\mathrm{Tr}\,B \quad \text{and} \quad (A,B) \mapsto c\langle A, (\mathbb{J}_\rho^f)^{-1}B\rangle$$

satisfy conditions (ii) and (iii) with constant $c > 0$. The essential point is to check

$$b(\mathrm{Tr}\,\rho)\mathrm{Tr}\,A^*\,\mathrm{Tr}\,A + c\langle A, (\mathbb{J}_\rho^f)^{-1}A\rangle \ge 0.$$

In the case of 1×1 matrices this is

$$b(x)|z|^2 + \frac{c}{x}|z|^2 \ge 0$$

which gives the condition $xb(x) + c > 0$. If this is true, then

$$\begin{aligned}
&\Big(\sum_i \lambda_i\Big) b\Big(\sum_i \lambda_i\Big) |\sum_i A_{ii}|^2 + c\Big(\sum_i \lambda_i\Big)\sum_{ij}\frac{1}{m_f(\lambda_i,\lambda_j)}|A_{ij}|^2 \\
&\ge -c\Big|\sum_i A_{ii}\Big|^2 + c\Big(\sum_i \lambda_i\Big)\sum_{ij}\frac{1}{m_f(\lambda_i,\lambda_j)}|A_{ij}|^2 \\
&\ge -c\Big|\sum_i A_{ii}\Big|^2 + c\Big(\sum_i \lambda_i\Big)\sum_i \frac{1}{\lambda_i}|A_{ii}|^2.
\end{aligned}$$

The positivity is the inequality

$$\Big(\sum_i \lambda_i\Big)\sum_i \frac{1}{\lambda_i}|A_{ii}|^2 \ge \Big|\sum_i A_{ii}\Big|^2$$

which is a consequence of the Schwarz inequality.

4. Skew information

The Wigner-Yanase-Dyson skew information is the quantity

$$I_p(D,A) := -\frac{1}{2}\mathrm{Tr}\,[D^p, A][D^{1-p}, A] \qquad (0 < p < 1).$$

Actually, the case $p = 1/2$ is due to Wigner and Yanase[47] and the extension was proposed by Dyson. The convexity of $I_p(D,A)$ in A is a famous result of Lieb[30]

It was observed in[39] that the Wigner-Yanase-Dyson skew information is connected to the Fisher information which corresponds to the function (28). For this function we have

$$\gamma_D(\mathrm{i}[D,A], \mathrm{i}[D,A]) = \frac{1}{2\beta(1-\beta)}\mathrm{Tr}\left([\rho^\beta, A][\rho^{1-\beta}, A]\right). \tag{28}$$

Apart from a constant factor this expression is the skew information proposed by Wigner and Yanase.[47] In the limiting cases $p \to 0$ or 1 we have the function (13) corresponding to the Kubo-Mori-Boguliubov case.

Let f be a standard function and $A = A^* \in \mathbf{M}_n$. The quantity

$$I_D^f(A) := \frac{f(0)}{2}\gamma_D^f(\mathrm{i}[D,A], \mathrm{i}[D,A])$$

was called **skew information** in[16] in this general setting. So the skew information is nothing else but the Fisher information restricted to $\mathcal{M}_D^c$, but it is parametrized by the commutator. Skew information appeared twenty years before the concept of quantum Fisher information. Skew information appears in a rather big literature, for example, connection with uncertainty relations.[3,9,10,13,25,31,32]

If $D = \mathrm{Diag}\,(\lambda_1, \ldots, \lambda_n)$ is diagonal, then

$$\gamma_D^f(\mathrm{i}[D,A], \mathrm{i}[D,A]) = \sum_{ij} \frac{(\lambda_i - \lambda_j)^2}{\lambda_j f(\lambda_i/\lambda_j)}|A_{ij}|^2.$$

This implies that the identity

$$I_D^f(A) = \mathrm{Cov}_D(A,A) - \mathrm{Cov}_D^{\tilde{f}}(A,A) \tag{29}$$

holds if $\mathrm{Tr}\,DA = 0$ and

$$\tilde{f}(x) := \frac{1}{2}\left((x+1) - (x-1)^2\frac{f(0)}{f(x)}\right). \tag{30}$$

It was proved in[8] that for a standard function $f : \mathbb{R}^+ \to \mathbb{R}$, $\tilde{f}$ is standard as well. Another proof is in[45] which contains the following theorem.

Theorem 4.1 *Assume that* $X = X^* \in \mathcal{M}$ *and* $\mathrm{Tr}\, DX = 0$. *If* f *is a standard function such that* $f(0) \neq 0$, *then*

$$\frac{\partial^2}{\partial t \partial s} S_F(D + t\mathrm{i}[D,X], D + s\mathrm{i}[D,X])\Big|_{t=s=0} = f(0)\gamma_D^f(\mathrm{i}[D,X], \mathrm{i}[D,X])$$

for the standard function $F = \tilde{f}$.

All skew informations are obtained from an f-divergence (or quasi-entropy) by differentiation.

Example 10 The function

$$f(x) = \left(\frac{1+\sqrt{x}}{2}\right)^2 \tag{31}$$

gives the Wigner-Yanase skew information

$$I^{WY}(D,A) = I_{1/2}(D,A) = -\frac{1}{2}\mathrm{Tr}\,[D^{1/2}, A]^2.$$

The skew information coming from the minimal Fisher information and it is often denoted as $I^{SLD}(D,A)$. The simple mean inequalities

$$\left(\frac{1+\sqrt{x}}{2}\right)^2 \leq \frac{1+x}{2} \leq 2\left(\frac{1+\sqrt{x}}{2}\right)^2$$

imply

$$I^{WY}(D,A) \leq I^{SLD}(D,A) \leq 2I^{WY}(D,A).$$

□

Acknowledgement

This work is supported by the Hungarian Research Grant OTKA 68258.

References

1. S. Amari, *Differential-geometrical methods in statistics*, Lecture Notes in Stat. **28** (Springer, Berlin, Heidelberg, New York, 1985)
2. S. Amari and H. Nagaoka, *Methods of information geometry*, Transl. Math. Monographs **191**, AMS, 2000.
3. A. Andai, Uncertainty principle with quantum Fisher information, J. Math. Phys. **49**(2008), 012106.
4. S. L. Braunstein and C. M. Caves, Statistical distance and the geometry of quantum states, Phys. Rev. Lett. **72**(1994), 3439–3443.
5. L.L. Campbell, An extended Centcov characterization of the information metric, Proc. Amer. Math. Soc. **98** (1986), 135–141.

6. J. Dittmann, On the Riemannian geometry of finite dimensional state space, Seminar Sophus Lie **3**(1993), 73–87
7. E. Fick and G. Sauermann, *The quantum statistics of dynamic processes* (Springer, Berlin, Heidelberg) 1990.
8. P. Gibilisco, D. Imparato and T. Isola, Uncertainty principle and quantum Fisher information II, J. Math. Phys. **48**(2007), 072109.
9. P. Gibilisco, D. Imparato and T. Isola, A volume inequality for quantum Fisher information and the uncertainty principle, J. Statist. **130**(2007), 545–559.
10. P. Gibilisco and T. Isola, Uncertainty principle and quantum Fisher information, Ann. Inst. Stat. Math, **59** (2007), 147–159.
11. P. Gibilisco, F. Hiai and D. Petz, Quantum covariance, quantum Fisher information and the uncertainty principle, IEEE Trans. Inform. Theory **55**(2009), 439–443.
12. P. Gibilisco, D. Imparato and T. Isola, Inequalities for quantum Fisher information, Proc. Amer. Math. Soc. **137**(2009), 317–327.
13. P. Gibilisco, D. Imparato and T. Isola, A Robertson-type uncertainty principle and quantum Fisher information, Lin. Alg. Appl. **428**(2008), 1706–1724.
14. M. Grasselli and R.F. Streater, Uniqueness of the Chentsov metric in quantum information theory, Infin. Dimens. Anal. Quantum Probab. Relat. Top., **4** (2001), 173-182.
15. F. Hansen, Characterizations of symmetric monotone metrics on the the state space of quantum systems, Quantum Inf. Comput., **6**(2006), 597–605.
16. F. Hansen, Metric adjusted skew information, Proc. Natl. Acad. Sci. USA. **105**(2008), 9909–9916.
17. M. Hayashi, Two quantum analogues of Fisher information from a large deviation viewpoint of quantum estimation, J. of Physics A: Mathematical and General, **35**(2002), 7689-7727.
18. M. Hayashi and K. Matsumoto, Asymptotic performance of optimal state estimation in quantum two level system, J. Math. Phys. **49**(2008), 102101.
19. M. Hayashi, *Quantum information. An introduction*, Springer-Verlag, Berlin, 2006.
20. C. W. Helstrom, *Quantum detection and estimation theory*, Academic Press, New York, 1976.
21. F. Hiai and D. Petz, Riemannian geometry on positive definite matrices related to means, Lin. Alg. Appl. **430**(2009), 3105–3130.
22. A. S. Holevo, *Probabilistic and statistical aspects of quantum theory*, North-Holland, Amsterdam, 1982.
23. A. Jencová, Geodesic distances on density matrices, J. Math. Phys. **45** (2004), 1787–1794.
24. O. Johnson, *Information theory and the central limit theorem*, Imperial College Press, 2004.
25. H. Kosaki, Matrix trace inequality related to uncertainty principle, Internat. J. Math. **16**(2005), 629–645.
26. W. Kumagai, A characterization of extended monotone metrics, to be published in Lin. Alg. Appl.

27. F. Kubo and T. Ando, Means of positive linear operators, Math. Ann. **246**(1980), 205–224.
28. S. Kullback and R.A. Leibler, On information and sufficiency, Ann. Math. Statistics, **22**(1951), 79–86.
29. S. Kullback, *Information theory and statistics*, John Wiley and Sons, New York; Chapman and Hall, Ltd., London, 1959.
30. E. H. Lieb, Convex trace functions and the Wigner-Yanase-Dyson conjecture, Advances in Math. **11**(1973), 267–288.
31. S. Luo and Z. Zhang, An informational characterization of Schrödinger's uncertainty relations, J. Stat. Phys. **114**(2004), 1557–1576.
32. S. Luo and Q. Zhang, On skew information, IEEE Trans. Inform. Theory, **50**(2004), 1778–1782.
33. S. Luo, Covariance and quantum Fisher information, Theory Probab. Appl. **53**(2009), 329–334.
34. H. Nagaoka, On Fisher information on quantum statistical models, in *Asymptotic Theory of Quantum Statistical Inference*, 113–124, ed. M. Hayashi, World Scientific, 2005.
35. M. Ohya and D. Petz, *Quantum entropy and its use*, Springer-Verlag, Heidelberg, 1993. Second edition 2004.
36. D. Petz, Quasi-entropies for states of a von Neumann algebra, Publ. RIMS. Kyoto Univ. **21**(1985), 781–800.
37. D. Petz, Quasi-entropies for finite quantum systems, Rep. Math. Phys., **23**(1986), 57-65.
38. D. Petz, Geometry of canonical correlation on the state space of a quantum System, J. Math. Phys. **35**(1994), 780–795.
39. D. Petz and H. Hasegawa, On the Riemannian metric of α-entropies of density matrices, Lett. Math. Phys. **38**(1996), 221–225
40. D. Petz, Monotone metrics on matrix spaces, Linear Algebra Appl. **244**(1996), 81–96.
41. D. Petz and Cs. Sudár, Geometries of quantum states, J. Math. Phys. **37**(1996), 2662–2673.
42. D. Petz and Cs. Sudár, Extending the Fisher metric to density matrices, in *Geometry of Present Days Science*, eds. O.E. Barndorff-Nielsen and E.B. Vendel Jensen, 21–34, World Scientific, 1999.
43. D. Petz, Covariance and Fisher information in quantum mechanics, J. Phys. A: Math. Gen. **35**(2003), 79–91.
44. D. Petz, *Quantum information theory and quantum statistics*, Springer, Berlin, Heidelberg, 2008.
45. D. Petz and V.E.S. Szabó, From quasi-entropy to skew information, Int. J. Math. **20**(2009), 1421–1430.
46. K. Temme, M. J. Kastoryano, M. B. Ruskai, M. M. Wolf and F. Verstraete, The χ^2-divergence and mixing times of quantum Markov processes, arXiv:1005.2358.
47. E.P. Wigner and M.M. Yanase, Information content of distributions, Proc. Nat. Acad. Sci. USA **49**(1963), 910–918.

COMPLEMENTARY SUBALGEBRAS IN FINITE QUANTUM SYSTEMS

D. PETZ and A. SZÁNTÓ

Department for Mathematical Analysis,
Budapest University of Technology and Economics
H-1521 Budapest XI., Hungary

While the notion of complementarity in quantum systems goes back to the beginning of quantum theory, the concept of complementary subalgebras is quite new. Here we overview the recent developments of the field, characterizations of complementarity and complementary decompositions are reviewed, and we point to some still open questions.

Keywords: complementarity, mutually unbiased bases, subalgebra, subsystem, quantum information, qubit

1. Complementary subalgebras

Complementarity of measurements and mutually unbiased bases (MUBs) are an interesting topic with a vast literature and several applications.[7,14] The term was first introduced by Schwinger,[17] and the first applications are due to Ivonovic.[4] Two von Neumann measurements are complementary, if the eigenbases $\{e_i\}_i$ and $\{f_j\}_j$ of the corresponding observables are mutually unbiased, that is

$$|\langle e_i|f_j\rangle|^2 = \frac{1}{n}$$

The term 'unbiased' comes from the fact that any measurement prepared in one of the bases is completely unrelated to any measurement made in another. Also, if a system is prepared in a state from one basis, then a measurement in another basis is unsubstantial, as the results are uniformly distributed. The related measurements are said to be 'complementary', because the information provided by them are non-overlapping.

The operators diagonal in a given base form a maximal Abelian subalgebra (MASA) of the matrix algebra $M_n(\mathbb{C})$. It is easy to see, that two bases are mutually unbiased if and only if the associated MASAs, $\mathcal{A}_1$ and

$\mathcal{A}_2$ are quasi-orthogonal, that is the linear subspaces $\mathcal{A}_1 \ominus \mathbb{C}I$ and $\mathcal{A}_2 \ominus \mathbb{C}I$ are orthogonal.

Let us denote the normalized trace with $\tau(\cdot) = \frac{1}{n}\operatorname{Tr}(\cdot)$. The following theorem[11,15] extends the notion of complementarity to any pair of subalgebras:

Theorem 1. *Let $\mathcal{A}_1$ and $\mathcal{A}_2$ be subalgebras of $M_n(\mathbb{C})$. Then the following conditions are equivalent:*

(i) The subspaces $\mathcal{A}_1 \ominus \mathbb{C}I$ and $\mathcal{A}_2 \ominus \mathbb{C}I$ are orthogonal.
(ii) If $P \in \mathcal{A}_1$ and $Q \in \mathcal{A}_2$ are minimal projections, then $\tau(PQ) = \tau(P)\tau(Q)$.
(iii) $\tau(A_1A_2) = \tau(A_1)\tau(A_2)$ if $A_1 \in \mathcal{A}_1$, $A_2 \in \mathcal{A}_2$.
(iv) If $E_1 : \mathcal{A} \to \mathcal{A}_1$ is the trace preserving conditional expectation, then E_1 restricted to $\mathcal{A}_2$ is a linear functional (times I).

While a MASA is related to a von Neumann measurement, as its minimal projections give a partition of unity, a subfactor is related to a subsystem of the original system, with the reduced density defined by the conditional expectation. Heuristically, a subfactor (representing a subsystem) is associated with quantum information, and a MASA (representing a measurement) is associated with classical information. Local measurement of a subsystem is associated with a MASA of the corresponding subfactor, so local measurements of complementary subsystems are complementary.

2. Characterization of complementarity

Two isomorphic subalgebras are related by a unitary transformation. Although given two subalgebras, this unitary is not unique, still, the complementarity of the two subalgebras is characterized with the properties of these unitaries.

In the case of MASAs, the subalgebras $\mathcal{A}$ and $U\mathcal{A}U^*$ are complementary if and only if the unitary U is a normalized *Hadamard matrix*, that is a unitary matrix with complex entries satisfying $|u_{ij}| = 1$. As the maximal number of mutually unbiased bases in arbitrary dimensions is an open problem, so are some related problems concerning Hadamard matrices.[18]

In the case of subfactors, the following theorem[6] characterizes the unitaries:

Theorem 2. *Let $\{E_i : 1 \leq i \leq n^2\}$ be an orthonormal basis in $M_n(\mathbb{C})$ and* *let*

$$W = \sum_{i=1}^{n^2} E_i \otimes W_i \in M_n(\mathbb{C}) \otimes M_m(\mathbb{C})$$

be a unitary. The subalgebra $W(\mathbb{C}I_n \otimes M_m(\mathbb{C}))W^*$ is complementary to $\mathbb{C}I_n \otimes M_m(\mathbb{C})$ if and only if

$$\frac{m}{n} \sum_k |W_k\rangle\langle W_k|$$

is the identity mapping on $M_m(\mathbb{C})$.

It is worth to note, that the condition of the theorem cannot hold if $m < n$, and in the case $n = m$ the condition means that

$$\{W_k : 1 \leq k \leq n^2\}$$

is an orthonormal basis in $M_m(\mathbb{C})$. Choda showed recently,[3] that if instead of $\mathbb{C}I \otimes M_m(\mathbb{C})$ we take $M_n(\mathbb{C}) \otimes \mathbb{C}I_m$, then the condition is

$$\tau(W_i^* W_j) = \frac{1}{n}\delta_{ij}, \tag{1}$$

which is quite similar.

A somewhat different characterization is also given[3] by Choda using the von Neumann entropy of positive matrices. The theorem is rephrased to match the notation used here:

Theorem 3. *Let $W = \sum_i E_i \otimes W_i$ be a unitary and $\rho[W] \in M_m(\mathbb{C})$ is the state corresponding to the operational partition of unity induced by W with elements $(\rho[W])_{ij} = \frac{1}{n}\tau(W_i^* W_j)$. The subalgebra $M_n(\mathbb{C}) \otimes \mathbb{C}I_m$ is complementary to $W(M_n(\mathbb{C}) \otimes \mathbb{C}I_m)W^*$ if and only if*

$$S(\rho[W]) = 2\log n$$

where S is the von Neumann entropy.

Proof. Assume first, that the two subalgebras are complementary. Then using (1) yields, that $\rho[W]$ is the diagonal matrix with entries $\frac{1}{n^2}$ in the diagonal, therefore the proposition holds.

For the converse, let us assume, that $\sum_{i=1}^{n^2} \lambda_i P_i$ is the spectral decomposition of $\rho[W]$. Now

$$\log n^2 = S(\rho[W]) = \sum_{i=1}^{n^2} \eta(\lambda_i),$$

so all the eigenvalues of $\rho[W]$ must be $\frac{1}{n^2}$, that is

$$(\rho[W])_{ij} = \frac{1}{n^2}\delta_{ij}.$$

The use of condition (1) finishes the proof. □

Weiner defined[19] a measure of complementarity for subalgebras as follows. If $E_{\mathcal{A}}$ and $E_{\mathcal{B}}$ are the trace preserving conditional expectations to the subalgebras $\mathcal{A}$ and $\mathcal{B}$, then let

$$c(\mathcal{A},\mathcal{B}) = \mathrm{Tr}(E_{\mathcal{A}}E_{\mathcal{B}})$$

Now $c(\mathcal{A},\mathcal{B}) \geq 1$, and equality holds if and only if the subalgebras are complementary.

3. Complementary decompositions of $M_n(\mathbb{C})$

Suppose that the MASAs $\{\mathcal{C}_i\}_{i=1}^m$ and the subfactors $\{\mathcal{A}_i\}_{i=1}^\ell$ are pairwise complementary subalgebras of $M_d(\mathbb{C})$, and $M_d(\mathbb{C})$ decomposes as a direct sum of these subalgebras. What can we say about m and ℓ? If we assume that all the subfactors isomorphic to $M_n(\mathbb{C})$, then comparing the dimensions of the subalgebras and $M_d(\mathbb{C})$ yields the simple upper bound $\ell \leq \frac{d^2-1}{n^2-1}$.

This question of complementary decompositions are well studied and understood in the smallest non-trivial case of $M_2(\mathbb{C}) \otimes M_2(\mathbb{C})$. The problem first arisen as a state determination problem.[10] It was first conjectured there, and later was proved,[12] that the maximal number of pairwise complementary subfactors isomorphic to $M_2(\mathbb{C})$ in $M_4(\mathbb{C})$ is four. It was shown,[6,13] that the complementary subalgebras have a special structure:

Theorem 4. *Let $\mathcal{A} \simeq M_2(\mathbb{C})$ be a subfactor of $M_4(\mathbb{C})$, $\mathcal{A}'$ its commutant, and $\mathcal{B}$ a subalgebra of $M_4(\mathbb{C})$ complementary to $\mathcal{A}$.*

a) If $\mathcal{B}$ is a MASA, then it is complementary to $\mathcal{A}'$.
b) If $\mathcal{B} \simeq M_4(\mathbb{C})$ is a subfactor, then either $\mathcal{B} = \mathcal{A}'$, or $\dim(\mathcal{B} \cup \mathcal{A}') = 2$

With the help of the above theorem, all possible complementary decompositions of M_4 can be described:[13]

Theorem 5. *Let $\mathcal{A}_k$ $(0 \leq k \leq 4)$ be pairwise complementary subalgebras of $M_4(\mathbb{C})$ such that all of them is either a subfactor isomorphic to $M_2(\mathbb{C})$ or a MASA. If k is the number of subfactors, then $k \in \{0, 2, 4\}$, and constructions for all those values are given.*

It is interesting to note, that the orthocomplement of the subspace generated by four pairwise complementary subfactors is always a MASA, and any choice of three complementary subfactors determines the fourth one. Theorem 4. also implies that in the case of two qubits measurements in the Bell basis are complementary to any local measurements of one of the qubits.

While the problem of determining the maximal number $\ell(2,n)$ of pairwise complementary subfactors of $M_{2^n}(\mathbb{C})$ isomorphic to $M_2(\mathbb{C})$ is still open for $n \geq 3$, it is conjectured that that the upper bound $\ell(2,n) \leq \frac{2^{2n}-1}{3}$ cannot be achieved, and a construction is given[6] with $\frac{2^{2n}-1}{3} - 1$ subfactors. Ohno constructed[8] full sets of subfactors in the case of $M_p(\mathbb{C})^n$, where $p > 2$ is a prime:

Theorem 6. *Assume that $p \geq 2$ is a prime, $n \geq 2$, and $k \geq 0$. Then there are*

$$\frac{p^{2kn}-1}{p^{2k}-1}$$

pairwise complementary subalgebras of $M_{p^{kn}}(\mathbb{C})$ isomorphic to $M_{p^k}(\mathbb{C})$.

Recently[19] the existence of some specific decompositions of $M_n^2(\mathbb{C})$ are disproved: there are no decompositions consisting of MASAs and of one or three subfactors isomorphic to $M_n(\mathbb{C})$.

References

1. L. Accardi, Some trends and problems in quantum probability, *Quantum probability and applications to the quantum theory of irreversible processes*, eds. L. Accardi, A. Frigerio and V. Gorini, Lecture Notes in Math. **1055**, pp. 1–19. Springer, 1984.
2. P. Busch and P.J. Lahti, The complementarity of quantum observables: theory and experiment, *Riv. Nuovo Cimento* **18**(1995), 27 pp.
3. M. Choda, Von Neumann entropy and relative position between subalgebras, preprint (2010)
4. I. D. Ivonovic, Geometrical description of quantal state determination, *J. Phys. A: Math. Gen.* **14** 3241 (1981)
5. K. Kraus, Complementary observables and uncertainty relations, *Phys. Rev. D* (3) **35**(1987), 3070–3075.
6. H. Ohno, D. Petz and A. Szántó, Quasi-orthogonal subalgebras of 4 × 4 matrices, *Linear Alg. Appl.* **425** (2007), 109–118.
7. H. Ohno and D. Petz, Generalized Pauli channels, *Acta Math. Hungar* **124** (2009), 165–177.
8. H. Ohno, Quasi-orthogonal subalgebras of matrix algebras, Linear Algebra Appl. **429** (2008), 2146–2158.

9. J. Oppenheim, K. Horodecki, M. Horodecki, P. Horodecki and R. Horodecki, A new type of complementarity between quantum and classical information, *Phys. Rev. A* **68**, 022307, 2003.
10. D. Petz, K. M. Hangos, F. Szöllősi, A. Szántó, State tomography for two qubits using reduced densities, *J. Phys. A: Math. Gen.* **39** 10901–10907. (2006)
11. D. Petz, Complementarity in quantum systems, *Rep. Math. Phys.* **59**(2007), 209–224.
12. D. Petz, J. Kahn, Complementary reductions for two qubits, J. Math. Phys. **48** (2007) 012107
13. D. Petz, A. Szántó and M. Weiner, Complementarity and the algebraic structure of 4-level quantum systems, *J. Infin. Dim. Anal. Quantum Probability and Related Topics* **12** (2009), 99–116.
14. D. Petz, Algebraic complementarity in quantum theory, *J. Math. Phys.* **51**, 015215 (2010)
15. S. Popa, Orthogonal pairs of *-subalgebras in finite von Neumann algebras, *J. Operator Theory* **9** (1983) 253-268
16. M. Rédei, Complementary observables in finite dimensions (in Hungarian), *Fizikai Szemle*, 1989, 338-343.
17. J. Schwinger, Unitary operator bases, *Proc. Natl. Acad. Sci. USA.* **46** (4)(1960)
18. W. Tadej and K. Zyczkowski, A concise guide to complex Hadamard matrices, *Open Systems & Infor. Dyn.* **13** 133–177 (2006)
19. M. Weiner, On orthogonal systems of matrix algebras, *Linear Alg. Appl.* **433** (2010), 520–533
20. H. Weyl, *Theory of groups and quantum mechanics*, Methuen, 1931. (Reprint: Dover, 1950)
21. W.K. Wooters and B.D. Fields, Optimal state determination by mutually unbiased measurements, *Annals of Physics*, **191**, 363–381, 1989.

ON THE LONG-TIME ASYMPTOTICS OF QUANTUM DYNAMICAL SEMIGROUPS

G. A. RAGGIO* and P. R. ZANGARA
FaMAF, Universidad Nacional de Córdoba,
Córdoba, Córdoba X5000, Argentina
**E-mail: raggio@famaf.unc.edu.ar*

We consider semigroups $\{\alpha_t : t \geq 0\}$ of normal, unital, completely positive maps α_t on a von Neumann algebra $\mathcal{M}$. The (predual) semigroup $\nu_t(\rho) := \rho \circ \alpha_t$ on normal states ρ of $\mathcal{M}$ leaves invariant the face $\mathcal{F}_p := \{\rho : \rho(p) = 1\}$ supported by the projection $p \in \mathcal{M}$, if and only if $\alpha_t(p) \geq p$ (i.e., p is sub-harmonic). We complete the arguments showing that the sub-harmonic projections form a complete lattice. We then consider r_o, the smallest projection which is larger than each support of a minimal invariant face; then r_o is sub-harmonic. In finite dimensional cases $\sup \alpha_t(r_o) = \mathbf{1}$ and r_o is also the smallest projection p for which $\alpha_t(p) \to \mathbf{1}$. If $\{\nu_t : t \geq 0\}$ admits a faithful family of normal stationary states then $r_o = \mathbf{1}$ is useless; if not, it helps to reduce the problem of the asymptotic behaviour of th e semigroup for large times.

Keywords: Quantum dynamical semigroups; sub-harmonic projections; long-time asymptotics.

1. Introduction and preliminaries

We consider a von Neumann algebra $\mathcal{M}$ and denote its normal state space by $\mathcal{S}$. A *quantum dynamical semigroup* $\{\alpha_t : t \geq 0\}$ is a family of normal, unital, positive, linear maps $\alpha_t : \mathcal{M} \to \mathcal{M}$ with the property $\alpha_t \circ \alpha_s = \alpha_{t+s}$ where α_0 is the identity. Then, the map $\nu_t : \mathcal{S} \to \mathcal{S}$ defined by $\nu_t(\rho) = \rho \circ \alpha_t$ is affine, ν_0 is the identity, and $\nu_t \circ \nu_s = \nu_{t+s}$. Conversely, given a semigroup $\{\nu_t : t \geq 0\}$ of affine maps on $\mathcal{S}$, the dual maps are a positive quantum dynamical semigroup.

One often demands on physical grounds, that α_t be completely positive. When $\mathcal{M}$ is the algebra of all bounded linear operators on a Hilbert space, if the dynamical semigroup is strongly continuous in t, and each α_t is completely positive, the generator has the canonical GKS-Lindblad form.

The long-time asymptotics of such semigroups has been studied in the 1970's and in the 1980's, after pioneering papers of E.B. Davies[1] , culminating with the work of Frigerio[2–4] , and U. Groh[5] . More recent studies are due to Fagnola & Rebolledo[6–8] , Umanitá[9] , Mohari[11,12] and Baumgartner & Narnhofer[13] . We refer to Ref. 8 for a recent overview. In pertinent cases, the asympotics can be studied via the GKS-Lindblad generator.

In this note, all projections are ortho-projections (self-adjoint equal to its square). $\mathbf{1}$ denotes the identity operator and for a projection p, $p^\perp = \mathbf{1} - p$. Limits in $\mathcal{M}$ are invariably in the w^*-topology. All states (positive linear functionals of unit norm) are normal. Limits of states are with respect to the distance induced by the norm. But recall that the norm-closure of a convex set of states coincides with its weak-closure. The *support* of a state ρ –written s_ρ– is the smallest projection $p \in \mathcal{M}$ such that $\rho(p) = 1$.
We will consider a quantum dynamical semigroup $\{\alpha_t : t \geq 0\}$ and will always explicitly mention any additional positivity hypotheses. In particular, if each α_t is completely positive, we say that the semigroup is CP. A state ω is *stationary* if $\nu_t(\omega) = \omega \circ \alpha_t = \omega$.

2. Invariant faces and sub-harmonic projections

A *face* is a convex subset $\mathcal{F}$ of $\mathcal{S}$ which is stable under convex decomposition: if $t\rho+(1-t)\mu \in \mathcal{F}$ for $0 < t < 1$ with $\rho, \mu \in \mathcal{S}$ then $\rho, \mu \in \mathcal{F}$. If $p \in \mathcal{M}$ is a projection then $\mathcal{F}_p := \{\rho \in \mathcal{S} : \rho(p) = 1\}$ is a closed face; we say it is the face *supported* by p. It is not so obvious but true[14,15] , that every closed face is of this form; i.e. it is the face supported by some projection. Clearly, $\mathcal{F}_p \subset \mathcal{F}_q$ if and only if $p \leq q$.

The following result is implicit or partially explicit in the work of Fagnola & Rebolledo and Umanitá.

Proposition 2.1. *Suppose ν is an affine map of $\mathcal{S}$ into itself and let α be the dual normal, linear, positive map of $\mathcal{M}$ into itself. For a projection $p \in \mathcal{M}$ the following conditions are equivalent: (1) the face $\mathcal{F}_p$ supported by p is ν-invariant; (2) $\alpha(p) \geq p$; (3) $p\alpha(a)p = p\alpha(pap)p$ fore every $a \in \mathcal{M}$; (4) $\alpha(p^\perp a p^\perp) = p^\perp \alpha(p^\perp a p^\perp) p^\perp$ for every $a \in \mathcal{M}$.*

Proof. We first prove the chain (2) $\Rightarrow$ (1) $\Rightarrow$ (3) $\Rightarrow$ (2). If $\alpha(p) \geq p$ then, for any state ρ one has $\nu(\rho)(p) = \rho(\alpha(p)) \geq \rho(p)$. Thus, $\rho(p) = 1$ implies

$\nu(\rho)(p)) = 1$, i.e. $\nu(\mathcal{F}_p) \subset \mathcal{F}_p$. If $\nu(\mathcal{F}_p) \subset \mathcal{F}_p$, we show that

$$(*) \quad \omega(p\alpha(pap)p) = \omega(p\alpha(a)p) \text{ for every } \omega \in \mathcal{S} .$$

Since every normal linear functional is the linear combination of at most four states, this then implies that $p\alpha(pap)p = p\alpha(a)p$. To prove (*) observe that, by the Cauchy-Schwarz inequality for states, the claim is trivially valid if $\omega(p) = 0$. Otherwise, consider the state $\omega_p(a) := \omega(pap)/\omega(p)$. Clearly $\omega_p \in \mathcal{F}_p$; thus,

$$\omega(p)^{-1}\omega(p\alpha(pap)p) = \omega_p(\alpha(pap)) = \nu(\omega_p)(pap) = \nu(\omega_p)(a)$$

$$= \omega_p(\alpha(a)) = \omega(p)^{-1}\omega(p\alpha(a)p) ;$$

which is (*). Finally, if $p\alpha(pap)p = p\alpha(a)p$, then $p - p\alpha(p)p = p\alpha(p^\perp)p = 0$ and Lemma 3.1 of the Appendix implies $\alpha(p) \geq p$.
If $0 \leq x = p^\perp x p^\perp \leq \mathbf{1}$ then, by Lemma 3.1 of the Appendix, $x \leq p^\perp$ and $\alpha(x) \leq \alpha(p^\perp)$; when (2) is the case $\alpha(p^\perp) \leq p^\perp$ so that $\alpha(x) \leq p^\perp$ which by the aforementioned Lemma, implies $p^\perp \alpha(x) p^\perp = \alpha(x)$. For general $0 \leq x = p^\perp x p^\perp$ we consider $x/\|x\|$ and obtain $p^\perp \alpha(x) p^\perp = \alpha(x)$. Since every $a \in \mathcal{M}$ is a linear combination of at most four positive elements, we conclude that (2) implies (4). But (4) implies $\alpha(p^\perp) = p^\perp \alpha(p^\perp) p^\perp$ which, by the same Lemma, gives $\alpha(p^\perp) \leq p^\perp$ which is equivalent to $\alpha(p) \geq p$. □

In the context of quantuym dynamical semigroups, a projection p satisfying $\alpha_t(p) \geq p$ has been termed *sub-harmonic*[6]. We say the projection p is *sub-harmonic* for the linear, normal, unital and positive map α on $\mathcal{M}$ if $\alpha(p) \geq p$. The previous proposition relates the sub-harmonic property of a projection to the more geometric notion of invariance of the supported face. This relationship can be immediately put to use:

Proposition 2.2. *If a family of projections is sub-harmonic for a linear, normal, unital and positive map α on $\mathcal{M}$, then the infimum of the family is sub-harmonic for α.*

Proof. If $\{\mathcal{F}_\iota : \iota \in I\}$ is a family of closed faces $\mathcal{F}_\iota$ of $\mathcal{S}$ then $\bigcap_\iota \mathcal{F}_\iota$ is a closed face and it is the largest closed face contained in each $\mathcal{F}_\iota$. The support of $\bigcap_\iota \mathcal{F}_\iota$ is exactly $\inf\{p_\iota : \iota \in I\}$, where p_ι is the support of $\mathcal{F}_\iota$. Moreover, if each $\mathcal{F}_\iota$ is ν-invariant then so is the intersection. □

The corresponding statement for the supremum of such a family has been observed and proved directly (Ref. 9).

For projections p that are *super-harmonic*, i.e. $\alpha(p) \leq p$ or equivalently $p^\perp$ is sub-harmonic, we have (in reply to a question posed in Ref. 10):

Corollary 2.1. *If a family of projections is super-harmonic for a linear, normal, unital and positive map α on $\mathcal{M}$, then the supremum of the family is super-harmonic for α.*

Proof. $\sup\{p : p \in \mathcal{F}\} = (\inf\{p^\perp : p \in \mathcal{F}\})^\perp$ and $\inf\{p^\perp : p \in \mathcal{F}\}$ is sub-harmonic by the previous proposition. □

The corresponding statement for the infimum of a super-harmonic family follows from the result for the supremum of a sub-harmonic family by orthocomplementation as above. Thus,

Theorem 2.1. *The set of sub-harmonic and the set of super-harmonic projections with respect to a linear, normal, unital and positive map on $\mathcal{M}$ are both complete lattices.*

A *minimal invariant face* is a closed ν_t-invariant face which does not properly contain another non-empty closed ν_t-invariant face. Equivalently, it is a face whose support is a minimal sub-harmonic projection, i.e. a sub-harmonic projection that is not larger than a non-zero sub-harmonic projection other than itself. One can prove, and this goes back to –at least– Davies (see Ref. 1, Theorem 3.8 of Sect. 6.3), that if the minimal invariant face admits a stationary state then it is unique and its support is the support of the face. Moreover (Ref. 5, Proposition 3.4) the restriction of ν_t to the face is ergodic (the Cesàro means converge to the stationary state).

A "recurrent" projection

We define the *minimal recurrent* projection r_o as the smallest projection which is larger than every minimal sub-harmonic projection. Equivalently, r_o is the support of the smallest ν_t-invariant face which contains every minimal ν_t-invariant face. By virtue of its definition and the result mentioned above –to the effect that the supremum of a family of sub-harmonic projections is sub-harmonic– it follows that the minimal recurrent projection is sub-harmonic. Hence the directed family $\alpha_t(r_o)$ which is bounded above by $\mathbf{1}$ has a lowest upper bound in $\mathcal{M}$ denoted by x which is positive and below $\mathbf{1}$. Since $x = \lim_{t\to\infty} \alpha_t(r_o)$ it follows that $\alpha_t(x) = x$ for every $t \geq 0$. Let $s[x]$ denote the support of x, that is the smallest projection $p \in \mathcal{M}$ with $xp = x$. The following treatment follows the lines of work by Mohari[12] .

Lemma 2.1. *If* $\{\alpha_t : t \geq 0\}$ *is CP, then* $s[x] = \mathbf{1}$.

Proof. $s[x]^\perp$ is the largest projection q with $xq = 0$, and it is sub-harmonic by a result of Ref. 12 quoted in the appendix. Assume that $s[x] \neq \mathbf{1}$; then there is a minimal sub-harmonic non-zero projection q with $q \leq s[x]^\perp$. One has $xq = 0$. By the definition of r_o, we have $q \leq r_o$ and thus $q = qr_oq \leq q\alpha_t(r_o)q \leq qxq = 0$, which contradicts the assumption. □

Let $\mathcal{J} := \{a \in \mathcal{M} : \lim_{t\to\infty} \alpha_t(a^*a) = 0\}$. Since for each state ρ, one has the Cauchy-Schwarz inequality

$$|\rho(\alpha_t(a^*b^*))| = |\rho(\alpha_t(ba))| = |\nu_t(\rho)(ba)|$$

$$\leq \sqrt{\nu_t(\rho)(bb^*)\nu_t(\rho)(a^*a)} \leq \|b\| \sqrt{\rho(\alpha_t(a^*a))}\ ;$$

we infer that $\mathcal{J}$ is a linear subspace of $\mathcal{M}$. If $c \in \mathcal{M}$ and $a \in \mathcal{J}$, the same inequality applied to $b = a^*c^*c$ shows that $ca \in \mathcal{J}$; thus $\mathcal{J}$ is a left-ideal.

If $\mathcal{M}$ is finite-dimensional (that is *-isomorphic to the direct sum of finitely many full matrix algebras) then, on the one hand $s[x] = \mathbf{1}$ implies that x is invertible, and Mohari[12] has shown that if x is invertible then $x = \mathbf{1}$; and –on the other hand– $\mathcal{J}$ is closed and there is a projection such that $\mathcal{J} = \mathcal{M} \cdot z$. Then

Theorem 2.2. *If* $\mathcal{M}$ *is finite dimensional and* $\{\alpha_t : t \geq 0\}$ *is CP, then* $\sup\{\alpha_t(r_o) : t \geq 0\} = \mathbf{1}$. *Moreover* $\mathcal{J} = \mathcal{M} \cdot r_o^\perp$ *and* r_o *is the smallest projection* $p \in \mathcal{M}$ *with* $\lim_{t\to\infty} \alpha_t(p) = \mathbf{1}$.

Proof. There is[14] a projection $z \in \mathcal{M}$ with $\mathcal{J} = \mathcal{M} \cdot z$. Lemma 3.1 implies that x is invertible and Theorem 2.5 of Ref. 12 gives $x = \mathbf{1}$. Hence $r_o^\perp \in \mathcal{J}$ and thus $r_o^\perp \leq z$ or $r_o \geq z^\perp$. Suppose p is a minimal sub-harmonic projection; there is a stationary state ω in the minimal invariant face supported by p and it follows (see the introduction) that it is unique and $s(\omega) = p$. Since $\omega(z) = \omega(\alpha_t(z)) \to 0$, we have $\omega(z^\perp) = 1$ and thus $p \leq z^\perp$. But then, by the definition of r_o, $r_o \leq z^\perp$. Thus $r_o = z^\perp$. □

Note. Despite the claim in Ref. 13, p. 8, one cannot conclude from $\lim_{t\to\infty} \alpha_t(p) = \mathbf{1}$ for a projection p, that p is sub-harmonic. Simple examples can be given[16] .

It follows from the Cauchy-Schwarz inequality for states that $\lim_{t\to\infty} \alpha_t(ar_o^\perp) = \lim_{t\to\infty} \alpha_t(r_o^\perp a) = 0$ for every $a \in \mathcal{M}$ so that $\alpha_t(a) \asymp$

$\alpha_t(r_o a r_o)$ for large t and every $a \in \mathcal{M}$. If $\{\nu_t : t \geq 0\}$ admits a faithful family of stationary states, then the minimal recurrent projection is the identity. This happens because for a stationary state ω, one has $\omega(r_o^\perp) = \omega(\alpha_t(r_o^\perp)) \downarrow 0$ and thus $\omega(r_o^\perp) = 0$. However in this case there are results[4,5,11] on the asymptotic behaviour of the semigroup.
Other recurrent projections have been considered. For example (Ref. 5, p. 407; Ref. 9), the supremum r of the supports of the stationary states (if any are available), which is then sub harmonic and above r_o.

There is no reason to expect that the above theorem holds in infinite dimension.

Acknowledgements

We thank the organizers of the 30th Conference on Quantum Probability and Related Topics, in Santiago, Chile. The support of CONICET (PIP 112200801017410) is acknowledged. The first author is grateful to M.E. Martín Fernández for generous support.

3. Appendix

We collect here two technical results used in the above proofs.

Lemma 3.1. *For $x \in \mathcal{M}$ satisfying $\mathbf{1} \geq x \geq 0$ and $p \in \mathcal{M}$ a projection one has:*

a) the following five conditions are equivalent: (1) $x \geq p$; (2)$pxp = p$; (3) $x = p + p^\perp x p^\perp$; (4)$xp = p$; (5) $px = p$.

b) the following four conditions are equivalent: (1) $p \geq x$; (2) $pxp = x$; (3) $x = xp$; (4) $x = px$.

Proof. a): Given $\mathbf{1} \geq x \geq p$, multiplication from left and right by p gives $p \geq pxp \geq p$ and thus $pxp = p$.
If $pxp = p$ then $p(\mathbf{1} - x)p = 0$ which implies $(\mathbf{1} - x)^{1/2}p = 0$ and thus $(\mathbf{1} - x)p = 0$ or $xp = p$; taking adjoints $p = px$.
And $xp = p$ or $px = p$ implies $pxp = p$.
Finally either of the equivalent conditions (4) or (5) imply that $x - p = p^\perp x p^\perp \geq 0$.

b): $p \geq x$ if and only if $p^\perp \leq \mathbf{1} - x$. Apply a). □

The following crucial observation and the proof, repeated here for convenience, are due to Mohari[12] .

Proposition 3.1 (Mohari). *Suppose* $\alpha : \mathcal{M} \to \mathcal{M}$ *is linear, unital, normal and completely positive and* $x \in \mathcal{M}$ *is positive with* $\alpha(x) = x$*. Then the support of* x *is super-harmonic.*

Proof. We may assume $\mathcal{M}$ is a von Neumann algebra on a Hilbert space $\mathcal{K}$. By the Stinespring Representation Theorem there is a normal $*$-homomorphism π of $\mathcal{M}$ into $\mathcal{B}(\mathcal{H})$ (the algebra of bounded linear operators on a Hilbert space $\mathcal{H}$) and an isometry $V : \mathcal{K} \to \mathcal{H}$ such that $\alpha(a) = V^*\pi(a)V$ for all $a \in \mathcal{M}$. Recall that the support of a self-adjoint element a is the smallest projection p of $\mathcal{M}$ such that $pa = a$ (equivalently $ap = a$). If $\mathcal{M} \subset \mathcal{B}(\mathcal{K})$ then the support coincides with the smallest projection $q \in \mathcal{B}(\mathcal{K})$ such that $qa = a$ (Proposition 1.10.4 of[14]). Now if x satisfies the hypothesis, s is its support and $z = s^\perp$, then $0 = zxz = z\alpha(x)z = zV^*\pi(x)Vz = (yVz)^*(yVz)$ where $y = \sqrt{\pi(y)}$. Thus $yVz = 0$ and hence $\pi(x)Vz = 0$. The support of $\pi(x)$ is $\pi(s)$ and since Vz maps $\mathcal{K}$ into the kernel of $\pi(x)$, we conclude that $\pi(s)Vz = 0$. But then, $\alpha(s)z = V^*\pi(s)Vz = 0$ or $\alpha(s) = \alpha(s)s$ which by the Lemma above implies $\alpha(s) \leq s$. □

References

1. E.B. Davies: *Quantum Theory of Open Systems.* (Academic Press, London 1976).
2. A. Frigerio: *Quantum dynamical semigroups and approach to equilibrium. Lett. Math. Phys.* **2**, 79–87 (1977/78).
3. A. Frigerio: *Stationary states of quantum dynamical semigroups. Comm. Math. Phys.* **63**, 269–276 (1978).
4. A. Frigerio, and M. Verri: *Long-time asymptotic properties of dynamical semigroups on W^*-algebras. Math. Z.* **180**, 275–286 (1982).
5. U. Groh: *Positive semigroups on C^*- and W^*-Algebras.* In *One-parameter Semigroups of Positive Operators*, edited by R. Nagel. Lecture Notes in Mathematics 1184, (Springer-Verlag, Berlin, 1986); pp. 369–425.
6. F. Fagnola, and R. Rebolledo: *On the existence of stationary states for quantum dynamical semigroups. J. Math. Phys.* **42**, 1296–1308 (2001).
7. F. Fagnola, and R. Rebolledo: *Subharmonic projections for quantum Markov semigroup. J. Math. Phys.* **43**, 1074–1082 (2002).
8. F. Fagnola, and R. Rebolledo: *Notes on the Qulitative Behaviour of Quantum Markov semigroups.* In *Open Quantum Systems III. Recent Developments*, edited by S. Attal, A. Joye, and C.–A. Pillet. Lecture Notes in Mathematics 1882. (Springer-Verlag, Berlin, 2006); pp. 161-206.

9. V. Umanitá: *Classification and decomposition of Quantum Markov Semigroups. Probab. Theory Relat. Fields* **134**, 603–623 (2006).
10. F. Fagnola: *Quantum Markov semigroups: structure and asymptotics. Rend. Circ. Mat. Palermo serie II Suppl. No.* **73**, 35–51 (2004).
11. A. Mohari: *Markov shift in non-commutative probability. J. Funct. Anal.* **199**, 189–209 (2003).
12. A. Mohari: *A resolution of quantum dynamical semigroups.* Preprint arXiv:math/0505384v1, May 2005.
13. B. Baumgartner, and H. Narnhofer: *Analysis of quantum semigroups with GKS–Lindblad generators: II. General. J. Phys. A: Math. Theor.* **41**, 395303 (2008).
14. S. Sakai: *C^*-algebras and W^*-algebras.* (Springer-Verlag, Berlin, 1971).
15. L. Asimow, and A.J. Ellis: *Convexity Theory and its Applications in Functional Analysis.* (Academic Press, London, 1980).
16. P.R. Zangara: *Evolución asintótica de sistemas cuánticos abiertos.* Trabajo Especial de Licenciatura en Física, FaMAF-UNC, Diciembre 2009.

HILBERT MODULES—SQUARE ROOTS OF POSITIVE MAPS

MICHAEL SKEIDE*

Dipartimento S.E.G.e S., Università degli Studi del Molise, Via de Sanctis, 86100 Campobasso, Italy, E-mail: skeide@unimol.it, Homepage: http://www.math.tu-cottbus.de/INSTITUT/lswas/_skeide.html

We reflect on the notions of positivity and square roots. We review many examples which underline our thesis that square roots of positive maps related to $*$–algebras are Hilbert modules. As a result of our considerations we discuss requirements a notion of positivity on a $*$–algebra should fulfill and derive some basic consequences.

*K*eywords: Quantum dynamics, quantum probability, positivity, Hilbert modules, product systems, E_0–semigroups. 2000 AMS-Subject classification: 46L53; 46L55; 46L08; 60J25; 81S25; 12H20.

1. Introduction

Let S denote a set, and let $\mathfrak{k}$ denote a map $S \times S \to \mathbb{C}$. Everybody knows that such a ***kernel*** $\mathfrak{k}$ ***over*** S is called ***positive definite*** if

$$\sum_{i,j} \bar{z}_i \mathfrak{k}^{\sigma_i,\sigma_j} z_j \geq 0 \tag{1.1}$$

for all finite choices of $\sigma_i \in S$ and $z_i \in \mathbb{C}$.

What is is the best way to show that some thing x is positive? The best way is writing x as a ***square***! It would, then, be justified to call an object y the positive thing x's ***square root***, if by writing down the object y's square we get back x. By *square*, of course, we mean a ***complex square*** like $\bar{y}y$ (y a complex number) or y^*y (y in a C^*–algebra) or $\langle y, y\rangle$ (y being in a Hilbert space).

Of course, for each choice $\sigma_i \in S$ and $z_i \in \mathbb{C}$ we may calculate the positive number in (1.1) and write down its positive square root $p(\sigma_1, \ldots, \sigma_n, z_1, \ldots, z_n)$ (or any other complex square root) and that's it.

*MS is supported by research funds of the University of Molise and the Italian MIUR (PRIN 2007).

Although, the collection of all p contains the full information about $\mathfrak{k}$ (for instance by suitable polarization procedures or by differentiation with respect to the parameters z_i), it is uncomfortable to do that. Also, the knowledge of some $p(\sigma_1,\ldots,\sigma_n,z_1,\ldots,z_n)$ for a fixed choice, does not at all help computing $p(\sigma_1,\ldots,\sigma_{n-1},z_1,\ldots,z_{n-1})$ for the same choice. We gain a bit but not very much, if we calculate for each choice $\sigma_1,\ldots,\sigma_n$ the positive (or some other) square root $P(\sigma_1,\ldots,\sigma_n) \in M_n$ of the positive matrix $\left(\mathfrak{k}^{\sigma_i\sigma_j}\right)_{i,j} \in M_n$. Still, the knowledge of some $P(\sigma_1,\ldots,\sigma_n)$ for a certain choice does not help computing $P(\sigma_1,\ldots,\sigma_{n-1})$ for the same choice. (Exercise: Try it and explain why it does not help!)

We wish something that allows easily to recover the function $\mathfrak{k}$ and that still gives evidence of positivity of the expressions in (1.1) by writing them as square. The solution to that problem is the well-known ***Kolmogorov decomposition***.

1.1 Theorem. *For every positive definite kernel $\mathfrak{k}$ over a set S with values in $\mathbb{C}$ there exist a Hilbert space H and a map $i\colon S \to H$ such that*

$$\mathfrak{k}^{\sigma,\sigma'} = \langle i(\sigma), i(\sigma')\rangle$$

for all $\sigma, \sigma' \in S$.

Proof. On the vector space $S_{\mathbb{C}} := \bigoplus_{\sigma\in S}\mathbb{C} = \left\{ \left(z_\sigma\right)_{\sigma\in S} \mid \#\{\sigma\colon z_\sigma \neq 0\} < \infty \right\}$ we define a sesquilinear form

$$\left\langle \left(z_\sigma\right)_{\sigma\in S}, \left(z'_\sigma\right)_{\sigma\in S}\right\rangle := \sum_{\sigma,\sigma'\in S} \bar{z}_\sigma \mathfrak{k}^{\sigma,\sigma'} z'_{\sigma'}.$$

Since $\mathfrak{k}$ is positive definite, this form is positive. Denote $e_\sigma := \left(\delta_{\sigma,\sigma'}\right)_{\sigma'\in S}$. Then $\langle e_\sigma, e_{\sigma'}\rangle = \mathfrak{k}^{\sigma,\sigma'}$. Denote by H the ***Hausdorff completion*** of $\bigoplus_{\sigma\in S}\mathbb{C}$ (that is, quotient out the subspace $\mathcal{N}$ of length-zero elements and complete that pre-Hilbert space). Then H with the function i defined by $i(\sigma) := e_\sigma + \mathcal{N}$ has the claimed properties. ■

Note that the subset $i(S)$ of H as constructed in the proof is total. Therefore, the pair (H,i) has the following universal property: If (G,j) is another Kolmogorov decomposition of $\mathfrak{k}$, then there is a unique bounded linear operators $v\colon H \to G$ such that $vi(\sigma) = j(\sigma)$ for all $\sigma \in S$. Note that v is isometric so that (H,i) is determined by that universal property up to unique unitary equivalence. This is just the same as the square root p of a positive number k, which is determined up to a unitary operator $e^{i\varphi}$ on the one-dimensional Hilbert space $\mathbb{C}$.

We like to think of the ***minimal Kolmogorov*** construction (H, i) as **the** square root of the kernel $\mathfrak{k}$. Obviously, every Hilbert space arises in that way. (Simply take the kernel $\mathfrak{k}^{h,h'} := \langle h, h' \rangle$. Then $(H, i\colon h \mapsto h)$ has the universal property.)

It is the scope of these notes to establish the idea of inner product spaces (like Hilbert modules) as square roots of maps that are positive in some sense. Apart from many instances of this interpretation, we intend also to discuss the just mentioned uniqueness issue for square roots, and to present the rudiments of what we consider a "good" notion of positivity in $*$–algebras:

In a "good notion of positivity" it should be a theorem that all positive things have a sort of square root.

Another scope is to point out the following insight about composition of positive things.

In the noncommutative world, if one wishes to compose positive things to get new ones, then these positive things must be maps on $*$–algebras, not elements in $*$–algebras.

Many of our examples have to do with product systems. We should mention that we systematically omit mentioning any relationship that has to do with commutants of von Neumann correspondences (Skeide[Ske03]). We refer the interested reader to the survey Skeide.[Ske08]

2. Kernels with values in a C^*–algebra

If $\mathfrak{k}\colon S \times S \to \mathcal{B}$ is a kernel ***over*** S with values in a C^*–algebra $\mathcal{B}$, then everything goes precisely as in the scalar-valued case, just that now the space emerging by Kolmogorov decomposition is a Hilbert $\mathcal{B}$–module.

A kernel $\mathfrak{k}$ is ***positive definite*** (or a ***PD-kernel***) if

$$\sum_{i,j} b_i^* \mathfrak{k}^{\sigma_i, \sigma_j} b_j \geq 0 \tag{2.1}$$

for all finite choices of $\sigma_i \in S$ and $b_i \in \mathbb{C}$. Let us equip the right $\mathcal{B}$–module $E_0 := S_{\mathbb{C}} \otimes \mathcal{B} = \bigoplus_{\sigma \in S} \mathcal{B} = \left\{ \left(b_\sigma\right)_{\sigma \in S} \mid \#\{\sigma\colon b_\sigma \neq 0\} < \infty \right\}$ with the sesquilinear map $\langle \bullet, \bullet \rangle\colon E_0 \times E_0 \to \mathcal{B}$

$$\left\langle \left(b_\sigma\right)_{\sigma \in S}, \left(b'_\sigma\right)_{\sigma \in S} \right\rangle := \sum_{\sigma \in S} b_\sigma^* \mathfrak{k}^{\sigma, \sigma'} b'_{\sigma'}.$$

Equation (2.1) is born to to make $\langle \bullet, \bullet \rangle$ ***positive***: $\langle x, x \rangle \geq 0$ for all $x \in E_0$. It also is ***right $\mathcal{B}$–linear***: $\langle x, yb \rangle = \langle x, y \rangle b$ for all $x, y \in E_0$ and $b \in \mathcal{B}$.

In other words, $\langle \bullet, \bullet \rangle$ is a ***semiinner product*** and E_0 is a ***semi-Hilbert $\mathcal{B}$–module***. By making appropriate use of ***Cauchy-Schwarz inequality***

$$\langle x, y \rangle \langle y, x \rangle \ \le \ \|\langle y, y \rangle\| \, \langle x, x \rangle$$

(Paschke[Pas73]), the function $x \mapsto \sqrt{\|\langle x, x \rangle\|}$ is a seminorm. So, we may divide out the right submodule of length-zero elements $\mathcal{N}$. In other words, $E_0/\mathcal{N}$ is a ***pre-Hilbert $\mathcal{B}$–module***, that is, it is a semi-Hilbert $\mathcal{B}$–module where $\langle x, x \rangle = 0$ implies $x = 0$ for all $x \in E_0/\mathcal{N}$. Moreover, $\|xb\| \le \|x\| \, \|b\|$ so that we may complete the quotient. In other words, $E := \overline{E_0/\mathcal{N}}$ is a ***Hilbert $\mathcal{B}$–module***, that is, E is a complete pre-Hilbert $\mathcal{B}$–module.

Recall that $e_\sigma \otimes b = \big(\delta_{\sigma,\sigma'} b\big)_{\sigma' \in S}$. If $\mathcal{B}$ is unital, then $i(\sigma) := e_\sigma \otimes \mathbf{1} + \mathcal{N}$ fulfills $\langle i(\sigma), i(\sigma') \rangle \ = \ \mathfrak{k}^{\sigma,\sigma'}$ and $\overline{\operatorname{span}}\, i(S)\mathcal{B} \ = \ E$. If $\mathcal{B}$ is nonunital, then choose an approximate unit $\big(u_\lambda\big)_{\lambda \in \Lambda}$ for $\mathcal{B}$, and verify that $\big(e_\sigma \otimes u_\lambda + \mathcal{N}\big)_{\lambda \in \Lambda}$ is a Cauchy net in E. Define $i(\sigma) := \lim_\lambda e_\sigma \otimes u_\lambda + \mathcal{N}$. In conclusion:

2.1 Theorem. *If $\mathfrak{k}$ is a $\mathcal{B}$–valued PD-kernel over S, then there is a pair (E, i) of a Hilbert $\mathcal{B}$–module E and map $i \colon S \to E$ satisfying*

$$\langle i(\sigma), i(\sigma') \rangle \ = \ \mathfrak{k}^{\sigma,\sigma'}$$

for all $\sigma, \sigma' \in S$ and $\overline{\operatorname{span}}\, i(S)\mathcal{B} = E$. Moreover, if (F, j) is another pair fulfilling $\langle j(\sigma), j(\sigma') \rangle = \mathfrak{k}^{\sigma,\sigma'}$, then the map $i(\sigma) \mapsto j(\sigma)$ extends to a unique ***isometry*** *(that is, an inner product preserving map) $E \to F$.*

By the universal property, it follows that the pair (H, i) is determined up to unique unitary equivalence. (A ***unitary*** is a surjective isometry.) We refer to it as the ***minimal Kolmogorov decomposition*** of $\mathfrak{k}$.

Once more, every Hilbert module E arises in that way, as the Kolmogorov decomposition (E, id_E) of the PD-kernel $(x, y) \mapsto \langle x, y \rangle$ over E. We, therefore, like to think of Hilbert modules as square roots of PD-kernels.

2.2 Example. For a positive element $b \in \mathcal{B}$ we may define the PD-kernel $\mathfrak{k} \colon (\omega, \omega) \mapsto b$ over the one-point set $S = \{\omega\}$. If we choose an element $\beta \in \mathcal{B}$ such that $\beta^*\beta = b$, then the right ideal $E = \overline{\beta\mathcal{B}}$ generated by β with inner product $\langle x, y \rangle := x^*y$ is a Hilbert $\mathcal{B}$–module. Moreover, the map $i \colon \omega \mapsto \beta$ fulfills $\langle i(\omega), i(\omega) \rangle = \mathfrak{k}^{\omega,\omega}$ and $\overline{\operatorname{span}}\, i(\omega)\mathcal{B} = E$.

If β' is a another square root, then Theorem 2.1 tells us that $\beta \mapsto \beta'$ extends as a unitary from $E = \overline{\beta\mathcal{B}}$ to $E' = \overline{\beta'\mathcal{B}}$. But more cannot be said about different choices of square roots. For instance, if $b = \mathbf{1}$, then every isometry $v \in \mathcal{B}$ is a square root. But as subsets of $\mathcal{B}$ the sets $v\mathcal{B} (= \overline{v\mathcal{B}})$ can be quite different. It can be all $\mathcal{B}$. (This happens if and only if v is a unitary.) But if v and v' fulfill $v^*v' = 0$, then they are even orthogonal to each other. If $\mathcal{B}$ is unital and $\overline{\beta\mathcal{B}} = \mathcal{B}$, then β is necessarily invertible. (Exercise!) If $\beta' \in \mathcal{B}$ fulfills $\beta^*\beta = \beta'^*\beta'$, then it is easy to show that $\beta'\beta^{-1}$ is a unitary. $(\beta^{-1})^*\beta'^*\beta'\beta^{-1} = (\beta^*)^{-1}\beta^*\beta\beta^{-1} = \mathbf{1}$.

Only, the picture of Kolmogorov decomposition for the kernel on a one-point set allows to make a precise statement.

2.3 Note. It seems that the concept of PD-kernels with values in an abstract C^*–algebra has not been considered before Barreto, Bhat, Liebscher, and Skeide.[BBLS04] The classical Stinespring theorem[Sti55] for CP-maps with values in a concrete C^*–algebra $\mathcal{B} \subset \mathscr{B}(G)$ is proved by using the Kolmogorov decomposition for a $\mathbb{C}$–valued PD-kernel; see Remark 4.4. However, analogue constructions for CP-maps with values in $\mathscr{B}^a(F)$ (the algebra of adjointable operators on a Hilbert $\mathcal{C}$–modules F) by Kasparov[Kas80] and Lance[Lan95] use proofs similar to Paschke's GNS-construction[Pas73] for CP-maps; see Note 5.9. Closest is Murphy's result in[Mur97] for $\mathscr{B}^a(F)$–valued kernels, whose proof uses techniques like reproducing kernels (Aronszajin[Aro50]); see Szafraniec's survey.[Sza09]

3. Composing PD-kernels?

It is well known that two positive definite $\mathbb{C}$–valued kernels $\mathfrak{l}$ and $\mathfrak{k}$ over the same set S may composed by ***Schur product***, that is, by the pointwise product

$$(\mathfrak{l}\mathfrak{k})^{\sigma,\sigma'} := \mathfrak{l}^{\sigma,\sigma'}\mathfrak{k}^{\sigma,\sigma'},$$

and the result is again a PD-kernel over S. Note that this Schur product of $\mathbb{C}$–valued kernels is commutative.

Of course, we may define the Schur product of $\mathcal{B}$–valued kernels by the same formula. But now the product, in general, depends on the order. However, if $\mathfrak{l}\mathfrak{k} \neq \mathfrak{k}\mathfrak{l}$, then neither of the two products is PD. (Note that by Kolmogorov decomposition, a PD-kernel is necessarily ***hermitian***: $\mathfrak{k}^{\sigma,\sigma'^*} = \langle i(\sigma), i(\sigma')\rangle^* = \langle i(\sigma'), i(\sigma)\rangle = \mathfrak{k}^{\sigma',\sigma}$.)

Does it help if we try to compose the square roots? Let us choose two positive elements $b = \beta^*\beta$ and $c = \gamma^*\gamma$, and, as in Example 2.2, consider the two PD-kernels $\mathfrak{k}\colon (\omega,\omega) \mapsto b$ and $\mathfrak{l}\colon (\omega,\omega) \mapsto c$ over the one-point set $S = \{\omega\}$. We may take the two square roots β and γ, multiply them, and use their product $\beta\gamma$ to define a PD-kernel $(\omega,\omega) \mapsto (\beta\gamma)^*(\beta\gamma) = \gamma^*\beta^*\beta\gamma$ on S.

There are two things to be noted. Firstly, if β and γ do note commute, then the "composed" kernel depends on the order. This as such is not too disturbing in a noncommutative context.

3.1 Note. Bercovici[Ber05] and Franz[Fra09] use such a procedure of a product in the definition of multiplicative monotone convolution of probability measures.

Secondly, and much more crucial, the kernel $\mathfrak{l}$ alone does not allow to determine that "composition". Or the other way round, different square roots γ of $\mathfrak{l}$ do, in general, not give rise to the same composition. What we know about the kernel is equivalently coded in its minimal Kolmogorov decomposition. However, as pointed out in Example 2.2, different square roots γ are indistinguishable both from the point of view of Kolmogorov decomposition and from the point of view of the kernel itself.

The puzzle is resolved, if we observe that, actually, we have to compute the map $\gamma^* \bullet \gamma \colon b \mapsto \gamma^* b \gamma$ — a map with strong positivity properties. If we wish to compose $\mathfrak{l}$ with arbitrary kernels $\mathfrak{k}$, then we need the entire information about that map. That information is encoded in the left ideal generated by γ. Doing also here a Kolmogorov type construction, we end up with the two-sided ideal, that is, the Hilbert $\mathcal{B}$–bimodule, generated by γ; see Example 4.5 below.

What we just discussed for a one-point set, for general sets S gives rise to the notion of completely positive definite (CPD) kernels. The Kolmogorov decomposition for CPD-kernels will result in a Hilbert bimodule rather than in an Hilbert module.

CPD-kernels may be composed, and the Kolmogorov decomposition for the composition of two CPD-kernels is reflected by the Kolmogorov decompositions of the factors. CPD-kernels are, therefore, the "correct" generalization of $\mathbb{C}$–valued PD-kernels.

This will be subject of the next section.

4. CPD-kernels

In a noncommutative context we have seen that, if we wish to compose kernels fulfilling some positivity condition in a way that preserves positivity, then it is practically forced to switch from kernels with values in $\mathcal{B}$ to kernels with values in the bounded maps on $\mathcal{B}$. Once we have map-valued kernels, there is no longer a reason that domain and codomain must coincide.

4.1 Definition. Let S be set and let $\mathfrak{K} \colon S \times S \to \mathfrak{B}(\mathcal{A}, \mathcal{B})$ be a kernel over S with values in the bounded maps from a C^*–algebra $\mathcal{A}$ to a C^*–algebra

$\mathcal{B}$. We say $\mathfrak{K}$ is a ***completely positive definite kernel*** (or ***CPD-kernel***) over S from $\mathcal{A}$ to $\mathcal{B}$ if

$$\sum_{i,j} b_i^* \mathfrak{K}^{\sigma_i,\sigma_j}(a_i^* a_j) b_j \ \geq\ 0 \tag{4.1}$$

for all finite choices of $\sigma_i \in S, a_i \in \mathcal{A}, b_i \in \mathcal{B}$. If $\mathcal{A} = \mathcal{B}$, then we also say a kernel ***on*** $\mathcal{B}$.

There are two possibilities to find the appropriate Kolmogorov decomposition for CPD-kernels. As pointed out in the end of the last section, it is no surprise that we obtain a Hilbert bimodule or, more fashionably, a *correspondence*. Recall that a ***correspondence*** from $\mathcal{A}$ to $\mathcal{B}$ is a Hilbert $\mathcal{B}$–module E with a nondegenerate(**!**) left action of $\mathcal{A}$ such that $\langle x, ay\rangle = \langle a^* x, y\rangle$ for all $x, y \in E$ and $a \in \mathcal{A}$.

4.2 Theorem. *If $\mathfrak{K}$ is a CPD-kernel over S from a unital C^*–algebra $\mathcal{A}$ to a C^*–algebra $\mathcal{B}$, there is pair (E, i) consisting of a correspondence E from $\mathcal{A}$ to $\mathcal{B}$ and a map $i \colon S \to E$ such that*

$$\langle i(\sigma), ai(\sigma')\rangle \ =\ \mathfrak{K}^{\sigma,\sigma'}(a)$$

for all $\sigma, \sigma' \in S$ and $a \in \mathcal{A}$, and such that $E = \overline{\operatorname{span}}\, \mathcal{A}i(S)\mathcal{B}$. Moreover, if (F, j) is another pair fulfilling $\langle j(\sigma), aj(\sigma')\rangle = \mathfrak{K}^{\sigma,\sigma'}(a)$, then the map $i(\sigma) \mapsto j(\sigma)$ extends to a unique bilinear isometry $E \to F$.

We refer to (E, i) as the ***Kolmogorov decomposition*** of the CPD-kernel $\mathfrak{K}$. By the universal property stated in the theorem, it is uniquely determined up to bilinear unitary equivalence. We also shall refer to E as the ***GNS-correspondence*** and to i as the ***cyclic map***; see Note 4.3.

Proof of Theorem 4.2. First possibility: By (4.1) it immediately follows that the kernel $\mathfrak{k}^{(a,\sigma),(a',\sigma')} := \mathfrak{K}^{\sigma,\sigma'}(a^* a')$ over $\mathcal{A} \times S$ is positive definite. Denote by $(E, \tilde{i})$ its Kolmogorov decomposition according to Theorem 2.1. On the subset $\tilde{i}(\mathcal{A} \times S)$ define a left action by $a\tilde{i}(a', \sigma') := \tilde{i}(aa', \sigma')$. This action fulfills $\langle \tilde{i}(a', \sigma'), a\tilde{i}(a'', \sigma'')\rangle = \langle a^*\tilde{i}(a', \sigma'), \tilde{i}(a'', \sigma'')\rangle$. Recall that the set $\tilde{i}(\mathcal{A} \times S)$ generates E as Hilbert $\mathcal{B}$–module. It is easy to prove that an action fulfilling the $*$–condition on a generating subset of E extends well-defined and uniquely to a left action on all of E. The pair (E, i) satisfies the stated properties.

Second possibility: Instead of appealing to Theorem 2.1, we imitate its proof. Indeed, if we equip the $\mathcal{A}$–$\mathcal{B}$–bimodule $E_0 := \mathcal{A} \otimes S_{\mathbb{C}} \otimes \mathcal{B}$ with the sesquilinear map defined by setting

$$\langle a \otimes e_\sigma \otimes b, a' \otimes e_{\sigma'} \otimes b'\rangle \ :=\ b^* \mathfrak{K}^{\sigma,\sigma'}(a^* a') b',$$

then the condition in Equation (4.1) is born to make it an semiinner product, which also fulfills $\langle a' \otimes e_{\sigma'} \otimes b', aa'' \otimes e_{\sigma''} \otimes b'' \rangle = \langle a^*a' \otimes e_{\sigma'} \otimes b', a'' \otimes e_{\sigma''} \otimes b'' \rangle$. We divide out the $\mathcal{A}$–$\mathcal{B}$–submodule of length-zero elements and complete, obtaining that way an $\mathcal{A}$–$\mathcal{B}$–correspondence E. The map $i(\sigma) := \lim_\lambda \mathbf{1}_\mathcal{A} \otimes e_\sigma \otimes u_\lambda + \mathcal{N}$ completes the construction. ■

4.3 Note. For one-point sets $S = \{\omega\}$ we get back the definition of CP-maps between C^*–algebras, and the second proof of Theorem 4.2 is just Paschke's ***GNS-construction*** for CP-maps; see.[Pas73]

4.4 Remark. Why did we present two proofs for Theorem 4.2? The first proof is more along classical lines: From the input data write down some kernel (in classical applications almost always $\mathbb{C}$–valued), show it is positive definite, and do the Kolmogorov decomposition. Only then start working in order to show that this Hilbert module (or, usually, Hilbert space in classical applications) has the desired structure.

For instance, classical proofs of the Stinespring construction for a CP-map $T\colon \mathcal{A} \to \mathcal{B}$ work approximately like that (cf. also Example 5.8): Represent your C^*–algebra $\mathcal{B}$ faithfully on a Hilbert space G and define a $\mathbb{C}$–valued kernel $\mathfrak{k}$ over the set $\mathcal{A} \times G$ as $\mathfrak{k}^{(a,g),(a',g')} := \langle g, T(a^*a')g' \rangle$. Work in order to prove it is positive definite. Do the Kolmogorov decomposition to get the pair $(H, \tilde{i})$. Work again in order to show that $\rho(a)\colon \tilde{i}(a', g) \mapsto \tilde{i}(aa', g)$ determines a representation ρ of $\mathcal{A}$ on H. Work still more in order to show that $v\colon g \mapsto i(\mathbf{1}, g)$ defines a bounded map such that $v^*\rho(a)v = T(a)$ for all $a \in \mathcal{A}$.

But how much work is this, is'nt it!

The second proof is different. We want an $\mathcal{A}$–$\mathcal{B}$–bimodule? That proof starts by writing down the $\mathcal{A}$–$\mathcal{B}$–bimodule E_0. It is immediate from the input data how to define a semiinner product that turns it into a semicorrespondence from $\mathcal{A}$ to $\mathcal{B}$. Apply the generalities from the theory of correspondences (essentially, Cauchy-Schwarz inequality; cf. Remark 7.7) that tells you that one may divide out kernels of semiinner products and complete. Identify the elements $i(\sigma)$, which, by definition of the inner product, fulfill the stated property. (Needless to say that representing the C^*–algebra $\mathcal{B}$ is neither necessary nor useful. Anyway, never represent an abstract C^*–algebra that it is not given as a concrete operator algebra from the beginning, unless you are going to prove a theorem that is explicitly about representations of $\mathcal{B}$ on a Hilbert space!)

After this deviation let us return to our subject: Square roots of positive things. Let us note that the Kolmogorov decomposition is a "good" square root of $\mathfrak{K}$. It allows easily to get $\mathfrak{K}$ back as $\mathfrak{K}^{\sigma,\sigma'}(a) = \langle i(\sigma), ai(\sigma')\rangle$. It puts into immediate evidence why $\mathfrak{K}$ is completely positive definite if it can be recovered by a pair (E,i). Indeed,

$$\sum_{i,j} b_i^* \mathfrak{K}^{\sigma_i,\sigma_j}(a_i^* a_j) b_j \;=\; \Big\langle \sum_i a_i i(\sigma_i) b_i, \sum_i a_i i(\sigma_i) b_i \Big\rangle \;\geq\; 0.$$

And it is unique up to suitable unitary equivalence of correspondences.

4.5 Example. Let us return to the situation with the PD-kernels $\mathfrak{k}\colon (\omega,\omega) \mapsto b = \beta^*\beta$ and $\mathfrak{l}\colon (\omega,\omega) \mapsto c = \gamma^*\gamma$ over the one-point set $S = \{\omega\}$ as discussed in Section 3. We have noted that in order to understand the composition defined as $(\omega,\omega) \mapsto (\beta\gamma)^*(\beta\gamma) = \gamma^*\beta^*\beta\gamma$ whatever b might be, we must know the map $\mathfrak{L}\colon b \mapsto \gamma^* b\gamma$ rather than just the kernel $\mathfrak{l}$. Of course, $\mathfrak{L}$ is a CPD-kernel over S, and its Kolmogorov decomposition is (F,j) with $F = \overline{\operatorname{span}}\, \mathcal{B}\gamma\mathcal{B}$ and $j(\omega) = \gamma$. Theorem 4.2 tells us that another γ' gives the same $\mathfrak{L}$ if and only if $\gamma \mapsto \gamma'$ extends as an isomorphism of correspondences over $\mathcal{B}$. This is precisely the case if $(\beta\gamma)^*(\beta\gamma)$ and $(\beta\gamma')^*(\beta\gamma')$ coincide whatever β is.

Having replaced $\mathfrak{l}\colon (\omega,\omega) \mapsto c$ with $\mathfrak{L}\colon (\omega,\omega) \mapsto \gamma^* \bullet \gamma$, it is only natural to do the same with $\mathfrak{k}\colon (\omega,\omega) \mapsto b$ and to replace it with $\mathfrak{K}\colon (\omega,\omega) \mapsto \beta^* \bullet \beta$. We can easily define $\mathfrak{L} \circ \mathfrak{K}\colon (\omega,\omega) \mapsto (\beta\gamma)^* \bullet (\beta\gamma)$. This brings us directly to the question of composition of kernels that motivated the definition of CPD-kernels.

4.6 Definition. Let S be a set, let $\mathfrak{K}$ be a kernel over S from $\mathcal{A}$ to $\mathcal{B}$, and let $\mathfrak{L}$ be a kernel over S from $\mathcal{B}$ to $\mathcal{C}$. Then we define their ***composition*** or ***Schur product*** $\mathfrak{L} \circ \mathfrak{K}$ over S from $\mathcal{A}$ to $\mathcal{C}$ by pointwise composition, that is, by

$$(\mathfrak{L} \circ \mathfrak{K})^{\sigma,\sigma'} \;:=\; \mathfrak{L}^{\sigma,\sigma'} \circ \mathfrak{K}^{\sigma,\sigma'}.$$

As the proof of the fact that the composition of CPD-kernels is CPD has to do with some of our considerations about positivity we wish to make in general, we postpone it to the next section. But once this is settled, it is clear what we will understand by a ***CPD-semigroup*** $\mathfrak{T} = \left(\mathfrak{T}_t\right)_{t\geq 0}$ of CPD-kernels $\mathfrak{T}_t$ over a set S from $\mathcal{B}$ to $\mathcal{B}$ (or ***on*** $\mathcal{B}$).

4.7 Note. A single CPD-kernel from $\mathcal{A}$ to $\mathcal{B}$ over the finite set $S = \{1,\ldots,n\}$ has been defined by Heo[Heo99] under the name of *completely*

multi-positive map. Special CPD-semigroups over $S = \{0, 1\}$ have been introduced in Accardi and Kozyrev.[AK01] This paper motivated the general definitions of CPD-kernels and of CPD-semigroups in Ref.[BBLS04] (One should note that the definition in Ref.[AK01] is slightly weaker, but this is compensated by that in Ref.[AK01] the algebra is $\mathcal{B}(H)$ and triviality of W^*–correspondences over $\mathcal{B}(H)$, where the weaker definition is equivalent to that in Ref.;[BBLS04] see Skeide,[Ske01] Lemma 5.27 and Remark 5.2.8, for an explanation.)
In Ref.,[BBLS04] Lemma 3.2.1, there are listed some properties of a kernel $\mathfrak{K}$ that are equivalent to that $\mathfrak{K}$ is CPD. The most interesting are:

(1) For each finite choice of $\sigma_i \in S, a_i \in \mathcal{A}, b_i \in \mathcal{B}$ the map

$$a \longmapsto \sum_{i,j} b_i^* \mathfrak{K}^{\sigma_i,\sigma_j}(a_i^* a a_j) b_j$$

is (completely) positive. (Apply positivity of these maps to $a = \mathbf{1} \geq 0$ to see that $\mathfrak{K}$ is CPD. The nontrivial direction follows directly from Kolmogorov decomposition, because $a \mapsto \langle x, ax \rangle$ is clearly CP.)

(2) For all choices $\sigma_1, \ldots, \sigma_n \in S$ ($n \in \mathbb{N}$) the map

$$\mathfrak{K}^{(n)}\colon (a_{ij}) \longmapsto (\mathfrak{K}^{\sigma_i,\sigma_j}(a_{ij}))$$

from $M_n(\mathcal{A})$ to $M_n(\mathcal{B})$ is (completely) positive.

For $\#S$ finite and fixed $n = \#S$, the second property with complete positivity is Heo's definition in Ref.[Heo99]

5. Positivity, tensor product, and Schur product

As we said in the introduction, a good possibility to check for positivity of something is to write that something as a square, that is, to find a square root. Looking at it from the other end, we can say a good notion of positivity is a notion that admits, as a theorem, the statement that every positive element has a whatsoever square root.

C^*–Algebras do fulfill that criterion. An element in b a C^*–algebra $\mathcal{B}$ is positive if and only if it can be written as $\beta^*\beta$ for some $\beta \in \mathcal{B}$. An equivalent condition (frequently used as definition of positive element) is that $b = b^*$ and the spectrum $\sigma(b)$ is contained in $\mathbb{R}_+$ (***spectral positivity***). Another equivalent condition is that $\varphi(b) \geq 0$ for all positive linear functionals φ on $\mathcal{B}$ (***weak positivity***), where φ is ***positive*** if $\varphi(b^*b) \geq 0$ for all $b \in \mathcal{B}$.

Spectral positivity of $b \in \mathcal{B}$ has the advantage that, apart from self-adjointness, it uses only the spectrum of b. And the spectrum of an element in a unital C^*–algebra depends only the unital C^*–subalgebra generated by that element. (The extra condition that a positive element must be self-adjoint, is something we gladly accept, looking at how powerful the other property is.) This means, no matter how big another C^*–algebra $\mathcal{A}$ is into which we embed $\mathcal{B}$, our element b, which is positive in $\mathcal{B}$, continues being positive also in $\mathcal{A}$.

Many other properties when dealing with positive elements are proved by spectral calculus. When constructing new Hilbert modules from given ones, it would be nice if checking on positivity of the new inner products in $\mathcal{B}$, one would no longer need manipulations involving spectral calculus. Instead, one would simply refer to known results about positivity in $\mathcal{B}$ or related C^*–algebras and for the rest has purely algebraic operations. We illustrate what we mean, by giving several algebraic constructions that result in a completely algebraic proof of positivity of the inner product in the tensor product and positivity of the Schur product.

The ***direct sum*** $E := E_1 \oplus \ldots \oplus E_n$ of Hilbert $\mathcal{B}$–modules with the inner product of $X = (x_1, \ldots, x_n)$ and $Y = (y_1, \ldots, y_n)$ defined as $\langle X, Y\rangle = \sum_i \langle x_i, y_i\rangle$ is a Hilbert $\mathcal{B}$–module. (Indeed, the sum of positive elements in a C^*–algebra is positive. The rest follows as for Hilbert spaces. The direct sum of an infinite family of Hilbert modules still has to be completed.)

By $\mathcal{B}^a(E,F)$ we denote the space of ***adjointable*** operators from a Hilbert module E to a Hilbert module F. It is easy to check that

$$\mathcal{B}^a(E) = \begin{pmatrix} \mathcal{B}^a(E_1,E_1) & \ldots & \mathcal{B}^a(E_n,E_1) \\ \vdots & & \vdots \\ \mathcal{B}^a(E_1,E_n) & \ldots & \mathcal{B}^a(E_n,E_n) \end{pmatrix},$$

acting in the obvious way on $X^n \in E$. Since $\mathcal{B}^a(E)$ is a C^*–algebra, for every $a \in \mathcal{B}^a(E_i, E_j)$ the *square* a^*a is a positive element of $\mathcal{B}^a(E_i)$, because it is positive in $\mathcal{B}^a(E)$.

5.1 Example. The ***linking algebra*** $\left(\begin{smallmatrix} \mathcal{B} & E^* \\ E & \mathcal{K}(E) \end{smallmatrix}\right) \subset \mathcal{B}^a(\mathcal{B} \oplus E)$ for a Hilbert module over a C^*–algebra $\mathcal{B}$ is the most important example. Here $E^* = \{x^* \colon x \in E\} \subset \mathcal{B}^a(E, \mathcal{B})$ where $x^* \colon y \mapsto \langle x, y\rangle$, and the C^*–subalgebra $\mathcal{K}(E) := \overline{\operatorname{span}}\, EE^*$ of $\mathcal{B}^a(E)$ is called the algebra of ***compact operators*** on E.

It follows that E^* is a correspondence from $\mathcal{B}$ to $\mathcal{K}(E)$ with inner product $\langle x^*, y^*\rangle = xy^*$. In particular, the element xx^* is a positive element of $\mathcal{K}(E) \subset \mathcal{B}^a(E)$. Note that the algebra of compact operators on E^* is $\mathcal{K}(E^*) = \overline{\operatorname{span}}\, E^*E = \overline{\operatorname{span}}\langle E, E\rangle =: \mathcal{B}_E$, the ***range ideal*** of the inner product of E.

Now we have direct sums (in particular, we have the ***column space*** E^n, the direct sum of n copies of E), and we have the ***dual correspondence*** E^*. So, nobody prevents us from combining these constructions. We define the ***row space*** $E_n := ((E^*)^n)^*$. Observe that E_n is a Hilbert module over the compact operators on $(E^*)^n$, that is, over $M_n(\mathcal{K}(E^*)) = M_n(\mathcal{B}_E)$. The Hilbert module structure of the ideal $M_n(\mathcal{B}_E)$ in $M_n(\mathcal{B})$, easily (and uniquely) extends to $M_n(\mathcal{B})$. After this, it is an easy exercise to verify that E_n consists of elements $X_n = (x_1, \ldots, x_n)$ with inner product and module action given by

$$\langle X_n, Y_n \rangle_{i,j} = \langle x_i, y_i \rangle \qquad (X_n B)_i = \sum_j x_j b_{j,i}.$$

(Exercise: $\mathcal{B}^a(E_n) = \mathcal{B}^a(E)$ with $aX_n = (ax_1, \ldots, ax_n)$. In particular, $X_n Y_n^* = \sum_i x_i y_i^* \in \mathcal{K}(E)$.)

5.2 Corollary. *For all choices of $n \in \mathbb{N}$ and $x_1, \ldots, x_n \in E$ the matrix $\big(\langle x_i, x_j \rangle\big)_{i,j}$ is a positive element of $M_n(\mathcal{B})$.*

Recall how this went. The inner product of direct sums (in particular, of column spaces) is positive, because the sum of positive elements in a $C*$–algebra is positive. xx^* is positive in $\mathcal{K}(E)$ because it is positive in the linking algebra. Therefore, E^* is Hilbert module over $\mathcal{K}(E)$. The rest is purely algebraic manipulation, iterating this dualization operation with direct sums.

5.3 Remark. The corollary is a key ingredient for proving positivity of tensor products and Schur products. Of course, both $xx^* \geq 0$ and the corollary can be proved in a different way, for instance, by making use of the well-known fact that $a \in \mathcal{B}^a(E)$ is positive if and only if $\langle x, ax \rangle$ is positive for all $x \in E$; see Ref.[Pas73] But, the proof of this fact requires considerably more spectral calculus. Even in the scalar case $\mathcal{B} = \mathbb{C}$, we think it will be difficult to find a simpler argument than ours above. But note that also in the scalar case we do have to recognize that the dual of a Hilbert space H, H^*, carries the structure of a Hilbert $\mathcal{K}(H)$–module.

To construct the ***tensor product*** $E \odot F$ (over $\mathcal{B}$) of a correspondence E from $\mathcal{A}$ to $\mathcal{B}$ and a correspondence F from $\mathcal{B}$ to $\mathcal{C}$ one tries to define a $\mathcal{C}$–valued semiinner product on $E \otimes F$ in the only reasonable way by

$$\langle x \otimes y, x' \otimes y' \rangle := \langle y, \langle x, x' \rangle y' \rangle$$

and sesquilinear extension. To show positivity we would be fine, if we had to show it only for elementary tensors. Indeed, the element $\langle x, x\rangle \in \mathcal{B}$ is positive, so that we may write it as $\beta^*\beta$. Hence, $\langle y, \langle x, x\rangle y\rangle = \langle y, \beta^*\beta y\rangle = \langle \beta y, \beta y\rangle \geq 0$. Now let $x_1, \ldots, x_n \in E$ and $y_1, \ldots, y_n \in F$ and put $X_n := (x_1, \ldots, x_n) \in E_n$ and $Y^n := (y_1, \ldots, y_n) \in F^n$. Observe that F^n is a correspondence from $M_n(\mathcal{B})$ to $\mathcal{C}$ with the obvious left action of $M_n(\mathcal{B})$. We find

$$\Big\langle \sum_i x_i \otimes y_i, \sum_i x_i \otimes y_i \Big\rangle = \sum_{i,j} \langle y_i, \langle x_i, x_j\rangle y_j\rangle = \langle X_n \otimes Y^n, X_n \otimes Y^n\rangle \geq 0.$$

Once this is established, we may divide out the length-zero elements $\mathcal{N}$, complete, and obtain $E \odot F$. We denote $x \odot y := x \otimes y + \mathcal{N}$. Note that $E \odot F$ is a correspondence from $\mathcal{A}$ to $\mathcal{C}$ which is determined up to bilinear unitary equivalence by the property that it is generated by elements $x \odot y$ having inner product $\langle x \odot y, x' \odot y'\rangle = \langle y, \langle x, x'\rangle y'\rangle$ and fulfilling $a(x \odot y) = (ax) \odot y$.

5.4 Corollary. *The composition $\mathfrak{L} \circ \mathfrak{K}$ of CPD-kernels $\mathfrak{K}$ and $\mathfrak{L}$ as in Definition 4.6, is CPD, too.*

Proof. Denote by (E, i) and (F, j) the Kolmogorov decompositions of $\mathfrak{K}$ and $\mathfrak{L}$, respectively. Then

$$\begin{aligned}(\mathfrak{L} \circ \mathfrak{K})^{\sigma,\sigma'} &= \mathfrak{L}^{\sigma,\sigma'} \circ \mathfrak{K}^{\sigma,\sigma'} = \langle j(\sigma), \langle i(\sigma), \bullet i(\sigma')\rangle j(\sigma')\rangle \\ &= \langle i(\sigma) \odot j(\sigma), \bullet i(\sigma') \odot j(\sigma')\rangle\end{aligned}$$

is, clearly, a CPD-kernel. ■

5.5 Observation. The preceding proof also shows that the GNS-correspondence of $\mathfrak{L} \circ \mathfrak{K}$ is the $\mathcal{A}$–$\mathcal{C}$–subcorrespondence of $E \odot F$ generated by all $i(\sigma) \odot j(\sigma)$, with $\sigma \mapsto i(\sigma) \odot j(\sigma)$ as cyclic map. Note that $E \odot F = (\overline{\operatorname{span}}\, \mathcal{A} i(S)\mathcal{B}) \odot (\overline{\operatorname{span}}\, \mathcal{B} j(S)\mathcal{C}) = \overline{\operatorname{span}} \{ a i(\sigma) \odot b j(\sigma')c \colon a \in \mathcal{A}; b \in \mathcal{B}; c \in \mathcal{C}; \sigma, \sigma' \in S\}$. So, $E \odot F$ is (usually much) bigger than the GNS-correspondence of $\mathfrak{L} \circ \mathfrak{K}$.

5.6 Example. A ***product system*** is a family $E^\odot = (E_t)_{t \geq 0}$ of $\mathcal{B}$–correspondences with an associative product $E_s \times E_t \ni (x_s, y_t) \mapsto x_s y_t \in E_{s+t}$ that extends as a bilinear unitary $E_s \odot E_t \to E_{s+t}$. If $\mathfrak{T} = (\mathfrak{T}_t)_{t \geq 0}$ is a CPD-semigroup over S on a unital C^*–algebra $\mathcal{B}$, then, by the preceding observation, the GNS-correspondences $\mathcal{E}_t$ of the $\mathfrak{T}_t$ fulfill

$$(\mathcal{E}_{s^n_{m_n}} \odot \ldots \odot \mathcal{E}_{s^n_1}) \odot \ldots \odot (\mathcal{E}_{s^1_{m_1}} \odot \ldots \odot \mathcal{E}_{s^1_1}) \supset \mathcal{E}_{s^n_{m_n} + \ldots + s^n_1} \odot \ldots \odot \mathcal{E}_{s^1_{m_1} + \ldots + s^1_1}.$$

For fixed $t > 0$, this gives rise to an inductive limit over tuples $(t_n, \ldots, t_1) \in (0, \infty)^n$ with $t_n + \ldots + t_1 = t$. For the resulting correspondences $E_t \supset \mathcal{E}_t$ the inclusion $\mathcal{E}_s \odot \mathcal{E}_t \supset \mathcal{E}_{s+t}$ becomes equality $E_s \odot E_t = E_{s+t}$. The elements $\xi_t^\sigma := i_t(\sigma) \in \mathcal{E}_t \subset E_t$ fulfill $\xi_s^\sigma \xi_t^\sigma = \xi_{s+t}^\sigma$, that is, for each $\sigma \in S$ the family $\xi^{\sigma\odot} = (\xi_t^\sigma)_{t \geq 0}$ is a ***unit***. Moreover, we have

$$\langle \xi_t^\sigma, \bullet \xi_t^{\sigma'} \rangle = \mathfrak{T}_t^{\sigma,\sigma'}$$

for all $\sigma, \sigma' \in S$, and the set $\{\xi^{\sigma\odot} : \sigma \in S\}$ of units generates $E^\odot$ as a product system. We refer to $E^\odot$ as the ***GNS-system*** of $\mathfrak{T}$ and to $\{\xi^{\sigma\odot} : \sigma \in S\}$ as the ***cyclic set of units***. We see:

The square root of a CPD-semigroup (in particular, of a CP-semigroup) is a product system.

5.7 Note. For CP-semigroups (that is, for a one-point set S), the preceding construction is due to Bhat and Skeide.[BS00] This seems to be the first publication where product systems of correspondences occur. The generalization to CPD-semigroup is from Ref.[BBLS04]
Meanwhile, structures like the family of GNS-correspondences $\mathcal{E}_t$ with inclusions such as $\mathcal{E}_{s+t} \subset \mathcal{E}_s \odot \mathcal{E}_t$ started to be investigated under the name of ***subproduct systems*** by Shalit and Solel,[SS09] and under the name of ***inclusion systems*** by Bhat and Mukherjee.[BM10] In Ref.,[BM10] which considers only subproduct systems of Hilbert spaces, it is proved (among other results) by the same inductive limit procedure that every subproduct system embeds into a proper product system. It is clear that this is true also for correspondences.

As a further application of the tensor product, we discuss a Stinespring construction for CPD-kernels.

5.8 Example. Let $\mathfrak{K}$ be CPD-kernel over a set S from $\mathcal{A} \ni \mathbf{1}_\mathcal{A}$ to $\mathcal{B}$, and denote by (E, i) its Kolmogorov decomposition. Suppose $\mathcal{B}$ is a concrete C^*–algebra of operators acting nondegenerately on a Hilbert space G. In other words, suppose G is a correspondence from $\mathcal{B}$ to $\mathbb{C}$. Put $H := E \odot G$. Observe that H is a correspondence from $\mathcal{A}$ to $\mathbb{C}$. In other words, H is a Hilbert space with a nondegenerate action of $\mathcal{A}$ that may be used to define the ***Stinespring representation*** ρ of $\mathcal{A}$ on H by $\rho(a) \colon x \odot g \mapsto a(x \odot g) = (ax) \odot g$. Note further that each $x \in E$ gives rise to an operator $L_x \colon g \mapsto x \odot g$. One readily verifies that

$$L_{i(\sigma)}^* \rho(a) L_{i(\sigma')} = \mathfrak{K}^{\sigma,\sigma'}(a)$$

for all $\sigma, \sigma' \in S$ and $a \in \mathcal{A}$.

5.9 Note. Note that this easily generalizes to the case when we replace $\mathcal{B} \subset \mathcal{B}(G)$ with $\mathcal{B} \subset \mathcal{B}^a(F)$ where F is some Hilbert $\mathcal{C}$–module. More precisely, F is a correspondence from $\mathcal{B}$ to $\mathcal{C}$. In this case, also $H = E \odot F$ is a Hilbert $\mathcal{C}$–module, but for the rest nothing changes. For a one-point set S, this is known as ***KSGNS-construction***; see Lance[Lan95] or Murphy.[Mur97]

For a one-point set S we recover the usual Stinespring construction for a CP-map from a unital C^*–algebra into a concrete operator C^*–algebra (cf. also Remark 4.4). However, neither the definition of CPD-kernel (or that of PD-kernel) nor the Kolmogorov decomposition for a CPD-kernel (or that of a PD-kernel) require that $\mathcal{B}$ is represented as an operator algebra. CPD-kernels $\mathfrak{K}$ and $\mathfrak{L}$ (in particular, CP-maps) may be composed, and the Kolmogorov decomposition of the composed CPD-kernel $\mathfrak{L} \circ \mathfrak{K}$ can easily be recovered inside the tensor product $E \odot F$ of those (E, i) and (F, j) of the factors $\mathfrak{K}$ and $\mathfrak{L}$, respectively, as ${}_{\mathcal{A}}i \odot j_{\mathcal{C}} := \overline{\operatorname{span}}\{ai(\sigma) \odot j(\sigma)c\colon a \in \mathcal{A}, \sigma \in S, c \in \mathcal{C}\}$. On the contrary, **never** will the Stinespring representation $\rho\colon a \mapsto a \odot \mathsf{id}_G$ of $\mathcal{A}$ on $H = E \odot G$ for $\mathfrak{K}$ help to determine the Stinespring representation of $\mathcal{A}$ for $\mathfrak{L} \circ \mathfrak{K}$!

Indeed, if $\mathcal{C} \subset \mathcal{B}(K)$, rather than the Stinespring representation of $\mathcal{A}$ associated with the identity representation of $\mathcal{B}$ on G, one would need the Stinespring representation of $\mathcal{A}$ associated with the Stinespring representation $\pi\colon b \mapsto b \odot \mathsf{id}_K$ of $\mathcal{B}$ on $L := F \odot K$ for $\mathfrak{L}$. (By this we mean the representation $a \mapsto a \odot \mathsf{id}_L$ of $\mathcal{A}$ on $E \odot L = E \odot F \odot K$.) The Stinespring representation of $\mathcal{A}$ for $\mathfrak{L} \circ \mathfrak{K}$ would, then, be the representation $a \mapsto a \odot \mathsf{id}_L$ on $E \odot F \odot K$ restricted to the invariant subspace $({}_{\mathcal{A}}i \odot j_{\mathcal{C}}) \odot K$. While this latter construction depends explicitly on $\mathfrak{L}$, or better on the Stinespring representation of $\mathcal{B}$ associated with $\mathfrak{L}$, the GNS-correspondence of $\mathfrak{K}$ is universal and works for composition with all $\mathfrak{L}$.

Conclusion:

Doing Stinespring representations for the individual members of a CP-semigroup on $\mathcal{B} \subset \mathcal{B}(G)$, is approximately as ingenious as considering a 2×2–system of complex linear equations as a real 4×4–system (ignoring all the structure hidden in the fact that certain 2×2–submatrices are very special) and applying the Gauß algorithm to the 4×4–system instead of trivially resolving the 2×2–system by hand.

5.10 Example. Let E and F be Hilbert modules over C^*–algebras $\mathcal{B}$ and

$\mathcal{C}$, respectively. Let φ be a map from $\mathcal{B}$ to $\mathcal{C}$. We say a linear map $T\colon E \to F$ is a ***φ–map*** if

$$\langle T(x), T(x')\rangle \;=\; \varphi(\langle x, x'\rangle)$$

for all $x, x' \in E$. Suppose $\mathcal{B}$ is unital and T is a φ–map for some CP-map φ from $\mathcal{B}$ to $\mathcal{C}$. Do the GNS-construction $(\mathcal{F}, \zeta)$ for φ. Then it is easy to verify that $v\colon x \odot b\zeta c \mapsto T(xb)c$ extends as an isometry $E \odot \mathcal{F} \to F$. Of course,

$$v(x \odot \zeta) \;=\; T(x).$$

Now let H_1, H_2 denote Hilbert spaces. Put $\mathcal{C} := \mathcal{B}(H_1)$, and put $F := \mathcal{B}(H_1, H_2)$. (This is a Hilbert $\mathcal{B}(H_1)$–module with inner product $\langle y, y'\rangle := y^*y'$.) Assume $\Phi\colon E \to F = \mathcal{B}(H_1, H_2)$ is a φ–map for the CP-map $\varphi\colon \mathcal{B} \to \mathcal{C} = \mathcal{B}(H_1)$, and construct the ingredients $\mathcal{F}, \zeta, v$ as before. Put $K_1 := \mathcal{F} \odot H_1$, and denote by $\rho\colon b \mapsto b \odot \mathsf{id}_{H_1}$ the Stinespring representation of $\mathcal{B}$ on K_1. Put $K_2 := E \odot K_1$, and denote by $\Psi\colon x \mapsto L_x = x \odot \mathsf{id}_{K_1}$ (the ***representation*** of E into $\mathcal{B}(K_1, K_2)$ ***induced*** by ρ). Denote $V := L_\zeta = \zeta \odot \mathsf{id}_{H_1} \in \mathcal{B}(H_1, K_1)$ and $W := (v \odot \mathsf{id}_{H_1})^* \in \mathcal{B}(F \odot H_1, E \odot \mathcal{F} \odot H_1) = \mathcal{B}(H_2, K_2)$. Then K_1; K_2; $V \in \mathcal{B}(H_1, K_1)$; $W \in \mathcal{B}(H_2, K_2)$; $\rho\colon \mathcal{B} \to \mathcal{B}(K_1)$; $\Psi\colon E \to \mathcal{B}(K_1, K_2)$ fulfill the following:

(1) $W^*\Psi(x)V = T(x)$ for all $x \in E$.
(2) $\Psi(x)^*\Psi(x') = \rho(\langle x, x'\rangle)$ for all $x, x' \in E$.
(3) ρ is a nondegenerate representation.
(4) W is a coisometry.

A sextuple with these properties is determined uniquely up to suitable unitary equivalence.

5.11 Note. Existence and uniqueness of a sextuple fulfilling these properties is proved in Bhat, Ramesh, Sumesh,[BRS10] Theorems 2.1 and 2.4. It should be noted that several additional conditions stated in Ref.,[BRS10] are automatic, once the four preceding conditions are fulfilled. (This result has been stated first by Asadi[Asa09] with an unnecessary condition and with an incorrect proof.) The construction of $\mathcal{F}, \zeta, v$ and the reduction of Ref.[BRS10] to it, is Skeide.[Ske10a]

6. More Examples

6.1 Example. Every $\mathbb{C}$–valued PD-kernel $\mathfrak{k}$ over a set S gives rise to a CPD-kernel $\mathfrak{K}$ on $\mathcal{B}(\mathbb{C}) = \mathbb{C}$ over S, if we interpret a complex number z as a map $z\colon w \mapsto wz$ on $\mathbb{C}$. Under this identification the Kolmogorov decompositions of $\mathfrak{k}$ and of $\mathfrak{K}$ coincide. (After all, every Hilbert space is also a correspondence over $\mathbb{C}$ in the only possible way.) And the composition of the CPD-kernels corresponds to the usual Schur product of the PD-kernels. Of course, the tensor product of the Kolmogorov decompositions is the usual tensor product of Hilbert spaces.

In the same way, a $\mathcal{B}$–valued PD-kernel $\mathfrak{k}$ over S can be interpreted as a CPD-kernel $\mathfrak{K}$ from $\mathbb{C}$ to $\mathcal{B}$, by interpreting $b \in \mathcal{B}$ as map $z \mapsto zb$.

About a CPD-kernel $\mathfrak{K}$ over $S$from $\mathcal{B}$ to $\mathbb{C}$ we cannot say much more than that it is, by Kolmogorov decomposition, a representation of $\mathcal{B}$ on a Hilbert space G which is generated by $\mathcal{B}$ from a set of $\#S$ vectors having certain inner products. Of course, for a one-point set $S = \{\omega\}$ we recover the GNS-construction for a positive linear functional $\varphi\colon \mathcal{B} \to \mathbb{C}$ with one vector $i(\omega)$ that is ***cyclic*** for $\mathcal{B}$ (that is, $\overline{\mathcal{B}i(\omega)} = G$) such that $\langle i(\omega), \bullet i(\omega)\rangle = \varphi$.

6.2 Example. The tensor product of the Kolmogorov decompositions of (C)PD-kernels on $\mathbb{C}$ is the usual tensor product of Hilbert spaces. But the inclusion of the Kolmogorov decomposition for the composition into that tensor product remains. In particular, the construction of a product system, as discussed in Example 5.6 for CP(D)-semigroups, works also here and the inductive limit is (almost always) proper. Just that now the result is a product system of Hilbert spaces (***Arveson system***, henceforth).

It is noteworthy that under very mild conditions (just measurability of the semigroups $\mathfrak{t}^{\sigma,\sigma'}$ in $\mathbb{C}$) the Arveson system of a (C)PD-semigroup (or Schur semigroup of positive definite kernels) $\mathfrak{t}$ can be computed explicitly. Indeed, under this condition we may define the generator $\mathfrak{l}$ of $\mathfrak{t}$ as the kernel

$$\mathfrak{l}^{\sigma,\sigma'} := \tfrac{d}{dt}\Big|_{t=0} \mathfrak{t}_t^{\sigma,\sigma'}.$$

One easily verifies that $\mathfrak{l}$ is ***conditionally positive definite***, that is,

$$\sum_{i,j} \overline{z}_i \mathfrak{l}^{\sigma_i,\sigma_j} z_j \ \geq 0$$

whenever $\sum_i z_i = 0$. Note, too, that $\mathfrak{l}$ is ***hermitian*** in the sense that $\overline{\mathfrak{l}^{\sigma,\sigma'}} = \mathfrak{l}^{\sigma',\sigma}$ for all $\sigma, \sigma' \in S$ (simply because $\mathfrak{t}_t$, obviously, is hermitian).

Note that for every choice of $\beta_\sigma \in \mathbb{C}$ ($\sigma \in S$) with $\mathfrak{t}$, also $\tilde{\mathfrak{t}}$ defined by setting $\tilde{\mathfrak{t}}_t^{\sigma,\sigma'} := e^{\overline{\beta}_\sigma t}\, \mathfrak{t}_t^{\sigma,\sigma'}\, e^{\beta_{\sigma'} t}$ is a PD-semigroup with generator $\tilde{\mathfrak{l}}^{\sigma,\sigma'} = \mathfrak{l}^{\sigma,\sigma'} + \beta_{\sigma'} + \overline{\beta}_\sigma$. Fix a $\sigma_0 \in S$. Then choose β_{σ_0} such that $\tilde{\mathfrak{t}}_t^{\sigma_0,\sigma_0} = 1$, that is, such that $\tilde{\mathfrak{l}}^{\sigma_0,\sigma_0} = \mathfrak{l}^{\sigma_0,\sigma_0} + \beta_{\sigma_0} + \overline{\beta}_{\sigma_0} = 0$. Further, for that β_{σ_0} choose the other β_σ such that $\tilde{\mathfrak{t}}_t^{\sigma_0,\sigma} = 1$, that is, such that $\tilde{\mathfrak{l}}^{\sigma_0,\sigma} = \mathfrak{l}^{\sigma_0,\sigma} + \beta_{\sigma_0} + \overline{\beta}_\sigma = 0$. Verify that the kernel $\tilde{\mathfrak{l}}$ we obtain in that way is not only conditionally positive definite, but really positive definite. Do the Kolmogorov decomposition (K, i) for that kernel. Define $E_t := \Gamma(L^2([0,t], K))$ (symmetric Fock space) and observe that the E_t form an Arveson system via

$$E_s \otimes E_t \ \to \ s_t E_s \otimes E_t \ = \ E_{s+t},$$

where the first step is the second quantized time-shift emerging from $[0, s] \ \mapsto \ [t, t+s]$, and where the second step is the usual factorization $\Gamma(H_1) \otimes \Gamma(H_2) = \Gamma(H_1 \oplus H_2)$. Note that for $\beta \in \mathbb{C}$ and $k \in K$ the elements $\xi_t(\beta, k) := e^{\beta t}\psi(\mathbb{I}_{[0,t]}k) \in E_t$ (ψ denoting exponential vectors and $\mathbb{I}$ indicator functions) form units. Finally,

$$\langle \xi_t(-\beta_\sigma, i(\sigma)), \xi_t(-\beta_{\sigma'}, i(\sigma'))\rangle \ = \ e^{-\overline{\beta}_\sigma t} e^{\mathfrak{l}^{\sigma,\sigma'} t} e^{-\beta_{\sigma'} t} \ = \ \mathfrak{t}_t^{\sigma,\sigma'}.$$

Since the $i(\sigma)$ generate K and since $i(\sigma_0) = 0$, these units generate the whole product system; see Skeide.[Ske00] In other words, we have just constructed the Kolmogorov decomposition of $\mathfrak{t}$. Moreover, a brief look at the construction shows, that it actually does not depend on that $\mathfrak{l}$ is *a priori* the generator of a PD-semigroup, but only on its properties to be hermitian and conditionally positive definite. We, thus, also have proved that every such kernel generates a PD-semigroup by exponentiation. This relation is called the ***Schönberg correspondence*** between PD-semigroups and conditionally positive definite hermitian $\mathbb{C}$–valued kernels.

6.3 Note. Without notions like product systems and units for them (these came not before Arveson[Arv89]), the possibility to realize PD-semigroups as inner products of suitably normalized exponential vectors has been discovered as early as Parthasarathy and Schmidt.[PS72] They applied it to characteristic functions (the Fourier transform) of the convolution semigroup of distributions of Lévy processes, and used it to prove representability of an arbitrary Lévy process (starting at 0) by (possibly infinite, for instance, for the Cauchy process) linear combinations of the usual creation, conservation, and annihilation processes on the Fock space; a result that Schürmann[Sch93] generalized to quantum Lévy processes.

6.4 Note. A natural question is if the a similar representation result holds for CPD-semigroups $\mathfrak{T}$ on $\mathcal{B}$, when we replace the symmetric Fock space by the time ordered Fock module; Ref.[BS00] The answer is: Yes, for von Neumann algebras, and if the CPD-semigroup is uniformly continuous; Ref.[BBLS04] The procedure is essentially the same. One has to find a σ_0 such that the semigroup $\mathfrak{T}^{\sigma_0,\sigma_0}$ can be normalized to give the trivial semigroup id on $\mathcal{B}$. However, as this is no longer possible in general,

one rather has to try add to a point σ_0 to S and to extend the CPD-semigroup over S to a CPD-semigroup over $S \cup \{\sigma_0\}$ in a consistent way. In the von Neumann case this is always possible. But unlike the scalar case, here this result is a hard to prove. In fact, it is equivalent to the deep result by Christensen and Evans[CE79] who found the form of the generator of a uniformly continuous CP-semigroup on a von Neumann algebra, and in Ref.[BBLS04] only equivalence to Ref.[CE79] is proved. From this point on, everything goes like the scalar case. The unit $\omega^\odot$ representing the value σ_0, actually already belongs to the GNS-system of $\mathfrak{T}$, and what is generated by all these units is a whole product system of time ordered Fock modules.
In the C^*–case the situation is worse. Not only is it possible that the GNS-system embeds into a product system of time ordered Fock modules, but is not isomorphic to one. It is even possible that a the GNS-system of a uniformly continuous CP-semigroup T (even an automorphism semigroup) does not embed at all into a product system of time ordered Fock modules. In fact, a the GNS-system embeds if and only if the CP-semigroup is ***spatial***. Spatial means that T dominates a CP-semigroup of the form $b \mapsto c_t^* b c_t$ for a norm continuous semigroup of elements c_t in $\mathcal{B}$. All this has been discussed for CP-semigroups in Bhat, Liebscher, and Skeide.[BLS10] The extension of the discussion to CPD-semigroup can be found in Skeide.[Ske10b]
It should be noted that in Skeide[Ske06] (preprint 2001) we called a product system $E^\odot$ ***spatial*** if it admits a unit $\omega^\odot$ that gives the trivial semigroup $\langle \omega_t, \bullet \omega_t \rangle = \mathsf{id}_\mathcal{B}$ on $\mathcal{B}$ (like the unit $\xi^{\sigma_0 \odot}$ corresponding to the index σ_0 of the extended CPD-semigroup above). Spatial product systems allow a classification scheme most similar to Arveson systems. They are type I (that is, Fock) and type II (non-Fock) depending on whether or not they are generated by a set of units that determines a uniformly continuous CPD-semigroup. And they have an index (the one-particle sector of the maximal Fock part) that behaves additive (direct sum) under a suitable spatial product of spatial product systems; see Ref.[Ske06] For Arveson systems (which, unlike general product systems, possess a tensor product) it may but need not coincide with the tensor product of Arveson systems,

6.5 Example. Of course, also a semigroup $\vartheta = (\vartheta_t)_{t \geq 0}$ of unital endomorphisms of $\mathcal{B}$ is a CP-semigroup. One readily verifies that its GNS-system consists of the correspondences $\mathcal{B}_t = \mathcal{B}$ as right Hilbert $\mathcal{B}$–module, but with left action $b.x_t := \vartheta_t(b)x_t$. The generating unit is $\mathbf{1}_t = \mathbf{1}$. (Note that each $\mathcal{B}_t$ is the GNS-correspondence of ϑ_t. No inductive limit is necessary, because $\mathcal{B}_s \odot \mathcal{B}_t = \mathcal{B}_{s+t}$.) On the other hand, if we have such a ***one-dimensional*** product system $\mathcal{B}_t$, then $\vartheta_t(b) := b.\mathbf{1}_t$ defines a unital endomorphism semigroup.

6.6 Note. This might appear to be a not so interesting product system. However, it turns out that every product system of correspondences over $\mathcal{B}$ arises from such a one-dimensional product system of correspondences over $\mathcal{B}^a(E)$. (Anyway, one should be alarmed by the fact that a one-dimensional product system of correspondences over $\mathcal{B}$ is isomorphic to the trivial one if and only if the endomorphism semigroup consists of inner automorphisms of $\mathcal{B}$.) The relation between these two product systems, is an operation of Morita equivelance; see the following Exampe 6.7 where the operation is scratched, and Note 6.8 where Morita equivalence is mentioned very briefly. This scratches the intimate relationship between E_0–semigroups (that is, unital endomorphism semigroups) on $\mathcal{B}^a(E)$ and product systems (first Arveson[Arv89,Arv90] for E_0–semigroups on $\mathcal{B}(H)$ and Arveson systems), and we have no space to give an account. Instead of a long list of papers, we mention Skeide[Ske09a] where the theory has been completed, and the references therein.

6.7 Example. In Example 5.8 and the remarks following it, we have emphatically underlined that we do not consider it too much of a good idea to tensor a GNS-correspondence (from $\mathcal{A}$ to $\mathcal{B}$, say) with a representation space G (or module F) of $\mathcal{B}$. This changes if actually $\mathcal{B} = \mathcal{B}(G)$ or $\mathcal{B} = \mathcal{B}^a(F)$.

To fix letters a bit more consistently, let E, F, and G be a Hilbert $\mathcal{B}$–module, a Hilbert $\mathcal{C}$–module, and a Hilbert $\mathcal{D}$–module, respectively. And suppose $\mathcal{E}$ and $\mathcal{F}$ are correspondences from $\mathcal{B}^a(E)$ to $\mathcal{B}^a(F)$, and from $\mathcal{B}^a(F)$ to $\mathcal{B}^a(G)$, respectively. If $\mathcal{E}$ is the GNS-correspondence of a CP-map (or even of a unital homomorphism), then nobody would be surprised that we require technical conditions for that CP-map (like normality in the case when E is a Hilbert space) that are reflected also by the left action of that correspondence. In the framework of C^*–modules this condition is ***strictness*** of the left action of $\mathcal{B}^a(E)$ (or, more precisely, strictness on bounded subsets). It is equivalent to that the compacts $\mathcal{K}(E)$ alone act already nondegenerately on $\mathcal{E}$. So, let us also suppose that $\mathcal{E}$ and $\mathcal{F}$ are strict correspondences in that sense. Let us compute $E^* \odot \mathcal{E} \odot F$ and $F^* \odot \mathcal{F} \odot G$. (Recall the definition of the dual correspondence E^* from Example 5.1.) Then

$$\begin{aligned}(E^* \odot \mathcal{E} \odot F) \odot (F^* \odot \mathcal{F} \odot G) &= E^* \odot \mathcal{E} \odot (F \odot F^*) \odot \mathcal{F} \odot G \\ &= E^* \odot \mathcal{E} \odot \mathcal{K}(F) \odot \mathcal{F} \odot G = E^* \odot (\mathcal{E} \odot \mathcal{F}) \odot G.\end{aligned}$$

Here $F \odot F^* = \mathcal{K}(F)$ via $y' \odot y^* \mapsto y'y^*$, and in the last step we made use of strictness of the left action of $\mathcal{F}$.

We see that "tensor-sandwiching" from the left and right with the respective representation modules is an operation that respects tensor products (and is, obviously, as associative as one may wish). Suppose $\mathcal{E}^{\odot}$ is a product system of strict correspondences over $\mathcal{B}^a(E)$. Then the correspondences $E_t := E^* \odot \mathcal{E}_t \odot E$ form a product system $E^{\odot}$ of correspondences over $\mathcal{B}$. If $\mathcal{E}^{\odot}$ is the one-dimensional product system of a strict E_0–semigroup on $\mathcal{B}^a(E)$ as in Example 6.5, then $E^{\odot}$ is indeed the product system of $\mathcal{B}$–correspondences associated with that E_0–semigroup. In the same way one obtains the product system of $\mathcal{B}$–correspondences of a strict CP(D)-semigroup group on $\mathcal{B}^a(E)$ from the GNS-system of that CP(D)-semigroup.

It is noteworthy that by this elegant and simple method, we recover for $\mathcal{B}(H)$ the constructions of Arveson systems from E_0–semigroups on $\mathcal{B}(H)$ or from CP-semigroups on $\mathcal{B}(H)$. However, even in that simple case it is indispensable to understand the dual correspondence of the Hilbert space H, H^*, as a fully qualified Hilbert $\mathcal{B}(H)$–module.

6.8 Note. For E_0–semigroups of $\mathcal{B}(H)$ this product system has been constructed in a different way by Bhat.[Bha96] It is anti-isomorphic to the Arveson system from Ref.[Arv89] (and need not be isomorphic to it; see Tsirelson[Tsi00]). We imitated Bhat's construction in Skeide;[Ske02] but since this construction requires existence of a unit vector in E, it is not completely general. The general construction above is from Skeide[Ske09b] (preprint 2004). The construction of the Arveson system of a CP-semigroup on $\mathcal{B}(H)$ is again due to Ref.;[Bha96] it goes via dilation of the CP-semigroup to an E_0–semigroup and, then, determining the Arveson system of that E_0–semigroup. In Skeide[Ske03] we gave a direct construction of that product system along the above lines. Bhat, Liebscher, and Skeide[BLS08] discuss the generalization from Ref.[Ske03] to CP-semigroups on $\mathcal{B}^a(E)$, Skeide[Ske10b] the case of CPD-semigroups. "Tensor-sandwiching", as we called it above, is actually *cum grano salis* an operation of Morita equivalence for correspondences. (For von Neumann algebras and modules, it is Morita equivalence. For C^*–algebras and modules it is Morita equivalence up to strict completion.) This has been explained in Ref.[Ske09b] and fully exploited in Ref.[Ske09a] to complete the theory of classification of E_0–semigroups by product systems.

7. Positivity in $*$–algebras

We said that if a notion of positivity is good, then positive things should have a square root. Positive elements of a C^*–algebra have a square root inside the C^*–algebra. In Skeide[Ske01] we worked successfully with the definition that an element in a pre-C^*–algebra $\mathcal{B}$ is positive if it is positive in the completion of $\mathcal{B}$, $\overline{\mathcal{B}}$. In other words, such an element has a square root not necessarily in $\mathcal{B}$, but it always has one in $\overline{\mathcal{B}}$.

7.1 Example. If polynomials behave nicely with respect to positivity, depends on where we wish to evaluate them. If p is a polynomial with complex coefficients and we consider functions $x \mapsto p(x)$ from $\mathbb{R}$ to $\mathbb{C}$, then p is positive in the sense that $p(x) \geq 0$ for all $x \in \mathbb{R}$ if and only if there is a polynomial q such that $p = \overline{q}q$. If we evaluate in $\mathbb{C}$ instead of $\mathbb{R}$ (requiring $p(z) \geq 0$ for all $z \in \mathbb{C}$), then Liouville's theorem tells us that p is constant. If we evaluate in a bounded interval $I = [a, b]$ only, then we get more positive polynomials. For instance, the polynomial $p(x) = -x$ is positive on the interval $[-1, 0]$. But no q will ever give back p as a square $\overline{q}q$. Note that such polynomials that are positive on bounded intervals occur (and are necessary!) in typical proofs of the spectral theorem, where they serve to approximate indicator functions of intervals from above. Of course, $-x$ has a square root in the C^*–algebra $C[-1, 0]$.

We already showed positivity of an element in $\mathcal{B}^a(E)$ by writing it as a^*a for some $a \in \mathcal{B}^a(E, F)$. So, it sometimes is convenient even to leave completely the algebra under consideration. Our thesis was also that Hilbert modules can be good square roots of positive-type maps.

It is our scope to propose here a new and flexible definition of positivity, that complements the algebraic notion of positivity from Accardi and Skeide[AS08] (applied successfully in Ref.[AS00] to the "square of white noise"). We will show that this notion allows for nice square roots of positive things.

For the balance of this section, $\mathcal{L}(V, W)$ stands for ***linear*** maps between vector spaces, and $\mathcal{L}^a(E, F)$ stands for ***adjointable*** maps between pre Hilbert $\mathcal{B}$–modules. The latter are linear automatically, but in general not necessarily bounded (unless at least one, E or F, is complete).

7.2 Definition. Let $\mathcal{B}$ be a unital $*$–algebra and let $\mathcal{S}$ be a set of positive linear functionals on $\mathcal{B}$. We say $b \in \mathcal{B}$ is ***$\mathcal{S}$–positive*** if $\varphi(c^*bc) \geq 0$ for all $\varphi \in \mathcal{S}, c \in \mathcal{B}$. We say $\mathcal{B}$ is ***$\mathcal{S}$–separated*** if the functionals in φ separate the points of $\mathcal{B}$ in the sense that $\varphi(c^*bc') = 0$ for all $\varphi \in \mathcal{S}$ and all $c, c' \in \mathcal{B}$ implies $b = 0$.

7.3 Observation. By polarization and since every $b \in \mathcal{B}$ is a linear combination of two self-adjoint elements, we may check separation only for $\varphi(c^*bc)$ and $b = b^*$.

7.4 Observation. $\varphi(c^*bc') = 0$ for all $c, c' \in \mathcal{B}$ means that the GNS-representation of the positive functional φ sends b to 0. In other words, if $\mathcal{B}$ is $\mathcal{S}$–separated, then the direct sum of all GNS-representations for $\varphi \in \mathcal{S}$ is a faithful representation. That is, we may and will interpret $\mathcal{B}$ as a concrete $*$–algebra of operators $\mathcal{B} \subset \mathcal{L}^a(G)$ where the pre-Hilbert space G is the (algebraic, of course) direct sum over the GNS-representations for all φ. By Observation 7.3, an element $b \in \mathcal{B}$ is $\mathcal{S}$–positive if and only if $\langle g, bg \rangle \geq 0$ for all $g \in G$.

One might be tempted to try the latter as a definition of positivity for an arbitrary $*$–algebra of operators. However, our representation has something special, namely, it is the direct sum of cyclic representations. It it this property that will allow us to work.

7.5 Remark. If $\mathcal{B}$ is not $\mathcal{S}$–separated, then we simply may quotient out the kernel of the representation on G. After all, $\mathcal{S}$ will typically contain those states that correspond to all possible measurements that the system described by our algebra $\mathcal{B}$ allows. If these measurements do not separate two points b_1 and b_2 from $\mathcal{B}$, it is pointless to consider them different.

In the sequel, we fix a unital $*$–algebra $\mathcal{B}$ that is $\mathcal{S}$–separated by a set $\mathcal{S}$ of positive functionals and represented faithfully as operator algebra $\mathcal{B} \subset \mathcal{L}^a(G)$ on the direct sum G of the GNS-spaces of all $\varphi \in \mathcal{S}$.

The following Kolmogorov decomposition is the generalization of Example 5.8 to the algebraic situation of this section.

7.6 Theorem. *Let $\mathcal{A}$ be a unital $*$–algebra. For some set S let $\mathfrak{K}\colon S \times S \to \mathcal{L}(\mathcal{A}, \mathcal{B})$ be a CPD-kernel over S from $\mathcal{A}$ to $\mathcal{B}$, in the sense that all sums in (4.1) are $\mathcal{S}$–positive. Then there exists a pre-Hilbert space H with a left action of $\mathcal{A}$, and a map $i\colon S \to \mathcal{L}^a(G, H)$ such that*

$$\mathfrak{K}^{\sigma,\sigma'}(a) = i(\sigma)^* a i(\sigma')$$

*for all $\sigma, \sigma' \in S$ and $a \in \mathcal{A}$. We refer to $E := \operatorname{span} \mathcal{A} i(S) \mathcal{B} \subset \mathcal{L}^a(G, H)$ as the **GNS-correspondence** of $\mathfrak{K}$ with **cyclic map** i, and to (E, i) its **Kolmogorov decomposition**.*

Proof. On $\mathcal{A} \otimes S_{\mathbb{C}} \otimes \mathcal{B}$ we define a sesquilinear map precisely as in the proof of Theorem 4.2. The map is right $\mathcal{B}$–linear, and the left action of $\mathcal{A}$ is a $*$–map. It is rather a tautology that $\langle \bullet, \bullet \rangle$ is positive for whatever notion of positivity we define on $\mathcal{B}$. So, we may call that map a ***semiinner product*** and $\mathcal{A} \otimes S_{\mathbb{C}} \otimes \mathcal{B}$ a ***semicorrespondence*** from $\mathcal{A}$ to $\mathcal{B}$.

The true work starts now when we wish to divide out the length-zero elements $\mathcal{N} := \{x \in \mathcal{A} \otimes S_{\mathbb{C}} \otimes \mathcal{B}\colon \langle x,x\rangle = 0\}$. Obviously, $x \in \mathcal{N}$ implies $xb \in \mathcal{N}$. Next we show that

$$x \in \mathcal{N} \implies \langle x,y\rangle = 0 \quad \text{for all} \quad y \in \mathcal{A} \otimes S_{\mathbb{C}} \otimes \mathcal{B}. \tag{7.1}$$

Indeed, $|\varphi(c^*\langle x,y\rangle c')|^2 = |\varphi(\langle xc,yc'\rangle)|^2 \leq \varphi(\langle xc,xc\rangle)\varphi(\langle yc',yc'\rangle) = 0$, so $\langle x,y\rangle = 0$. From this, we immediately conclude that $x,y \in \mathcal{N}$ implies $x+y \in \mathcal{N}$, so $\mathcal{N}$ is a right $\mathcal{B}$ submodule. Suppose again $x \in \mathcal{N}$. Then $\langle ax,ax\rangle = \langle x,a^*ax\rangle = 0$, by (7.1), so $\mathcal{N}$ is an $\mathcal{A}$–$\mathcal{B}$–submodule. We, therefore, may define the $\mathcal{A}$–$\mathcal{B}$–module $E := (\mathcal{A} \otimes S_{\mathbb{C}} \otimes \mathcal{B})/\mathcal{N}$. Now, if $x,y \in \mathcal{A} \otimes S_{\mathbb{C}} \otimes \mathcal{B}$ and $n,m \in \mathcal{N}$. Then $\langle x+n,y+m\rangle = \langle x,y\rangle$, again by (7.1). Therefore, by $\langle x+\mathcal{N},y+\mathcal{N}\rangle := \langle x,y\rangle$ we well-define a semiinner product on E which is ***inner*** (that is $\langle x,x\rangle = 0$ implies $x = 0$ for all $x \in E$). We, thus, may call E a ***precorrespondence*** from $\mathcal{A}$ to $\mathcal{B}$. Of course, the elements $j(\sigma) := \mathbf{1} \otimes e_\sigma \otimes \mathbf{1} + \mathcal{N}$ fulfill $\langle j(\sigma), aj(\sigma')\rangle = \mathfrak{K}^{\sigma,\sigma'}(a)$.

Next we wish to construct $E \odot G$. To that goal, we define as usual a sesquilinear map on $E \otimes G$ by setting $\langle x \otimes g, x' \otimes g'\rangle := \langle g, \langle x,x'\rangle g'\rangle$. We have to face the problem to show that

$$\sum_{i,j}\langle x_i \otimes g_i, x_j \otimes g_j\rangle \;=\; \sum_{i,j}\langle g_i, \langle x_i,x_j\rangle g_j\rangle \;\geq\; 0$$

for all finite choices of $g_i \in G$ and $x_i \in E$. Recall that G decomposes into a direct sum of pre-Hilbert subspaces G_φ with cyclic vectors g_φ, say, such that $\langle g_\varphi, bg_{\varphi'}\rangle = 0$ for all $\varphi \neq \varphi'$ and such that $g = \sum_\varphi b_\varphi g_\varphi$ for suitable b_φ (different from 0 only for finitely many φ). It follows that

$$\begin{aligned}\sum_{i,j}\langle g_i, \langle x_i,x_j\rangle g_j\rangle &= \sum_\varphi\sum_{i,j}\langle b^i_\varphi g_\varphi, \langle x_i,x_j\rangle b^j_\varphi g_\varphi\rangle \\ &= \sum_\varphi\Big\langle g_\varphi, \langle \textstyle\sum_i x_i b^i_\varphi, \sum_i x_i b^i_\varphi\rangle g_\varphi\Big\rangle \;\geq\; 0.\end{aligned}$$

Now it is clear that every $x \in E$ defines an operator $L_x\colon g \mapsto x \odot g$ with adjoint $L^*_x\colon y \odot g \mapsto \langle x,y\rangle g$. We end the proof, by putting $i(\sigma) := L_{j(\sigma)}$. ■

7.7 Note. Observe that (7.1) functions as substitute for Cauchy-Schwarz inequality. We wish to underline that at least as important as the estimates that follow from Cauchy-Schwarz inequality for pre-Hilbert modules (for instance, that the norm of a Hilbert modules is a norm, or that the operator norm of $\mathcal{B}^a(E)$ is a C^*–norm), is that Cauchy-Schwarz inequality is true also for semiinner products. Note, too, that unlike for Hilbert spaces, for semi-Hilbert modules the case $\langle x,x\rangle = 0 = \langle y,y\rangle$ in the proof Cauchy-Schwarz inequality require a small amount of additional work; see, for instance, Ref.,[Ske01] Proposition 1.2.1.

We should note that when we wish to compose such kernels, so that also the algebra to the left needs a positivity structure, then the CPD-condition must be supplemented with a compatibility condition for that positivity. (Compare this with the solution for the algebraic definition of positivity in Ref.[AS08]) We leave this to future work and close with a corollary about existence of square roots for $\mathcal{S}$–positive elements of $\mathcal{B}$.

7.8 Corollary. *(1) Suppose that* $\mathcal{A} = \mathbb{C}$ *and* $S = \omega$ *and choose an* $\mathcal{S}$*–positive* $b \in \mathcal{B}$. *Then* $\beta = i(\omega)$ *is an adjointable operator* $G \to H$ *such that* $b = \beta^* \beta$.

(2) By Friedrichs' theorem and spectral calculus for self-adjoint operators, there exists a positive self-adjoint operator $\sqrt{b}\colon \overline{G} \supset D \to \overline{G}$ *with* $G \subset D$. *Then* $\beta := \sqrt{b} \upharpoonright G$ *and* $H := \sqrt{b}G$ *is a possible choice for (1).*

Proof. Only for (2) there is something to show. Indeed, after having chosen $\sqrt{b}$ and defined β and H as stated, let us choose some β' and H' that exist according to (1). If necessary replace H' with $\beta' G$, so that β' is surjective. Then $v\colon \beta' g \mapsto \sqrt{b} g = \beta g$ defines an isometry into $\overline{G}$ and a unitary onto H. Since β' has an adjoint $H' \to G$, it follows that $\beta\colon G \to H$ has the adjoint $\beta^* = \beta'^* v^*$. Of course, $\beta^* \beta g = \beta'^* v^* \beta g = \beta'^* \beta' g = bg$. ■

References

AK01. L. Accardi and S. Kozyrev, *On the structure of Markov flows*, Chaos Solitons Fractals **12** (2001), 2639–2655.

Aro50. N. Aronszajn, *The theory of reproducing kernels*, Trans. Amer. Math. Soc. **68** (1950), 337–404.

Arv89. W. Arveson, *Continuous analogues of Fock space*, Mem. Amer. Math. Soc., no. 409, American Mathematical Society, 1989.

Arv90. ______, *Continuous analogues of Fock space IV: essential states*, Acta Math. **164** (1990), 265–300.

AS00. L. Accardi and M. Skeide, *Hilbert module realization of the square of white noise and the finite difference algebra*, Math. Notes **86** (2000), 803–818, (Rome, Volterra-Preprint 1999/0384).

AS08. ______, *Interacting Fock space versus full Fock module*, Commun. Stoch. Anal. **2** (2008), 423–444, (Rome, Volterra-Preprint 1998/0328, revised 2000).

Asa09. M.B. Asadi, *Stinespring's theorem for Hilbert C^*–modules*, J. Operator Theory **62** (2009), 235–238.

BBLS04. S.D. Barreto, B.V.R. Bhat, V. Liebscher, and M. Skeide, *Type I product systems of Hilbert modules*, J. Funct. Anal. **212** (2004), 121–181, (Preprint, Cottbus 2001).

Ber05. H. Bercovici, *Multiplicative monotonic convolution*, Illinois J. Math. **3** (2005), 929–951.

Bha96. B.V.R. Bhat, *An index theory for quantum dynamical semigroups*, Trans. Amer. Math. Soc. **348** (1996), 561–583.

BLS08. B.V.R. Bhat, V. Liebscher, and M. Skeide, *A problem of Powers and the product of spatial product systems*, Quantum Probability and Infinite Dimensional Analysis — Proceedings of the 28th Conference (J.C. Garcia, R. Quezada, and S.B. Sontz, eds.), Quantum Probability and White Noise Analysis, no. XXIII, World Scientific, 2008, (arXiv: 0801.0042v1), pp. 93–106.

BLS10. ———, *Subsystems of Fock need not be Fock: Spatial CP-semigroups*, Proc. Amer. Math. Soc. **138** (2010), 2443–2456, electronically Feb 2010, (arXiv: 0804.2169v2).

BM10. B.V.R. Bhat and M. Mukherjee, *Inclusion systems and amalgamated products of product systems*, Infin. Dimens. Anal. Quantum Probab. Relat. Top. **13** (2010), 1–26, (arXiv: 0907.0095v1).

BRS10. B.V.R. Bhat, G. Ramesh, and K. Sumesh, *Stinespring's theorem for maps on Hilbert C^*-modules*, Preprint, arXiv: 1001.3743v1, 2010, To appear in J. Operator Theory.

BS00. B.V.R. Bhat and M. Skeide, *Tensor product systems of Hilbert modules and dilations of completely positive semigroups*, Infin. Dimens. Anal. Quantum Probab. Relat. Top. **3** (2000), 519–575, (Rome, Volterra-Preprint 1999/0370).

CE79. E. Christensen and D.E. Evans, *Cohomology of operator algebras and quantum dynamical semigroups*, J. London Math. Soc. **20** (1979), 358–368.

Fra09. U. Franz, *Monotone and Boolean convolutions for non-compactly supported probability measures*, Indiana Univ. Math. J. **3** (2009), 1151–1185.

Heo99. J. Heo, *Completely multi-positive linear maps and representations on Hilbert C^*-modules*, J. Operator Theory **41** (1999), 3–22.

Kas80. G.G. Kasparov, *Hilbert C^*-modules, theorems of Stinespring & Voiculescu*, J. Operator Theory **4** (1980), 133–150.

Lan95. E.C. Lance, *Hilbert C^*-modules*, Cambridge University Press, 1995.

Mur97. G.J. Murphy, *Positive definite kernels and Hilbert C^*-modules*, Proc. Edinburgh Math. Soc. **40** (1997), 367–374.

Pas73. W.L. Paschke, *Inner product modules over B^*-algebras*, Trans. Amer. Math. Soc. **182** (1973), 443–468.

PS72. K.R. Parthasarathy and K. Schmidt, *Positive definite kernels, continuous tensor products, and central limit theorems of probability theory*, Lect. Notes Math., no. 272, Springer, 1972.

Sch93. M. Schürmann, *White noise on bialgebras*, Lect. Notes Math., no. 1544, Springer, 1993.

Ske00. M. Skeide, *Indicator functions of intervals are totalizing in the symmetric Fock space* $\Gamma(L^2(\mathbb{R}_+))$, Trends in contemporary infinite dimensional analysis and quantum probability (L. Accardi, H.-H. Kuo, N. Obata, K. Saito, Si Si, and L. Streit, eds.), Natural and Mathematical Sciences Series, vol. 3, Istituto Italiano di Cultura (ISEAS), Kyoto, 2000, Volume in honour of Takeyuki Hida, (Rome, Volterra-Preprint 1999/0395), pp. 421–424.

Ske01. ______, *Hilbert modules and applications in quantum probability*, Habilitationsschrift, Cottbus, 2001, Available at `http://www.math.tu-cottbus.de/INSTITUT/lswas/\_skeide.html`.

Ske02. ______, *Dilations, product systems and weak dilations*, Math. Notes **71** (2002), 914–923.

Ske03. ______, *Commutants of von Neumann modules, representations of* $\mathcal{B}^a(E)$ *and other topics related to product systems of Hilbert modules*, Advances in quantum dynamics (G.L. Price, B .M. Baker, P.E.T. Jorgensen, and P.S. Muhly, eds.), Contemporary Mathematics, no. 335, American Mathematical Society, 2003, (Preprint, Cottbus 2002, arXiv: math.OA/0308231), pp. 253–262.

Ske06. ______, *The index of (white) noises and their product systems*, Infin. Dimens. Anal. Quantum Probab. Relat. Top. **9** (2006), 617–655, (Rome, Volterra-Preprint 2001/0458, arXiv: math.OA/0601228).

Ske08. ______, *Product systems; a survey with commutants in view*, Quantum Stochastics and Information (V.P. Belavkin and M. Guta, eds.), World Scientific, 2008, (Preferable version: arXiv: 0709.0915v1!), pp. 47–86.

Ske09a. ______, *Classification of* E_0*–semigroups by product systems*, Preprint, arXiv: 0901.1798v2, 2009.

Ske09b. ______, *Unit vectors, Morita equivalence and endomorphisms*, Publ. Res. Inst. Math. Sci. **45** (2009), 475–518, (arXiv: math.OA/0412231v5 (Version 5)).

Ske10a. ______, *A factorization theorem for* φ*–maps*, Preprint, arXiv: 1005.1396v1, To appear in J. Operator Theory, 2010.

Ske10b. ______, *The Powers sum of spatial CPD-semigroups and CP-semigroups*, Banach Center Publications **89** (2010), 247–263, (arXiv: 0812.0077).

SS09. O.M. Shalit and B. Solel, *Subproduct systems*, Documenta Math. **14** (2009), 801–868, (Preprint, arXiv: 0901.1422v2).

Sti55. W.F. Stinespring, *Positive functions on* C^**–algebras*, Proc. Amer. Math. Soc. **6** (1955), 211–216.

Sza09. F.H. Szafraniec, *Murphy's positive definite kernels and Hilbert* C^**–modules reorganized*, Preprint, arXiv: 0906.5408v1, 2009.

Tsi00. B. Tsirelson, *From random sets to continuous tensor products: answers to three questions of W. Arveson*, Preprint, arXiv: math.FA/0001070, 2000.

MULTIPARAMETER QUANTUM STOCHASTIC PROCESSES

WILLIAM J. SPRING

Quantum Information and Probability Group,
School of Computer Science, University of Hertfordshire,
College Lane, Hatfield. AL10 9AB, UK
e-mail: j.spring@herts.ac.uk

We extend the non-commutative constructions outlined in Ref.[1–3] to the general n parameter case, these being integrals over an n - dimensional parameter space, employing martingales as integrand. We extend previous results in type one,[1,4] type two[1,5] stochastic integrals and introduce new integrals for the Clifford sheet. These stochastic integrals are orthogonal, centred L^2 - martingales, obeying isometry properties. We develop the construction to consider general martingale representations for the Clifford sheet.

Keywords: **martingale, increments, type r integrals, representation.**

1. Introduction

Throughout this paper we work with a stochastic base $(X, \mathcal{A}, (\mathcal{A}_z), m, I)$, in which X will retain Hilbert space properties, $\mathcal{A}$ may represent a von Neumann Algebra, $C*$-Algebra, or Hilbert Space, $(\mathcal{A}_z)$ an associated filtration of $\mathcal{A}$, m represents a gage or state, a linear functional acting on $\mathcal{A}$ and I a parameter set, for us the set $\mathbb{R}^n_+$, with $z \in \mathbb{R}^n_+$.

2. Clifford Construction

For the Clifford construction X is taken to represent the anti-symmetric fermion Fock space $\mathcal{F}(L^2(\mathbb{R}^n_+))$ generated by tensor products of $L^2(\mathbb{R}^n_+)$. X is a Hilbert space of the form $\mathcal{F}(\mathcal{H}) = \bigoplus_{r=0}^{\infty}\mathcal{H}_r$, with $\mathcal{H} = L^2(\mathbb{R}^n_+)$ and $\mathcal{H}_r = \mathcal{H} \otimes \mathcal{H} \otimes \cdots \otimes \mathcal{H}$ the tensor product of r copies of $\mathcal{H}$.[6] $\mathcal{A}$ is taken to be a von Neumann algebra of operators generated by polynomials of the form $\sum_i^l \prod_j^m \psi(f_{i,j})$ in which each $\psi(f) = a(f) + a^*(f)$, is a sum of creation and annihilation operators acting on X with f real valued, $f \in L^2(\mathbb{R}^n_+)$. $\psi = \psi^*$, satisfies the CAR properties with $\{\psi(f), \psi(g)\} = 2(f, g)_{L^2(\mathbb{R}^n_+)}$.

$(\mathcal{A}_z)$ represents a filtration generated from $\mathcal{A}$ through the restriction on the fermion fields ψ to elements of the form $\chi_{R_z} g \in L^2(R_z) \subseteq L^2(\mathbb{R}_+^n)$. The gage $m(\circ) = (\Omega, \circ\,\Omega)$, establishes a faithful, central state on $\mathcal{A}$ in which $\Omega = 1 \in \mathbb{C} = \mathcal{H}_0$ in $\mathcal{F}(\mathcal{H})$. The triple $(X, \mathcal{A}, m)$ forms a probability gage space.[7] $m(|a|) = (\Omega, |a^p|^{1/p}\Omega)$ allow one to establish noncommutative $L^p(\mathcal{A})$ spaces, with $1 \leq p < \infty$, (for $L^\infty(\mathcal{A})$ we employ the usual operator norm). Conditional expectations may be defined on $\mathcal{A}$ and extended to $L^p(\mathcal{A})$.[8]

2.1. *Increments*

Let $z = (z_1, z_2, \ldots z_n) \in \mathbb{R}_+^n$ and R_z the n-dimensional cube in $\mathbb{R}_+^n$ based at the origin, by which we mean that $inf R_z$ is based at the origin and $sup R_z$ at z. We work with partially ordered sets, referred to as 'type r' increments, in which, for each r, there exists at least one set Δ_A with $n-r+1$ parameters forward of the corresponding parameters in z.

For the case $n = 1$,[9] we meet one form of increment, a type 1 parameter set, of the form $[z, z')$ with $z < z' \in \mathbb{R}$.

For the case $n = 2$,[1] we meet two types of increment; a type 1 parameter set and a type 2 parameter set. A type 1 set Δ_A, is one in which for each $z' \in \Delta_A$, each parameter z'_i in z' is forward of z_i found in z, (by which we mean $z_i \leq z'_i$ for $i = 1, 2$). Type 2 parameter sets Δ_B and Δ_C are sets in which for all $z' \in \Delta_B$, just one parameter is forward of those found in $z = z' \vee z''$, and it is the same parameter for each $z' \in \Delta_B$. For the case $n = 2$, all elements $z'' \in \Delta_C$ are also forward of those found in z in the 'other' parameter. Δ_B and Δ_C may be thought of as mutually exclusive parameter sets.

For the three dimensional index set $\mathbb{R}_+^3$ a type 1 parameter set generalises in a natural way, to form a cube Δ_A in which all points z' are forward of $z = inf\Delta_A$. Following the classical case,[10] one of the two type 2 parameter sets Δ_B and Δ_C will be forward in 2 parameters whilst the 'other' parameter set may be forward in either one parameter or two parameters, (forward that is of $z = inf\Delta_B \vee inf\Delta_C$). For Δ_B and Δ_C both forward of z in two parameters it was felt that this could lead to a new type of classical stochastic integral.[10]

Proposition 2.1. *Let Δ_B and Δ_C both be forward of z in two parameters. Then any product $\psi(\chi_{\Delta_B})\psi(\chi_{\Delta_C})$ can be expressed to within epsilon of a* *sum of type 2 products in which one parameter set is forward of z in two parameters whilst the other is forward in just one.*

Proof. Without loss of generality we may assume that Δ_B is forward of $z = inf\Delta_B \vee inf\Delta_C$ in the z_1 and z_3 coordinates, whilst Δ_C is forward of z in the z_2 and z_3 coordinates. If we cut Δ_B and Δ_C in half parallel to the $z_1 - z_2$ plane to form Δ_{B_1} 'sitting on top of' Δ_{B_2} and Δ_{C_1} 'sitting on top of' Δ_{C_2}, then:

$$\begin{aligned}
&\psi(\chi_{\Delta_B})\psi(\chi_{\Delta_C})\\
=&\psi(\chi_{\Delta_{B_1}}) + \chi_{\Delta_{B_2}})\psi(\chi_{\Delta_{C_1}} + \chi_{\Delta_{C_2}})\\
=&\psi(\chi_{\Delta_{B_1}})\psi(\chi_{\Delta_{C_1}}) + \psi(\chi_{\Delta_{B_1}})\psi(\chi_{\Delta_{C_2}}) + \psi(\chi_{\Delta_{B_2}})\psi(\chi_{\Delta_{C_1}}) + \psi(\chi_{\Delta_{B_2}})\psi(\chi_{\Delta_{C_2}})
\end{aligned}$$

We note that $\psi(\chi_{\Delta_{B_1}})\psi(\chi_{\Delta_{C_2}})$ and $\psi(\chi_{\Delta_{B_2}})\psi(\chi_{\Delta_{C_1}})$ may be viewed as type 2 increments in which Δ_{B_1} is forward of $z' = inf\Delta_{B_1} \vee z$ in 2 parameters whilst Δ_{C_2} is forward of z' in just one parameter. The same is true for $\psi(\chi_{\Delta_{B_2}})\psi(\chi_{\Delta_{C_1}})$. Hence we may write

$$\begin{aligned}
&\psi(\chi_{\Delta_B})\psi(\chi_{\Delta_C}) - \psi(\chi_{\Delta_{B_1}})\psi(\chi_{\Delta_{C_2}}) - \psi(\chi_{\Delta_{B_2}})\psi(\chi_{\Delta_{C_1}})\\
&= \sum_{i=1}^{2} \psi(\chi_{\Delta_{B_i}})\psi(\chi_{\Delta_{C_i}})
\end{aligned}$$

Repeating this process leads us to a representation of $\psi(\chi_{\Delta_B})\psi(\chi_{\Delta_C})$ with norm to within ϵ of the required type. □

Following the proposition we work with type 2 increments Δ_B and Δ_C in which one increment is forward of $z = inf\Delta_B \vee inf\Delta_C$ in two parameters whilst the other is forward of z in just one parameter.

In moving from the parameter space $\mathbb{R}^2_+$ to $\mathbb{R}^3_+$, we meet a new type 3 increment, in which each of three parameter sets, Δ_D, Δ_E and Δ_F consists of points forward of $z = inf\Delta_D \vee inf\Delta_E \vee inf\Delta_F$ in just one of the three parameters available. Each of the three increments are situated on a face of R_z, one to each of the three forward faces in R_z. Again each of the parameter sets (increments) may be thought of as mutually exclusive. For a type 3 increment one may conjecture other possibilities, for example three increments, forward of z, each of the three increments situated on a forward edge of R_z, one to each of the three forward edges in R_z. However each of these may be seen to be within ϵ of a sum involving combinations of types already discussed.

3. Quantum Stochastic Integrals

Definition 3.1. Let $h : \mathbb{R}^n_+ \times \cdots \times \mathbb{R}^n_+ \longrightarrow L^2(\mathcal{A})$ by $(z_1, \ldots, z_r) \mapsto h(z_1, \ldots, z_r) = a \underset{i=1}{\overset{r}{\pi}} \chi_{\Delta_i}$ with $a \in \mathcal{A}_{inf \Delta_1 \vee \cdots \vee inf \Delta_r}$ and $\Delta_1 \hat{\wedge} \ldots \hat{\wedge} \Delta_r$.[1] Then h is said to be an elementary $\mathcal{A}$ valued, r adapted process.

Elementary adapted processes are extend to simple processes by linearity. We work in general, with type r integrals, in which $1 \leq r \leq n$, such that r increments are forward of the region R_z in at least one parameter $z_i \in \mathbb{R}$.

Definition 3.2. Let $a \in \mathcal{A}_z$, with $z \in \mathbb{R}^n_+$. Let Δ_i denote an increment forward of R_z. Let $h(z) = a\chi_{\Delta_1} \cdots \chi_{\Delta_r}$, the parameter set forming a type r parameter set.
We define a type r integral to be one of the form

$$\begin{aligned} \mathcal{S}_r(h, z, f_1, \ldots, f_r) &= \int_{R_z} \cdots \int_{R_z} h(z_1, \ldots, z_r) d\Psi_{z_1}(f_1) \ldots d\Psi_{z_r}(f_r) \\ &= a\Psi(\chi_{\Delta_1 \cap R_z} f_1) \ldots \Psi(\chi_{\Delta_r \cap R_z} f_r) \end{aligned}$$

These integrals extend by linearity to simple adapted processes.

The type r integrals are pairwise orthogonal, centered martingales satisfying isometry conditions which extend (via isometry) to appropriate completions in which the same properties hold.[11,12]

4. Representation

Following the development of representation theorems for the Clifford setting over a two dimensional parameter set it is natural to consider more general cases. Here we extend to three and general dimensional parameter settings.

Theorem 4.1. *Let $(X_z)_{z \in \mathbb{R}^3_+}$ denote an $L^2(\mathcal{A})$ valued martingale adapted to the family $(\mathcal{A}_z)_{z \in \mathbb{R}^3_+}$ of von Neumann subalgebras of $\mathcal{A}$. Then $\exists$ unique $f \in L^2_\psi$, $g \in L^2_{\psi\psi}$ and $h \in L^2_{\psi\psi\psi}$ (type 1, 2 and 3 resp.) s.t.*

$$X_z = X_0 + \mathcal{S}_1(f, z) + \mathcal{S}_2(g, z) + \mathcal{S}_3(h, z)$$

For the case $n = 2$ any product of ψ's is of the form type I, type II or a limit of type I and type II's.[1] A type I integral $a\psi(\chi_\Delta u)$ of an elementary adapted process is of the form $a \in \mathcal{C}_z$ for some $z \in \mathbb{R}^2_+$ with $z \prec z' \, \forall z' \in \Delta$. For a type II integral the elementary adapted processes were of the form $a\chi_{\Delta_1}\chi_{\Delta_2}$ with integral $a\psi(\chi_{\Delta_1}\chi_{\Delta_2}v)$ with $a \in \mathcal{C}_z$ with $z = inf\chi_{\Delta_1} \vee inf\chi_{\Delta_2}$

and $\Delta_1 \hat{\wedge} \Delta_2$ the so called 'cock-eyed' domains. The isometry property and linearity of the ψ's allowed us to extend the above to simple processes and then to appropriate completions.

For the case $n = 3$ we work with the three partially ordered sets, discussed above. We consider the general case.

Let $X = Y \overset{p_{z_1}}{\underset{i=1}{\pi}} \psi(\chi_{A_i}) \overset{p_{z_2}}{\underset{j=1}{\pi}} \psi(\chi_{B_j}) \ldots \overset{p_{z_{(k+1)}}}{\underset{t=1}{\pi}} \psi(\chi_{C_t})$ denote a product of ψ's arranged so that all increments proud of R_z are arranged to the right hand side of the product. Y is assumed to be the same type of product but within $\mathcal{A}_z$. Let p_{z_l} denote the number of 'proud domains' (increments) in the z_l-direction. Include in this product domains which are forward in more than one direction (so that type 1, 2, 3, ... type r increments are included). These may be associated, for the sake of order, with the z_l group corresponding to the minimum z_l direction in which the increment s proud. The approach taken here is to apply a double induction. The first of these will be in the parameter set $\mathbb{R}_+^{k+1}$, the second in $\mathbb{R}_+^k$. By linearity of the Ψ's we may assume that each of the proud domains are mutually exclusive in $\mathbb{R}_+^{k+1}$ and of the same size. If we cut these domains in half, parallel to the surface upon which they lie, the linearity of ψ's generates products with the same number of proud domains, and also products with less than $p_{z_1}p_{z_2}p_{z_3}$ proud domains. Our first inductive assumption is that any products obtained as a result of the 'cutting process' with less than $p_{z_1}p_{z_2}p_{z_3}$ proud domains is a sum of type r integrals, with $1 \leq r \leq k+1$. At each 'cut' we subtract all obtained type r integrals from X and focus on the remaining products left behind. These are of the same type as the original but with $k+1$ dimensional volume halved with each cut. The projections of these remaining 'proud domains' in $\mathbb{R}_+^{k+1}$ onto the surrounding $\mathbb{R}_+^k$ planes will be of the same form as was X but with again with their k dimensional volume halved. Our second hypothesis is that this type of projection to products with proud domains defined over $\mathbb{R}_+^k$, will have a norm tending to zero, as the number of cuts tends to infinity. It will follow that in $\mathbb{R}_+^{k+1}$ the norm of X - (sum of type r integrals) will tend to zero. and hence that X is a sum of type r integrals and/or a limit of a sum of such integrals.

To start the process we consider such products with parameter space $\mathbb{R}_+^3$. For just one proud domain Δ we move R_z back to $R_{inf\Delta}$. If Y produces new proud domains in this move then this is an example of more than one proud domain for a region $R_{z'}$, for some $z' \in \mathbb{R}_+^3$. If Y does not produce new proud domains then $\psi(\chi_{A_i})$ is a type 1 integral. For any configuration of two proud domains we can express the product as a sum of type 1 and type 2 integrals. For example, with one domain forward in each parameter of the other, we have a type 1 integral. With one domain on an edge and the other on a face with no edge in common, we have a type 2 integral. With two on different edges, we have seen that we can express this to within ϵ of a sum of type 2 integrals. With three proud domains we may have, for example, one domain on an edge and the other two on a face which has no edge in common with the first domain. If we cut this parallel to the face upon which the second and third proud domains are this leads to two type 2 integrals and two of the same kind. Further cuts lead to a sum of type 2 integrals. We note that one may proceed in this manner as for the two dimensional parameter sets[1] considering each case in turn and achieve an upper bound on X - (sum of type r integrals) for each case in terms of the volume, which again tends to zero as the number of cuts tends to infinity.

Alternatively we note that the *cutting and subtracting from X procedure* produces projections, to products over $\mathbb{R}^2_+$ that have been shown[1] to be dependent upon the number of cuts, resulting in a norm that tends to zero as the number of cuts tends to infinity. This occurs in each of the planes and hence in $\mathbb{R}^3_+$ the norm of X - (sum of type r integrals) tends to zero as the number of cuts tends to infinity. Repeating the process for three proud domains and then four proud domains allows us to verify the inductive claims for $k = 2$ and $k + 1 = 3$. Assuming the result is true for $n = k$ and verifying for $n = k+1$ follows now by cutting proud domains parallel to the faces, using the linearity of the ψ's and subtracting sums of type r integrals from X. The remaining products with same type of proud domain as X will then have a reduced volume whose norm is a function of the reduced volume, tending to zero as the number of cuts tends to infinity.

For the general case we therefore have:

Theorem 4.2. *Let $(X_z)_{z\in\mathbb{R}^n_+}$ denote an $L^2(\mathcal{A})$ valued martingale adapted to the family $(\mathcal{A}_z)_{z\in\mathbb{R}^n_+}$ of von Neumann subalgebras of $\mathcal{A}$. Then $\exists$ unique $f_i \in L^2_{\psi^n}$ s.t.*

$$X_z = X_0 + \sum_{i=1}^{n} \mathcal{S}_i(f_i, z)$$

The uniqueness of $f, g,$ and h for the case $n = 3$ and the f_i in the general case follows by application of the conditional expectation operator and isometry.[1]

5. Concluding Remarks

Further work has been carried out for the general case in both the Clifford and quasi-free setting. The details will appear elsewhere. The author would like to thank the organisers and participants for an extremely enjoyable and productive conference.

References

1. W. J. Spring and I. F. Wilde, Rep. Math. Phys. **42**, 389 (1998).
2. W. J. Spring, Foundations of Quantum Probability and Physics **4**, (2006).
3. W. J. Spring, Quantum Communication, Measurement and Computing **8**, (2007).
4. K. Itô, Memoirs of the American Mathematical Society **4**, 1 (1951).
5. E. Wong and M. Zakai, Z. Wahrscheinlichkeitstheorie und Verw. Gebiete **29**, 109 (1974).
6. O. Bratteli and D. W. Robinson, *Operator Algebras and Quantum Statistical Mechanics II* (Springer, New York, 1981).
7. I. E. Segal, Ann. of Math. **57/58**, (1953).
8. M. Terp, Lic. Scient. Thesis, University of Odense (1981)
9. C. Barnett, R. F. Streater, I. F. Wilde J. Funct. Anal. **48**, (1982)
10. P. Imkeller, Stoch. Proc. and their App. **20**, 1 (1985).
11. W. J. Spring, Int. Journ. Pure and App. Math. **49**, 3 (2008)
12. W. J. Spring, Quantum Communication, Measurement and Computing **9**, (2009).